Gengxiang Hu, Xun Cai, Yonghua Rong
**Materials Science**

# Also of interest

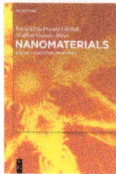

Gengxiang Hu, Xun Cai, Yonghua Rong

# Materials Science

Volume 2: Phase Transformation and Properties

Revised and translated by Zhuguo Li, Da Shu, Zhenghong Guo, Xiaodong Wang, and Guo He

**DE GRUYTER**

上海交通大學出版社
SHANGHAI JIAO TONG UNIVERSITY PRESS

**Authors**
Prof. Gengxiang Hu
School of Material Science and Engineering
Shanghai Jiao Tong University
No. 800 Dongchuan Road
Shanghai 200240, China

Prof. Xun Cai
Shanghai Key Laboratory of Materials Laser Processing and Modification
School of Material Science and Engineering
Shanghai Jiao Tong University
No. 800 Dongchuan Road
Shanghai 200240, China

Prof. Yonghua Rong
School of Material Science and Engineering
Shanghai Jiao Tong University
No. 800 Dongchuan Road
Shanghai 200240, China

ISBN 978-3-11-049515-7
e-ISBN (PDF) 978-3-11-049537-9
e-ISBN (EPUB) 978-3-11-049270-5

**Library of Congress Control Number: 2020949159**

**Bibliographic information published by the Deutsche Nationalbibliothek**
The Deutsche Nationalbibliothek lists this publication in the Deutsche Nationalbibliografie;
detailed bibliographic data are available on the Internet at http://dnb.dnb.de.

# Foreword

Shanghai Jiao Tong University pioneered in setting up the Department of Materials Science and Engineering in China in 1988, which began with the Department of Metallurgy in 1958, and renamed the School of *Materials Science* and Engineering in 1997. In the background of discipline development, Materials Science textbook, written by Professor Gengxiang Hu, Professor Xun Cai, and Professor Yonghua Rong, was published by Shanghai Jiao Tong University Press in 2000; then the second and third editions were successively published in 2006 and 2010, respectively. This textbook has been used in more than 30 universities or colleges in China so far. In the textbook, a trial has been performed, that is, the knowledge of metals, ceramics, and polymers is not individually described in chapters, but is distributed in knowledge structure of materials science: atomic structure and interatomic bonding, the structure of solids, crystal defects, deformation and recrystallization, phase transformation and phase diagrams, and diffusion. Such a consideration is to let students better understand the difference among metals, ceramics, and polymers. This textbook focuses on the knowledge of structural materials based on mechanical properties and also refers to functional materials based on physical properties. Although it is devoted to materials science, the basic knowledge of the synthesis, processing, and treatment of various materials in practical applications is also introduced because both the theoretical background and engineering common sense are a prerequisite for any engineer to be successful.

As a part of China–Germany cultural exchange, in 2015, De Gruyter invited Shanghai Jiao Tong University Press to publish the *Materials Science* as English edition in Germany. For this reason, Professor Zhuguo Li built a team in the School of Materials Science and Engineering; its members were selected from professors and associate professors who have been teaching this course for many years. This English edition of *Materials Science* adds several sections, cancels some content, corrects some mistakes, and rearranges chapters based on the Chinese edition textbook according to the development of materials science and teaching experience.

I read and revised the whole manuscript of *Materials Science* (English edition), and I hope that this textbook is also welcomed by the European undergraduates.

Yonghua Rong
Professor
School of Materials Science and Engineering
Shanghai Jiao Tong University
December, 2018

https://doi.org/10.1515/9783110495379-202

# Contents

# About the authors

Professor Gengxiang Hu graduated from the Mechanical Engineering Department of Shanghai Jiao Tong University (SJTU) in 1952 and worked as assistant professor at the Mechanical Engineering Department. From 1956 to 1979, he was a lecturer at the Department of Metallurgy at SJTU, and from 1979 to 1980, he was an associate professor in Materials Science Department at SJTU, and deputy director of the Institute of Superalloy. From 1980 to 1982, he was a visiting scholar at the University of California, Berkeley, and Lawrence National Laboratory for further study on "embrittlement of iron-based superalloys." In 1985, he was promoted to professor at the Department of Materials Science.

Professor Hu coauthored the textbook *Metal Science*, which has been widely praised by teachers and students since it was published in 1980. It is used as the undergraduate textbook in more than 50 universities in China. In 1987, it won the first prize as an excellent textbook of China National Machinery Commission, and in 1988, it won the National Excellent Textbook award of China National Higher Education Commission. He also received a number of awards for excellence in teaching at SJTU.

In terms of scientific research, Professor Hu is mainly engaged in the research of "superalloys" and "superplasticity" of metal materials. He has published more than 100 scientific research papers in domestic and foreign journals. In order to commend Professor Hu for his outstanding contribution to the development of higher education in China, he is entitled to a special allowance from the State Council.

Professor Xun Cai Born on 29 November 1943 in Zhejiang, China, he graduated from the Department of Materials Science and Engineering at the Harbin Institute of Technology in 1965 and received PhD in physics at the Technical University of Vienna in 1988. After college, he was employed at first as an engineer and then as a senior engineer at Shanghai Electric Micromotors Research Institute. In 1988, he was appointed as an associate professor and then promoted to university professor at the Department of Materials Science and Engineering of Shanghai Jiao Tong University. In 1992, he received the title of "Outstanding Scientist Who Made Notable Contributions to China" from the Chinese government.

Professor Cai's research concentrated on the areas of electric contact alloys, thin-film materials, and surface modification technologies. He has published 10 books and more than 300 research papers on materials science

Professor Yonghua Rong graduated from the University of Science and Technology, Beijing, in 1976, majoring in metal physics, then become an assistant professor at Shanghai Jiao Tong University (SJTU). From 1994 to 1995, he, as a senior visiting scholar, studied 718 superalloys using transmission electron microscope at Lehigh University, USA. In 1999, he was promoted to professor in the School of Materials Science and Engineering at SJTU.

He wrote the textbook *Introduction to Analytical Electron Microscopy* (Chinese version) for postgraduates and the monograph *Characterization of Microstructures by Analytical Electron Microscopy (AEM)* (English version). He also presided and wrote *Microstructural Characterization of Materials* for undergraduates.

https://doi.org/10.1515/9783110495379-203

Professor Rong was a director of "Phase Transformation Theories and Their Applications" group from 1998 to 2010. He is interested in phase transformation in nanomaterials and shape-memory alloys, especially advanced high strength steels, which reflect in his National Natural Science Foundation projects and one key project as well as in his monograph *Advanced High Strength Steels and Their Process Development.* He published more than 200 scientific research papers in domestic and foreign journals. He won the second prize of Natural Science Award of Chinese Universities in 2000, the second prize of National Teaching Achievement Award in 2001, the first prize of Shanghai Teaching Achievement Award in 2000, and the second prize of Shanghai Science and Technology Progress Award in 2004. He also received six awards for excellence in teaching at SJTU.

# About the translators

Professor Zhuguo LI
School of Materials Science and Engineering
Shanghai Jiao Tong University

**Professor Zhuguo LI** has obtained his bachelor degree in welding and automation in 1994 from Shanghai Jiao Tong University, master's degree in welding engineering in 1997 from the same university, and PhD in materials processing science from Osaka University in 2004. Since then, he joined Shanghai Jiao Tong University. He became a full-time professor since 2010 and a distinguished professor since 2020. Now he is the vice dean of School of Materials Science and Engineering, Shanghai Jiao Tong University, and the director of Shanghai Key Laboratory of Materials Laser Processing and Modification. He also serves as the deputy secretary general of China Welding Society. His research interests include laser welding, laser cladding, and laser additive manufacturing, and he is the coauthor of more than 150 SCI papers published on peer-review journals. He won the first prize of Science and Technology Award from CMES in 2016, and the first prize of Science and Technology Progress Award from Shanghai Municipality in 2015 and 2019.

**Professor Da Shu** is currently the deputy director of Shanghai Key Laboratory of Advanced High-Temperature Materials and Precision Forming, School of Materials Science and Engineering, Shanghai Jiao Tong University (SJTU). He received his BSc (1994) at Nanjing University of Aeronautics and Astronautics and his MSc (1997) and PhD (2001) at SJTU. His subsequent career has been a postdoc (2001), a lecturer (2003), an associate professor (2007), and a professor (2013) at SJTU. He was an academic visitor at the Department of Materials, Oxford University, from 2009 to 2010. His research focuses on processing of advanced metallic materials and control of solidified structure. He has published more than 100 peer-reviewed articles and patented 48 inventions (including one US patent). He has also won numerous awards, including (1) the second prize of National Technology Invention Awards in 2018 and 2006, (2) Rising Star of Shanghai Municipal Science and Technology Commission in 2007, (3) the second prize of Science and Technology Awards of China Non-Ferrous Metals Industry Association in 2008, and (4) the New-Century Talents of Ministry of Education in 2013.

**Zhenghong Guo** is an associate professor at the School of Materials Science and Engineering, Shanghai Jiao Tong University, where he has been a faculty member since 2001. From 2002, he has been taking charge of several courses for students, mainly national excellent course "The Fundamentals of Materials Science" for undergraduate students and school-level honor course "Principle of Phase Transformation" for graduate students.

Guo earned his doctoral degree in materials science at Shanghai Jiao Tong University. His research interests lie in the area of microstructural control by solid-state transformation, covering theoretical exploration to novel steel development.

https://doi.org/10.1515/9783110495379-204

**Xiaodong Wang** is an associate professor at the School of Materials Science and Engineering, Shanghai Jiao Tong University. For a long time, he taught the National Excellent Course "Fundamentals of Materials Science" for undergraduates and served as the responsible professor of the core course "Microscopy and Spectroscopic Characterization" for postgraduates. He studied modestly and was good at thinking. He actively explored the curriculum teaching reform and put it into practice, and finally formed his own unique teaching concept and teaching method, which is popular and well received by students. He won the second prize of individual award of "Teaching and Education Award," the fifth "Excellent Teaching Award" of Shanghai Jiao Tong University, and "the most popular teacher" award of School of Materials Science and Engineering. He cares for students and encourages them to be honest and innovative. He earnestly fulfills the primary duty of teaching and educating.

**Guo He** is currently a professor of materials science and engineering at Shanghai Jiao Tong University (Shanghai, China). He received his PhD in materials science from Northwestern Polytechnical University (Xian, China) in 1994, MS in materials science from Xi'an Jiao Tong University (Xian, China) in 1991, and BS in materials science and engineering from Hefei University of Technology (Hefei, China) in 1983. From 1994, he served as a lecturer and an associate professor successively at the University of Science and Technology Beijing (Beijing, China). In 2000, he moved to City University of Hong Kong (Hong Kong, China) as a visiting scientist at the Department of Physics and Materials Science. Later, he worked at Leibniz Institute for Solid State and Materials Research Dresden (Dresden, Germany) as an Alexander-von-Humboldt research fellow from 2001 to 2003, then, he moved to National Institute for Materials Science (Tsukuba, Japan) as a visiting scientist. In 2005, he joined the Faculty of Shanghai Jiao Tong University as a professor. His research interests focus on porous titanium, nanostructured titanium, and bioactive coating or surface modification of titanium for orthopedic applications. He has published hundreds of peer-reviewed journal papers and more than a dozen patents. His papers have been cited thousands of times. In recent years, he has devoted himself to the development of bioceramic materials and the application of the porous titanium technology that is being developed in his lab.

# Chapter 6
# Change from liquid or vapor phase to solid phase for one-component system

A system consisting of an element or a stable compound as a component is called one-component system. For one-component system, the constituted phase will change from a structure (phase) to another structure (phase) with variation in temperature or/and pressure, in which the conventional definition of a phase in the alloy is *the so-called phase refers to the homogeneous body with the uniform chemical composition, the identical crystal structure and properties and separated from each other by the interface.* The change between phases is called phase transformation. The change from liquid to solid is called solidification; furthermore, if the solid is crystal, it is also called crystallization. The change between solid phases is called solid–solid (solid status) transformation. The change from vapor to solid is called vapor–solid transformation (deposition). These phase transformations can be directly and clearly described by an equilibrium phase diagram for a corresponding system. This chapter starts with the derivation of Gibbs phase rule based on phase equilibria and its application in one-component phase diagram as an example, then clarifies the conditions of solidification from thermodynamic viewpoint, and derivates the critical radius and energy barrier of homogeneous and heterogeneous nucleation, respectively; summarizes the growth mechanisms of crystals and the influences of internal and external factors on the crystal growth morphology. The chapter also includes the deposition process of thin films because the vapor deposition technology is widely applied in the preparation of various functional thin film materials; finally refers to "the characteristics of polymer crystallization" in view of the particularity of single component polymer (homopolymer).

## 6.1 Phase equilibria and thermodynamics of phase transformation

### 6.1.1 Phase rule

For phase equilibria, it means that the partial Gibbs free energy of each component distributed in different phases, that is, chemical potential, is equal. Therefore, if the symbol $\mu$ is denoted to chemical potential and a system is composed of $P$ phases from $\alpha, \beta, \ldots$ to $P$ constituted by $C$ components from $1, 2, \ldots$ to $C$, the conditions of phase equilibria are

https://doi.org/10.1515/9783110495379-006

$$\mu_1^\alpha = \mu_1^\beta = \cdots = \mu_1^P$$

$$\mu_2^\alpha = \mu_2^\beta = \cdots = \mu_2^P$$

.

$$\qquad\qquad\qquad\qquad\qquad\qquad\qquad\qquad\qquad\qquad\qquad (6.1)$$

.

.

$$\mu_C^\alpha = \mu_C^\beta = \cdots = \mu_C^P$$

It can be found from eq. (6.1) that there are $C(P–1)$ equations.

The state of every phase depends on three variables: temperature, pressure, and composition. For each phase with $C$ components, the independent composition variable number is $(C–1)$, according to the restraining condition that the sum of fractions of all components in every phase is equal to 1. Obviously, for $P$ phases, there are $P(C–1)$ composition variables. Therefore, the state of a system can be described by $P(C–1) + 2$ variables, in which "2" includes the variables pressure and temperature together. However, these variables are all not independent because $C(P–1)$ equations from eq. (6.1) link composition variables, in other words, in $P(C–1) + 2$ variables the number of independent ones that are sufficient to determine phase equilibria, that is, the number of freedom degrees, $f$, is

$$f = [P(C-1) + 2] - C(P-1) = C - P + 2 \qquad\qquad (6.2)$$

This is phase rule, which was proposed by American physicist J. Willard Gibbs in the nineteenth century. Gibbs phase rule describes a criterion for the number of phases that can coexist within a system at equilibrium, and thus it is an important tool for the analysis of phase diagram.

## 6.1.2 One-component phase diagrams

For one-component system, it is obvious that $f = 3–P$. Then in a single phase region two freedom degrees occur: pressure and temperature can be selected independently. At a transition from one phase to another, only one freedom degree can be varied independently: either temperature or pressure. An equilibrium (or coexist) of three phases corresponds with fixed values for both pressure and temperature: that is, *triple point*. These conclusions can be clearly interpreted by $H_2O$ transformation for example, as shown in Fig. 6.1(a).

It may be noted from Fig. 6.1(a) that there are three regions for different phases: solid, liquid, and vapor. Each phase will exist under equilibrium conditions over the temperature–pressure ranges of its corresponding area $(f = 2)$. These regions are

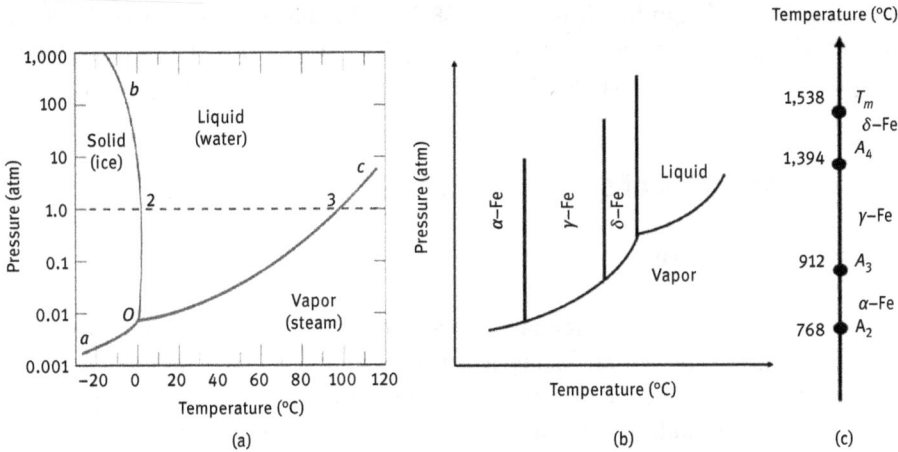

Fig. 6.1: Pressure–temperature phase diagram of $H_2O$ and pure iron, respectively.

separated by three curves, namely, phase boundaries ($f = 1$), shown on Fig. 6.1(a) (labeled as $aO$, $bO$, and $cO$, respectively). At any point on either curve, the two phases are in coexistence (or equilibrium) with one another. That is, along curve $aO$ means equilibrium between solid and vapor phases, likewise curve $bO$ for solid–liquid and curve $cO$ for liquid–vapor. Also, upon crossing a boundary, transformation from one phase to another occurs. For example, at one atmospheric pressure, during heating the transformation from solid to liquid phase occurs (i.e., melting occurs) at point 2 on Fig. 6.1(a) (i.e., the intersection of the dashed horizontal line with the solid–liquid phase boundary); this point is called the melting point at this pressure ($T = 0$ °C). Similarly, at point 3, the intersection with the liquid–vapor boundary represents the boiling reaction from liquid to vapor or vaporization (boiling point: $T = 100$ °C). Of course, upon cooling, the reverse transformation from liquid to solid (solidification) occurs at point 2 and from vapor to liquid phase (condensation) occurs at point 3. And, finally, solid (ice) sublimes or vaporizes upon crossing the curve $aO$. At triple point ($f = 0$), it is the equilibrium of three phases with fixed pressure ($6.04 \times 10^{-3}$ atm) and temperature (273.16 K).

In the one-component system, apart from the transformation between either two of gas, liquid, and solid states, some substances may also undergo isomeric transformation in the solid state. For example, Fig. 6.1(b) is a pure iron phase diagram, where $\delta$-Fe and $\alpha$-Fe are body-centered cubic structures with slightly different lattice constants and $\gamma$-Fe is face-centered cubic. There are two curves to separate them in the diagram. For metals, the temperature range of interest is generally below the boiling point, and the external pressure of interest is usually a standard atmospheric pressure. Therefore, this diagram can be simplified by omitting pressure axis, as shown in Fig. 6.1(c). In this figure, $T_m$(1,538 °C) is the melting point; $A_4$(1,394 °C) is the transformation temperature of $\delta$-Fe and $\gamma$-Fe; $A_3$(912 °C) is the transformation temperature of $\gamma$-Fe and $\alpha$-Fe; $A_2$(768 °C) is the magnetic transition temperature.

## 6.2 Solidification of single component

### 6.2.1 Structure of melt

Solidification is a phase transformation in which a melt (liquid) structure turns into a solid structure. Structural changes between solid and melt may be revealed clearly based on their radial distribution functions (RDFs) obtained by X-ray diffraction method. RDFs indicate quantitatively the feature of the average environment of atoms or molecules in a phase at local scales. It has been revealed by RDFs that the average atomic radii in liquid is slightly higher than that in solid. In addition, it is interesting to know that in many liquid phases, especially those from close-packed families, the average coordination number $Z_{melt} \approx 11$. This average coordination number for metallic melts is typically lower, by one, than the corresponding coordination number in the crystalline solids, for example, FCC and HCP. Table 6.1 lists the coordination numbers $(Z)$ and atomic radii $(r_{atom})$ in metallic crystals and their melts. On the other hand, when solids melt from nonclose-packed crystal structures, such as BCT–Sn, they usually form melts with a higher $Z_{melt}$ value than $Z_{cry}$ value in the crystalline state. The mean can either increase or decrease in bulk mass density during melting, depending on the change in coordination numbers. Besides, the most important structural information of melt is characterized by the presence of well-developed short-range order instead of long-range order in crystal, thereby minimizing the energy of melt consisting of small atom clusters. These clusters are not stable in size and always evolve at every moment by collapsing and restructuring. This phenomenon is called as structural fluctuation and is different from that in crystal with stable long-range order structure.

**Table 6.1:** Coordination numbers and ratio of atomic radii for crystals and melts.

| Metal | Lattice | $Z_{melt}$ | $Z_{cry}$ | $r_{atom}$ (Å) |
|-------|---------|------------|-----------|----------------|
| Al | FCC | 10–11 | 12 | (1.49/1.43) |
| Au | FCC | 11 | 12 | (1.43/1.43) |
| Pb | FCC | 12 | 12 | (1.86/1.75) |
| Ar | FCC | 10–11 | 12 | (1.98/1.92) |
| Zn | HCP | 11 | 12 | (1.47/1.40) |
| Sn | BCT | 11 | 8 | (1.60/1.53) |

## 6.2.2 Free energy change during solidification

In the most general sense, crystallization occurs at the pressure of 1 atm. In terms of the second law of thermodynamics, the spontaneous process always decreases free energy of the assembly under the isothermal and isobaric condition. Taking free energy $G$ as

$$G = H - TS \tag{6.3}$$

where $H$ is enthalpy, $T$ is temperature, and $S$ is entropy. Using Maxwell's equations, we can obtain

$$dG = Vdp - SdT \tag{6.4}$$

and therefore:

$$\frac{dG}{dT} = -S \tag{6.5}$$

under the isobaric condition ($dp = 0$). Since $S > 0$, the free energy shows the negative temperature dependence.

Figure 6.2 illustrates the free energy variation of melt and solid, respectively, as a function of temperature. During melting, the configuration entropy increases because the randomness of atomic movements increases by change of long-range order structure in crystal into short-range order structure in melt. At the same time, vibration entropy also increases slightly due to the increase in atomic vibration amplitude. As a result, $S_L > S_S$, that is, the slope of tangent to liquid $G \sim T$ curve ($S_L$) is bigger than that to the solid one ($S_S$). The intersection point of $G \sim T$ curves defines

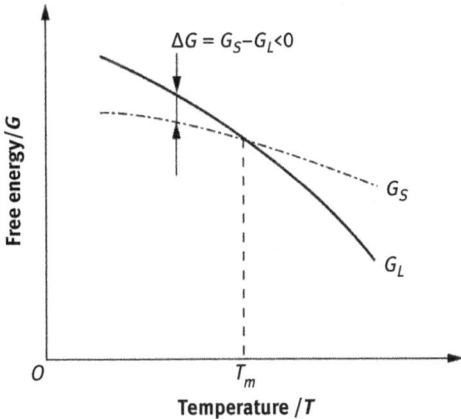

**Fig. 6.2:** Schematic illustration of the free energy variation as a function of temperature.

the coexistent temperature of liquid and solid with same free energy. This temperature is called theoretical temperature for solidification, or melting point $T_m$. However, at the melting point when the liquid and solid coexist in an equilibrium state, neither solidification nor melting process can be accomplished. In practice, the temperature of a liquid must be less than $T_m$ for crystallization to occur. Similarly, the temperature of a solid must be greater than $T_m$ for melting to occur. It means that additional driving force (free energy difference between liquid and solid) is required to trigger solidification or melting process.

The change in free energy corresponding to phase transformation at a given temperature $T$ is

$$\Delta G = \Delta H - T\Delta S \tag{6.6}$$

Let $\Delta G_v$ to be the change in free energy per unit volume corresponding to solidification, then:

$$\Delta G_v = G_S - G_L \tag{6.7}$$

where $G_S$ and $G_L$ are the free energies of solid and liquid per unit volume at $T$, respectively. From eq. (6.3), the following equations can be derived under the isobaric condition ($dp = 0$):

$$\Delta G_v = (H_S - H_L) - T(S_S - S_L) \tag{6.8}$$

and

$$\Delta H_P = H_S - H_L = -L_m \tag{6.9}$$

$$\Delta S_m = S_S - S_L = -L_m/T_m \tag{6.10}$$

where $L_m$, taking a positive value due to energy absorption from environment, is the latent heat of melting, and thus $L_m$ represents latent heat release in solidification and $\Delta S_m$ is entropy of freezing.

Substituting eqs. (6.9) and (6.10) into (6.8) gives

$$\Delta G_v = \frac{-L_m \Delta T}{T_m} \tag{6.11}$$

where $\Delta T = T_m - T$ is the difference between melting point and the given temperature $T$. The requirement of $\Delta G_v < 0$ indicates $\Delta T > 0$ in eq. (6.11), that is, $T_m > T$. Here, $\Delta T$ is called supercooling, meaning that somehow the liquid is cooled below the melting point by an amount $\Delta T$, a specific temperature range in which liquid phase can be maintained below $T_m$. The above thermodynamic conditions indicate that the actual solidification temperature should be lower than the melting point $T_m$, that is, supercooling is required for crystallization.

### 6.2.3 Nucleation

Crystallization is the (natural or artificial) process by which a solid forms, where the atoms or molecules are regularly arranged into a structure. Crystallization process consists of two major events according to time sequence. The first is nucleation, that is, the appearance of a small crystal from the melt. The second is called as crystal growth, corresponding to the size increase of particles by consuming residual liquid continuously and leading to a crystal state finally.

The nucleation modes can be divided into two types:

(1) Homogeneous nucleation: Nucleation without preferential nucleation sites is homogeneous nucleation. It occurs spontaneously and randomly in the supercooled melt, away from the surfaces and impurities in the assembly.

(2) Heterogeneous nucleation: Nucleation with preferential nucleation sites, such as at surfaces on vessel walls or impurities.

Although heterogeneous nucleation is much more common than homogeneous nucleation, the basic principle of nucleation is originated from the understanding of homogeneous nucleation and, therefore, the mechanism of homogeneous nucleation is discussed first.

### Homogeneous nucleation

#### a. Consideration of the critical radius and energy barrier

Since structural fluctuation occurs at any supercooled melt, there exists a statistical distribution of atom clusters or embryos with different sizes having the solid feature within the melt. The appearance probability of an embryo with a given size increases as the temperature decreases. Although the free energy of solid phase is lower than that of liquid one below $T_m$, the embryo may not be stable because a new barrier is created from the interfacial energy between embryo and melt. Nucleation occurs only when the supercooling reaches to a threshold that there are sufficient embryos with their radius larger than the critical value, which result in the decrease of free energy of system with increasing the sizes of embryos, and these embryos are named nuclei. It is worthy to point out that the appearance of an embryo will generate strain energy due to the density difference between embryo and melt, but the strain energy can fully release in parent phase–liquid, and thus the strain energy opposing transformation can be ignored in solidification, while this factor cannot be ignored in solid–solid transformation.

The change in total free energy of system, $\Delta G$, to form a spherical embryo with radius, $r$, from liquid of a pure component involves the variation of the volume (bulk) free energy (driving force, negative value) and the interfacial energy (resistance, positive value), and it is given by

$$\Delta G = \frac{4}{3}\pi r^3 \Delta G_v + 4\pi r^2 \sigma \qquad (6.12)$$

where $\sigma$ is the interfacial energy (surface tension) per unit area or specific interfacial energy between the embryo and its surroundings.

As $\Delta G_v$ and $\sigma$ belong to intensive properties, that is, for a given temperature they are constant, $\Delta G$ is only a function of $r$. Figure 6.3 draws a comprehensive picture of the free energy variation of an spherical embryo as a function of its radius. The critical radius, $r^*$, can be obtained (eq. (6.13)) when $\Delta G$ has a peak value given by the condition, $d(\Delta G)/dr = 0$:

$$r^* = -\frac{2\sigma}{\Delta G_v} \quad \text{or} \quad r^* = \frac{2\sigma T_m}{L_m \Delta T} \qquad (6.13)$$

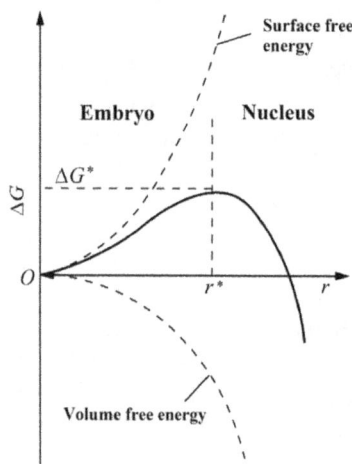

**Fig. 6.3:** Variation of $\Delta G$ as a function of $r$ during the formation of a nucleus.

When the value of $r$ is less than $r^*$, the square dependence of interfacial energy term dominates over volume free energy term and total free energy, $\Delta G$, increases monotonically with $r$. When the value of $r$ is larger than $r^*$, the cubic dependence of volume free energy term dominates over surface free energy term and $\Delta G$ passes through the peak value at the critical radius, $r^*$. It is thought that a size distribution of small clusters (embryos) of atoms exists in liquid at any time and these clusters continually gain and lose atoms. With the help of a thermal fluctuation, an embryo of these clusters continues to gain more atoms than it loses, and the embryo becomes a nucleus with radius larger than $r^*$; growth occurs spontaneously due to the decrease in total free energy. The free energy change to form a critical sized nucleus, that is, the critical activation energy $(\Delta G^*)$ for an embryo with radius $r^*$ can be found by substituting eqs. (6.13) into (6.12):

$$\Delta G^* = \frac{16\pi\sigma^3}{3(\Delta G_v)^2} \quad \text{or} \quad \Delta G^* = \frac{16\pi\sigma^3 T_m^2}{3(L_m\Delta T)^2} \tag{6.14}$$

Obviously, the value of $r^*$ is mainly affected by $\Delta T$ because other terms $\sigma$ and $L_m$ are not sensitive to the temperature change. The larger supercooling corresponds to both the smaller critical radius $(r^*)$ and the critical activation energy $(\Delta G^*)$, energy barrier (nucleation work), accompanying with the higher nucleation rate. At the melting point $T_m$, $\Delta T = 0$, hence $r^* = \infty$ and $\Delta G^* = \infty$ and no embryo is qualified as a nucleus and solidification is inhibited completely.

Since the interfacial area of critical nucleus is

$$A^* = 4\pi(r^*)^2 = \frac{16\pi\sigma^2}{\Delta G_v^2} \tag{6.15}$$

the following equation can be derived by substituting it into eq. (6.14):

$$\Delta G^* = \frac{1}{3}A^*\sigma \tag{6.16}$$

Thus, it is obvious that the total free energy is still positive when a critical nucleus is formed, which comes from about 1/3 of interfacial energy, in other words, the decrease of volume free energy due to nucleation can only compensate for 2/3 interfacial energy required for creating critical nucleus, and the remnant needs to be supplemented by thermal (energy) fluctuation in the melt. Energy fluctuation, usually associated with the structural fluctuation, is a phenomenon that the actual energy of every atom in the melt can deviate the mean value by fluctuation instantaneously. In general, the conditions of solidification for one-component system are supercooling, structural fluctuation, and energy fluctuation.

In the multicomponent assembly, in addition to energetic and structural fluctuations, the third one, constitutional fluctuation is also required for nucleation. For a binary alloy, $\Delta G_v$, in eq. (6.9), depends not only on the temperature but also on the composition of the liquid $(c_L)$ and of the solid nuclei $(c_s)$. Thus, for a given liquid composition, critical values of nucleus composition as well as size are required to determine $\Delta G^*$. For alloy nucleation, the appropriate expression for $\Delta G_v$ is obtained by dividing the expression given in the following eq. (6.17) by the molar volume of the solid:

$$\Delta G_V = \left[(\mu_S^A - \mu_L^A)(1 - c_S) - (\mu_S^B - \mu_L^B)c_S\right]/V_{AV} \tag{6.17}$$

where $\mu_S^A$ and $\mu_S^B$ are the chemical potentials for species $A$ and $B$ in the solid, $\mu_L^A$ and $\mu_L^B$ are the chemical potentials in the liquid, and $V_{AV}$ is the volume per unit mole atoms in solid.

It is apparent from eq. (6.17) that $\Delta G$ can be maximized for a composition of the solid when $(\mu_S^A - \mu_L^A) = (\mu_S^B - \mu_L^B)$, that is, by a parallel tangent construction as shown in Fig. 6.3: $\Delta\mu^A = \Delta\mu^B$. This maximum driving force condition has been proposed to find the favored nucleus composition for a given temperature and liquid composition (Fig. 6.4).

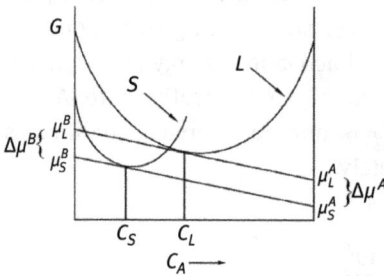

Fig. 6.4: The composition of a nucleus, $c_s$, that maximizes the free energy change at a given supercooling, is given by the parallel tangent construction for liquid with composition $c_L$.

The solidification mechanism of multicomponent assembly will be discussed in detail in the next chapter.

## b. Nucleation rate

The homogeneous nucleation rate, $N$, is defined as the nuclei number formed in per unit volume liquid per unit time. To evaluate the nucleation rate, it is necessary to find expressions for the nuclei number that have size and the attaching rate of atoms or molecules from embryos to the critical nuclei. By considering the entropy of mixing between a small number, $N_n^*$, of critical nuclei, each of which contains $n^*$ atoms, and in $N_L$ atoms of the liquid, a mathematical relationship for the equilibrium number of critical nuclei with $n^*$ atoms can be derived as

$$\frac{N_n^*}{N_L} = K_1 \exp\left(-\frac{\Delta G^*}{kT}\right) \tag{6.18}$$

where $k$ is Boltzmann's constant and $K_1$ is a constant. If one assumes that the critical embryo can transform to the stable nucleus by attaching at least one atom diffused from the neighboring liquid, the nucleation rate ($N$) can be obtained by multiplying the diffusion factor into right terms in eq. (6.18):

$$N = K \exp\left(\frac{-\Delta G^*}{kT}\right) \exp\left(\frac{-Q}{kT}\right) \tag{6.19}$$

where $K$ is the pre-exponential term, $Q$ is the activation energy for atomic diffusion crossing liquid–solid interface.

As shown in Fig. 6.5, $N$ goes through a maximum at a specific supercooling, which means nucleation rate is dominated by the factor of nucleation work mainly

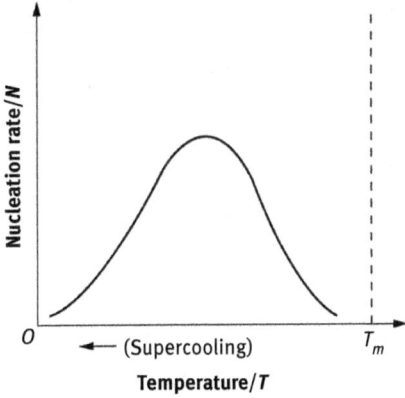

Fig. 6.5: Relationship between nucleation rate and temperature.

$(\Delta G^*)$ under the low supercooling, the decrease of temperature in this range results in the decrease of critical radius for nucleation, favoring the increase of nucleation rate rapidly until to maximum. However, a continued decrease of temperature causes the decrease of nucleation rate from maximum. This is because of the significant decrease in diffusivity, and this effect counteracts the decrease in critical radius. That is, the nucleation rate is dominated by the factor of atomic diffusion $(Q)$ at this temperature range and, in general, nucleation rate decreases from maximum with respect to the further decrease in temperature. According to eq. (6.19), $N$ increases very rapidly at a critical supercooling in the range of $0.2T_m$ to $0.4T_m$, when the temperature is decreased from $T_m$. Changes in the term $K$ in eq. (6.19) by orders of magnitude do not significantly affect the calculated supercooling for sensible nucleation rates. This rapid increase in nucleation rate with supercooling effectively defines a nucleation temperature in theory, and mostly falls into the range of $\Delta T^*/T_m \approx 0.15$–$0.25$, where $\Delta T^* = T_m - T^*$, from the experimental results (Table 6.2).

Table 6.2: Measured effective nucleation temperature.

|      | $T_m$ (K) | $T^*$ (K) | $T^*/T_m$ |
|------|-----------|-----------|-----------|
| Hg   | 234.3     | 176.3     | 0.247     |
| Sn   | 505.7     | 400.7     | 0.208     |
| Pb   | 600.7     | 520.7     | 0.133     |
| Al   | 931.7     | 801.7     | 0.140     |
| Ge   | 1,231.7   | 1,004.7   | 0.184     |
| Ag   | 1,233.7   | 1,006.7   | 0.184     |
| Au   | 1,336     | 1,106     | 0.172     |
| Cu   | 1,356     | 1,120     | 0.174     |
| Fe   | 1,803     | 1,508     | 0.164     |
| Pt   | 2,043     | 1,673     | 0.181     |

**Table 6.2** (continued)

|  | $T_m$ (K) | $T^*$ (K) | $T^*/T_m$ |
|---|---|---|---|
| $BF_3$ | 144.5 | 126.7 | 0.123 |
| $SO_2$ | 197.6 | 164.6 | 0.167 |
| $CCl_4$ | 250.2 | 200.2 ± 2 | 0.202 |
| $H_2O$ | 273.2 | 273.7 ± 1 | 0.148 |
| $C_5H_5$ | 278.4 | 208.2 ± 2 | 0.252 |
| $C_{10}H_8$ | 353.1 | 258.7 ± 1 | 0.267 |
| LiF | 1,121 | 889 | 0.21 |
| NaF | 1,265 | 984 | 0.22 |
| NaCl | 1,074 | 905 | 0.16 |
| KCl | 1,045 | 874 | 0.16 |
| KBr | 1,013 | 845 | 0.17 |
| KI | 958 | 799 | 0.15 |
| RbCl | 988 | 832 | 0.16 |
| CsCl | 918 | 766 | 0.17 |

In a low-viscosity liquid, the nucleation rate increases abruptly when the temperature drops to a certain value $T^*$. This temperature $T^*$ is called as the effective temperature for homogeneous nucleation. With further increase in supercooling, the nucleation rate continues to increase, leading to the end of crystallization before reaching to the maximum nucleation rate as shown in Fig. 6.6.

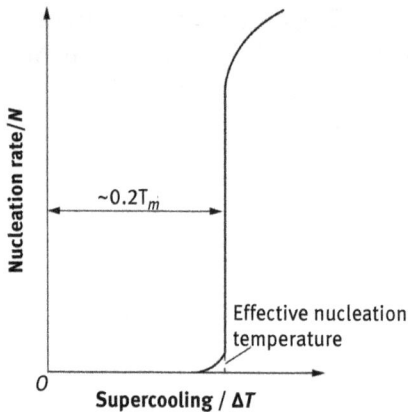

Fig. 6.6: Relationship between nucleation rate and supercooling in metals.

An example to calculate atomic numbers in a critical nucleus is mentioned here. Taking the $T_m = 1{,}356$ K, $\Delta T = 256$ K, $\sigma = 177 \times 10^{-3}$J/m$^2$ (Table 6.3), $L_m = 1{,}628 \times 10^6$J/m$^3$

**Table 6.3:** The maximum supercooling and surface tension in metal liquids.

| Metal | Maximum supercooling (K) | Surface tension $\sigma$ ($\times 10^{-3} \text{J} \cdot \text{m}^{-2}$) | Metal | Maximum supercooling (K) | Surface tension $\sigma$ ($\times 10^{-3} \text{J} \cdot \text{m}^{-2}$) |
|-------|--------------------------|------------------------|-------|--------------------------|------------------------|
| Al | 195 | 121 | Au | 230 | 132 |
| Mn | 308 | 206 | Ga | 76 | 56 |
| Fe | 295 | 204 | Ge | 227 | 181 |
| Co | 330 | 234 | Sn | 118 | 59 |
| Ni | 319 | 255 | Sb | 135 | 101 |
| Cu | 236 | 177 | Hg | 77 | 28 |
| Pd | 332 | 209 | Bi | 90 | 54 |
| Ag | 227 | 126 | Pb | 80 | 33 |
| Pt | 370 | 240 | | | |

and $a_0 = 3.615 \times 10^{-10}$m (lattice constant) of copper, for example, the critical radius is obtained by eq. (6.13):

$$r^* = \frac{2\sigma T_m}{L_m \Delta T} = \frac{2 \times 177 \times 10^{-3} \times 1,356}{1,628 \times 10^6 \times 236} = 1.249 \times 10^{-9}\text{m}.$$

So the critical nucleus volume is

$$V_c = \frac{4}{3}\pi r^{*3} = 8.157 \times 10^{-27}\text{m}^3$$

Comparing to the volume of unit cell:

$$V_L = (a_0)^3 = 4.724 \times 10^{-29}\text{m}^3$$

the cell number within a critical nucleus is

$$n = \frac{V_c}{V_L} \approx 173 \text{ cells}$$

That is, a critical nucleus contains 692 copper atoms due to fcc structure with four atoms within a cell. This result demonstrates that although experimental errors may affect the calculation accuracy, the probability that hundreds of atoms spontaneously aggregate together to form a nucleus is very low. So the homogeneous nucleation is difficult to occur in practice.

### Heterogeneous nucleation

As mentioned earlier, homogeneous nucleation is only possible for large supercooling (in the order of $0.2T_m$) under special experimental conditions. However, the small contamination particles exist in the melt and the oxides float on the melt surface inevitably. Including the vessel walls, these mediums may catalyze nucle-ation at a much smaller supercooling and with fewer atoms required to form the critical nucleus. This is named as heterogeneous nucleation.

In Fig. 6.7, heterogeneous nucleation is assumed to occur at a flat wall of vessel and with isotropic surface tensions. For example, the embryo is a spherical cap that has an angle $\theta$ with the substrate given by

$$\sigma_{Lw} = \sigma_{\alpha L}\cos\theta + \sigma_{\alpha w} \tag{6.20}$$

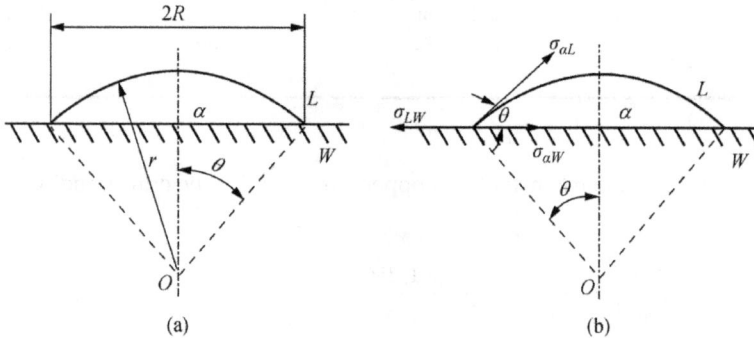

(a)           (b)

**Fig. 6.7:** Schematic illustration of heterogeneous nucleation ($\alpha$: nucleus; $L$: liquid phase).

where $\sigma_{Lw}$, $\sigma_{\alpha L}$, and $\sigma_{\alpha w}$ are the surface tension between wall–liquid, solid–liquid and solid–wall interface, respectively.

The change in the surface energy $\Delta G_s$ may be written as

$$\Delta G_s = A_{\alpha L}\sigma_{\alpha L} + A_{\alpha w}\sigma_{\alpha w} - A_{\alpha w}\sigma_{Lw} \tag{6.21}$$

where $A_{\alpha L}$ and $A_{\alpha w}$ are the surface areas of solid–liquid and solid–wall interface, respectively.

Because

$$A_{\alpha w} = \pi R^2 = \pi r^2\sin^2\theta \text{ and } A_{\alpha L} = 2\pi r^2(1-\cos\theta) \tag{6.22}$$

so

$$\Delta G_s = A_{\alpha L}\sigma_{\alpha L} - \pi r^2\sin^2\theta\cos\theta\sigma_{\alpha L}$$

$$= (A_{\alpha L} - \pi r^2\sin^2\theta\cos\theta)\sigma_{\alpha L.} \tag{6.23}$$

In addition, the volume of spherical cap, $V_\alpha$, is

$$V_\alpha = \pi r^3\left(\frac{2-3\cos\theta+\cos^3\theta}{3}\right) \tag{6.24}$$

and the change in volume free energy corresponding to form a nucleus $\Delta G_t$ is

$$\Delta G_t = V_a \Delta G_v = \pi r^3 \left( \frac{2 - 3\cos\theta + \cos^3\theta}{3} \right) \Delta G_v \qquad (6.25)$$

Therefore, the change in total free energy of assembly, $\Delta G = \Delta G_t + \Delta G_s$, is

$$\Delta G = \left( \frac{4}{3}\pi r^3 \Delta G_v + 4\pi r^2 \sigma_{aL} \right) \left( \frac{2 - 3\cos\theta + \cos^3\theta}{4} \right)$$

$$= \left( \frac{4}{3}\pi r^3 \Delta G_v + 4\pi r^2 \sigma_{aL} \right) f(\theta) \qquad (6.26)$$

Comparing eqs. (6.26) and (6.12) reveals that the only difference between them is the factor of contact angle $f(\theta)$. This factor has no relationship with respect to nucleus radius. Therefore, the critical radius for heterogeneous nucleation:

$$r^* = -\frac{2\sigma_{aL}}{\Delta G_v} \qquad (6.27)$$

which is obtained by setting $\partial\Delta G/\partial r = 0$ as well.

This result indicates that at identical supercooling, $\Delta T$, the critical radius of the spherical cap is given by eq. (6.27) again, however, the atomic number in the spherical cap (critical nucleus) is smaller than that for homogeneous nucleation due to the catalytic effect. Hence, the critical activation energy (energy barrier) for nucleation $G^*_{het}$ is reduced by a factor $f(\theta)$ to

$$\Delta G^*_{het} = \Delta G^*_{hom} \left( \frac{2 - 3\cos\theta + \cos^3\theta}{4} \right)$$

$$= \Delta G^*_{hom} f(\theta) \qquad (6.28)$$

For a planar surface, the reduction in the critical activation energy compared to homogeneous nucleation relies on the contact angle between 0° and 180°. When $\theta = 180°$, the surface loses catalytical effect and the nucleus does not interact with the substrate surface, $f(\theta) = 1$ and the homogeneous nucleation result is obtained. When $\theta = 0°$, the substrate "wets" the nucleus, $f(\theta) = 0$, and $G^*_{het} = 0$. As a result, nucleation can begin immediately when the liquid temperature decreases to the freezing point ($T_m$). When $0 < \theta < 180°$, the substrate partially "wets" the nucleus, and this is the most common case in practice.

The main difference between homogeneous and heterogeneous nucleation is the value of thermodynamic barrier $G^*$. The very low barrier for heterogeneous nucleation results in the maximum nucleation rate at very low supercooling, such as ~$0.02T_m$ shown in Fig. 6.7. Another feature of heterogeneous nucleation is the rate that switches from minimum to maximum, then decreases again because the most available nucleation sites have been occupied (Fig. 6.8).

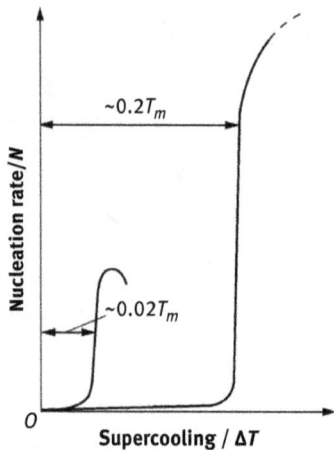

**Fig. 6.8:** Nucleation rate as the function of temperature for both homogeneous and heterogeneous nucleation. Supercooling for each is also shown.

Let us reconsider the example of copper again. The volume of a spherical cap is

$$V_{cap} = \frac{\pi h^2}{3}(3r - h) \tag{6.29}$$

where $h$ taken as $0.2\,r$ is the height of cap. In a critical nucleus, we can obtain $V_{cap} = 2.284 \times 10^{-28}\,\text{m}^3$ and $V_{cap}/V_L \approx 5$ (cells). Only about 20 atoms are required to form a critical nucleus. Obviously, this corresponds to very small structural fluctuation, and in turn to the very small supercooling.

## 6.2.4 Crystal growth

Crystal growth is the second stage of a crystallization process, namely, new atoms, ions, or polymer strings are arranged into the crystalline Bravais lattice continuously. The growth typically follows either homogeneous or heterogeneous nucleation, unless a "seed" crystal, is threw into the liquid in purpose to start the growth. Here, the crystal growth requires to move an interface between a crystal and surrounding liquid at a given velocity. The growth manner is related to the structure of liquid–solid interface.

### Structure of liquid–solid interface

Two distinct types of interfaces are observed: faceted interface (Figs. 6.9 and 6.11(a)) and nonfaceted interface (Figs. 6.10 and 6.11(b)). The faceted interface is thought to grow at the motion of small ledges. The atoms are added only at the ledges so that the interface advances only when a ledge passes along it. The nonfaceted interface is thought to advance by addition of atoms at all points of its surface. The nonfaceted

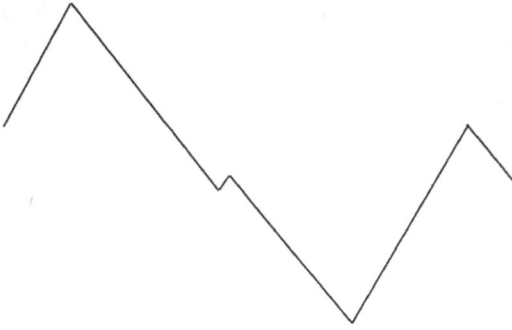

**Fig. 6.9:** Schematic diagram of the facet structure at grain boundary.

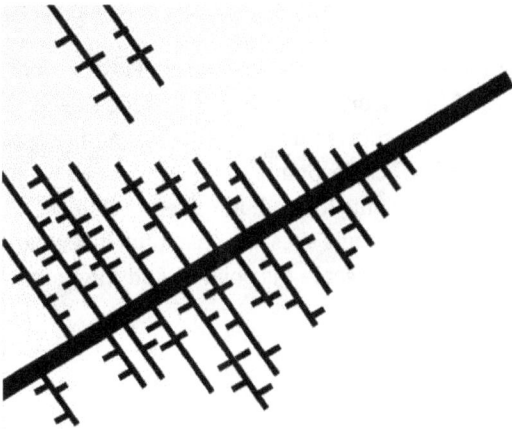

**Fig. 6.10:** Schematic diagram of the nonfaceted interface due to formation of dendritic structure.

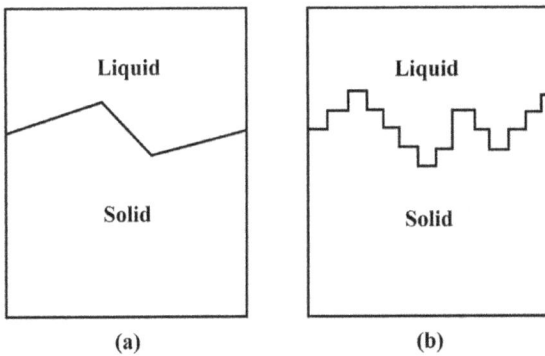

**Fig. 6.11:** Schematic illustration of liquid–solid interface (a) smooth interface; (b) rough interface.

interface is sometimes called a diffuse or rough interface. First, we describe how to differentiate two kinds of interfaces based on the work of K. A. Jackson.

Jackson proposed a sharp interface model and evaluates the conditions when a smooth interface becomes a rough one and then becomes a smooth one again by atoms in liquid randomly occupied on the smooth interface, and these atoms are called "solid atoms." Using a near-neighbor bond model, he derived a mathematical relationship between the change in free energy and the fraction, $x$, of the $N_T$ possible sites occupied by atoms as

$$\frac{\Delta G_s}{N_T k T_m} = \alpha x (1-x) + x \ln x - (1-x) \ln(1-x) \tag{6.30}$$

where

$$\alpha = \frac{\xi L_m}{k T_m} \tag{6.31}$$

Here, $\xi$, always less than 1 and being largest for close-packed planes, is a factor related to the interfacial crystallography. This theory has been successfully used to discriminate growth morphologies. Figure 6.12 shows plots of eq. (6.30) for different values of $\alpha$. When $\alpha < 2$, the valley value of $\Delta G_s$ corresponds to $x = 1/2$; that is, when half the sites are occupied by atoms. This represents a rough interface. Solidification then carries on by continuous growth since enough sites for easy attachment are available. It is worthy to note that in macroscopic scale the solid–liquid interface is

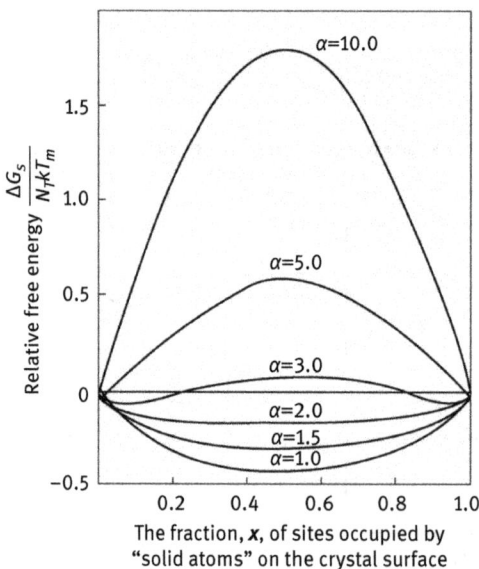

Fig. 6.12: Relationships between $\dfrac{\Delta G_s}{N_T k T_m}$ and $x$ with different $\alpha$.

nonfaceted but may exhibit flat plane approximately. When $\alpha > 2$, there are two valley values of the curve at both small and large values of $x$, indicating a smooth interface with nearly none (or all) of the sites occupied by atoms. This represents a smooth (faceted) interface. Solidification is then carried on by layer or lateral growth.

Experimental observations have found that: (1) the rough interface usually appears in the metals and some organic compounds with a small entropy change in melting because $\alpha \leq 2$; (2) while the faceted interface usually appears in some semi-metals such as bismuth, antimony, gallium, arsenic, in semiconductor materials such as germanium and in silicon and most inorganic compounds because $\alpha \geq 2$; (3) in polymers, the elementary structural unit for crystallization is molecular chain, which is different from above materials formed by the atomic migration. Therefore, this model is invalid for polymers.

Jackson's approach is limited to the equilibrium surface structure. Another approach to evaluate interface roughness focuses on effect of thermal vibrations on the surface tension of a step on an otherwise faceted interface. This corresponds to dynamic supercooling ($\Delta T_k$) in the assembly. The result shows that the step energy vanishes when $L_m/kT_m$ falls below a critical value of order unity. When the step energy closes to zero, there is hardly any or even no barrier to surface roughening. This analysis is consistent with the result as Jackson's approach quantitatively. The importance of the ratio $L_m/kT$ in eq. (6.31) was also emphasized by different statistical multilevel models of interface structure and Monte Carlo simulations. In a word, the roughness of the interface increases with the decrease in $L_m/kT$ in these models.

### Growth manner and growth velocity

The growth manner of crystal, including continuous growth, two-dimensional nucleation, and screw dislocation growth, is closely related to the structure of liquid–solid interface.

### a. Continuous growth

The motion of a rough interface is called continuous or normal growth because the interface immigrates itself in a continuous manner due to the enough sites for easy attachment of atoms. To establish the mathematical relationship between growth velocity and supercooling, the former is assumed as a product of function involving the driving force for crystallization and a kinetic factor involving the interface mobility:

$$v(T_i) = v_c(T_i)[1 - \exp(\Delta G/RT_i)] \tag{6.32}$$

where $T_i$ is the interface temperature, $v_c(T_i)$ is proportional to the diffusivity in the liquid, and $\Delta G$ is the free energy change of substance solidified (it is negative for solidification). The bracketed term in eq. (6.32) means a difference between the "forward flux" (liquid → solid) and the "backward flux." The kinetic factor, $v_c(T_i)$, corresponding to the hypothetical maximum growth velocity at infinite driving force, is the forward flux rate alone. Near equilibrium, a linear relation between velocity and supercooling can be obtained by Taylor expansion of the exponential term:

$$v(T) = v_c(T_m) \frac{L_m \Delta T_k}{RT_m^2} \qquad (6.33)$$

or briefly has the form:

$$v_g = \mu \Delta T_k \qquad (6.34)$$

where $\mu$ is kinetic coefficient, estimated as $\sim 10^{-2}$m/(s·K), so a relatively high growth velocity can be realized at a small supercooling. $\Delta T_k$ is called kinetic supercooling. The kinetic supercooling in nonfaceted growth is so small (0.01–0.05 °C) that to the present time there has been no accurate measurement of it. For inorganic compounds such as oxide and organic compounds with high viscosity, the growth velocity, however, will decrease after it reaches maximum with the increase in supercooling. Besides, growth velocity is also controlled by the efficiency of latent heat dissipation. When the latent heat released accompanying thickening of rough surface is very small, the growth velocity is very high.

### b. Two-dimensional nucleation-controlled growth

If the interface is smooth in atomic scale and without any defects, the growth rate is controlled by the heterogeneous nucleation on interface. That is, some atomic clusters form on the interface to create the necessary bulges (Fig. 6.13). Then the growth can be carried on by lateral spreading. The rate is assumed to have nearly same magnitude as the continuous growth manner described above. The growth law (assuming the case of cylindrical surface clusters) can be expressed as

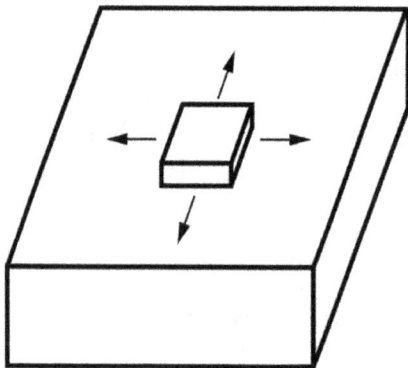

**Fig. 6.13:** Growth mechanism by two-dimensional nucleation.

$$v_g \sim \exp\left(\frac{-\pi\sigma_e^2 h T_m}{L_m k T \Delta T_k}\right) \qquad (6.35)$$

or briefly has the form:

$$v_g = \mu\exp\left(\frac{-b}{\Delta T_k}\right) \qquad (6.36)$$

where $\sigma_e$ is the surface tension per unit area at lateral interface and $h$ is the bulge height. According to eq. (6.36), the growth rate is close to zero at small supercooling and increases abruptly at some critical supercooling. If the nuclei number is extremely high, the growth is more proper to be described by the continuous mechanism.

### c. Growth by screw dislocation
If there emerges one or more screw dislocations at the solid–liquid interface, the sites for atomic attachment are naturally created at the lateral surface due to the formation of step with the height close to the magnitude of Burger vector (Fig. 6.14). Each step moves one plane at a time when it sweeps around the dislocation. Since the distance

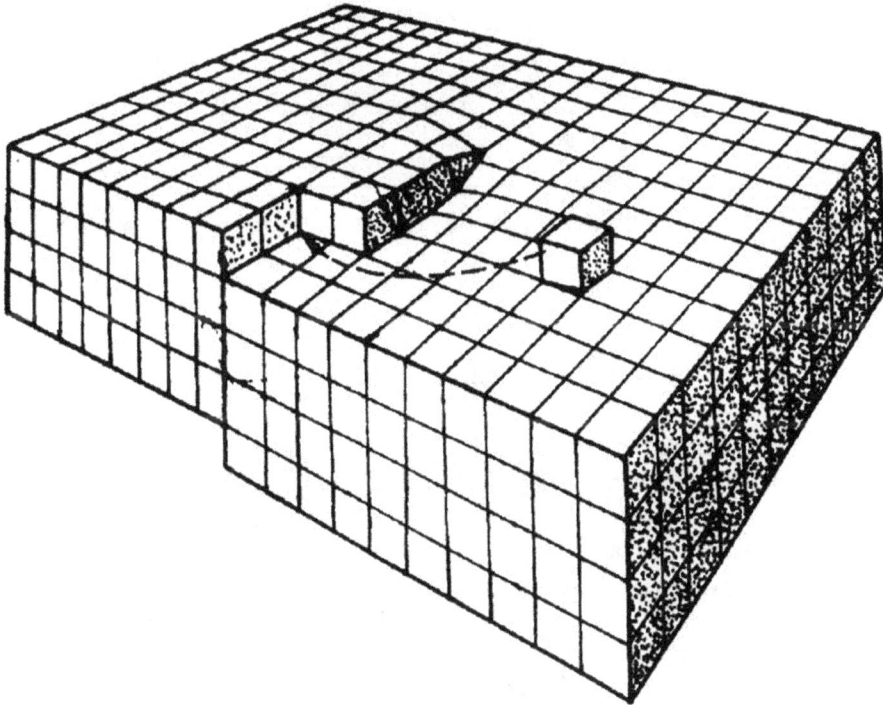

**Fig. 6.14:** Growth mechanism by screw dislocation.

between neighboring turns of the spiral is inversely proportional to $\Delta T_k$, the total length of step is directly proportional to $\Delta T_k$. Under the condition of small supercooling, the growth rate is

$$v_g \sim (\Delta T_k)^2 \tag{6.37}$$

because the growth rate per unit length of step is also proportional to $\Delta T_k$. Growth by screw dislocation has been observed in some nonmetal crystals. And by using single screw dislocation, a new technology has been developed to prepare crystal whisker in oxide, sulfide, halide, alkali, and some metals, which gives these materials very high yield strength and other excellent properties. Figure 6.15 schematically illustrates the laws for the abovementioned continuous, two-dimensional nucleation, and screw dislocation-assisted growth.

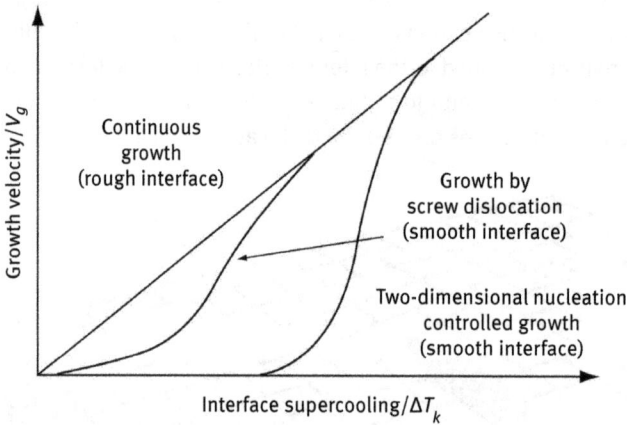

**Fig. 6.15:** Growth velocity versus interface supercooling according to the different interface kinetics.

### 6.2.5 Crystallization kinetics and solidified microstructures

#### Crystallization kinetics
Assuming that after a time $\tau$, a new nucleus (grain) forms in the melt and begins to grow. Frequently, this new nucleus will increase in size at a constant rate until it impinges upon a neighboring nucleus. The nucleation event occurs randomly and homogeneously over the entire remaining melt, the kinetic equation describing the fraction of crystallized materials as the function of holding time at a constant temperature can be derived. Since nucleus growth is isotropic, constant and uninterfered by previous nuclei, each of them will grow spherically with radius $R$

$$R = v_g(t - \tau) \tag{6.38}$$

where $v_g = dR/dt$ is growth velocity; $\tau$ is incubation period for nucleation, $t$ is the time of growth. So the volume of a nucleus is:

$$V_n = \frac{4}{3}\pi v_g^3 (t - \tau)^3 \tag{6.39}$$

The equation provides an expression for volume transformed per nuclei so that now we must determine the number of nuclei. We define a nucleation rate:

$$N = \frac{\text{number nuclei formed/time}}{\text{uncrystallized volume}} \tag{6.40}$$

The nuclei number formed in the time interval $dt$ is $NdtV_u$, where $V_u$ is the uncrystallized volume. The value of $V_u$ will be a function of time and thus difficult to be determined. Suppose that we consider the function $NdtV$ where $V$ is total volume of melt. This function counts the number of nuclei formed in both the uncrystallized and crystallized volumes of melt. Since nuclei may not form in the already crystallized volume, we call these *phantom nuclei*, as shown in Fig. 6.16. We then define an imaginary number of nuclei, $n_i$, as the sum of the real nuclei, $n_r$, and phantom nuclei, $n_p$,

$$n_i = n_r + n_p \tag{6.41}$$

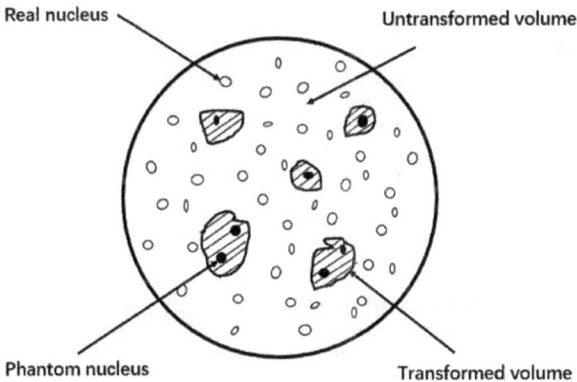

Fig. 6.16: The real and phantom nuclei in a crystallizing melt.

The volume associated with the imaginary nuclei in the time interval $dt$ may be expressed as follows:

$$V_s = \int_\tau^t \frac{4}{3}\pi v_g^3 (t-\tau)^3 NV_u dt \tag{6.42}$$

It turns out to be more convenient to work with the fraction of volume ($\varphi_s$):

$$\varphi_s = \frac{V_s}{V} = \int_\tau^t \frac{4}{3}\pi v_g^3 (t-\tau)^3 N dt \tag{6.43}$$

We would now like to relate $\varphi_s$ to the real volume fraction crystallized, $\varphi_r$. The real and phantom nuclei formed in any time interval $dt$ will have the same volume per nuclei, so that we may write

$$\frac{dn_r}{dn_i} = \frac{dV_r}{dV_i} = \frac{d\varphi_r}{d\varphi_i} \tag{6.44}$$

Let the number of nuclei formed per volume in a time $dt$ be called $dP$, so that we have $dn_r = V_u dP$ and $dn_i = V dP$. We now assume that the nuclei are formed randomly throughout the melt. This means $dP$ will be independent of position, so we may write

$$\frac{dn_r}{dn_i} = \frac{V_u}{V} = \frac{V - V_r}{V} = 1 - \varphi_r \tag{6.45}$$

Combining eqs. (6.44) and (6.45),

$$\frac{d\varphi_r}{d\varphi_i} = 1 - \varphi_r \tag{6.46}$$

This simple differential equation is solved to give

$$\varphi_r = 1 - e^{-\varphi_i} \tag{6.47}$$

If we assume that both $v_g$ and $N$ are constant and that $\tau$ is negligibly small we integrate eq. (6.43) to obtain

$$\varphi_i = \frac{\pi}{3} N v_g^3 t^4 \tag{6.48}$$

Combining eqs. (6.47) and (6.48) we obtain

$$\varphi_r = 1 - \exp\left(-\frac{\pi}{3} N v_g^3 t^4\right) \tag{6.49}$$

This is the famous Johnson–Mehl (J-M) equation, and it is applicable to any phase transformation subject to the four restrictions of random (homogeneous) nucleation, constant $N$, constant $v_g$, and small $\tau$. A plot of eq. (6.49) is presented in Fig. 6.17 for different values of the growth rate and the nucleation rate.

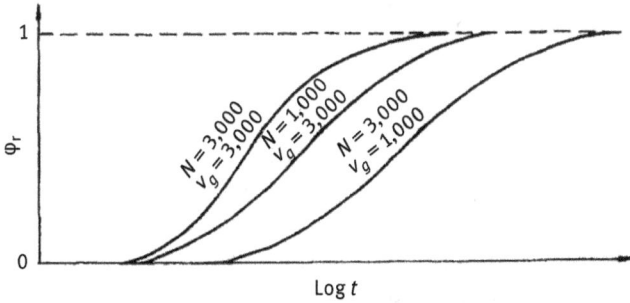

**Fig. 6.17:** A plot of the Johnson–Mehl equation for various values of $v_g$ and $N$.

These curves demonstrate a characteristic S-shaped or sigmoidal profile reflecting the low rates at the beginning and the end of the crystallization but the high rate in between them.

At the beginning, the low rate may be resulted from the time required for the formation and growth of a significant nuclei number. During the intermediate period, the rate is high because the existing nuclei grow by consuming the melt while the new ones form continuously in the remaining melt. Once the crystallization approaches completion, there is little melt for further nucleation and the nucleation rate begins to slow down. In addition, the existing particles begin to touch mutually, forming boundaries and stopping growth.

When temperature changes, the position of S-shaped curve moves and shape changes due to the changes of incubation period, nucleation rate and growth velocity together, such as shown in Fig. 6.18(a). However, the volume fraction crystallized corresponding to the maximum crystallization rate hardly changes according to following derivation:

$$\frac{d^2\varphi_r}{dt^2} = \left[ 4\pi N v_g^3 t^2 - \left( \frac{4\pi N v_g^3 t^3}{3} \right)^2 \right] \exp\left( -\frac{\pi N v_g^3 t^4}{3} \right) = 0 \qquad (6.50)$$

that is,

$$t^4 = 9/\left( 4\pi N v_g^3 \right) \qquad (6.51)$$

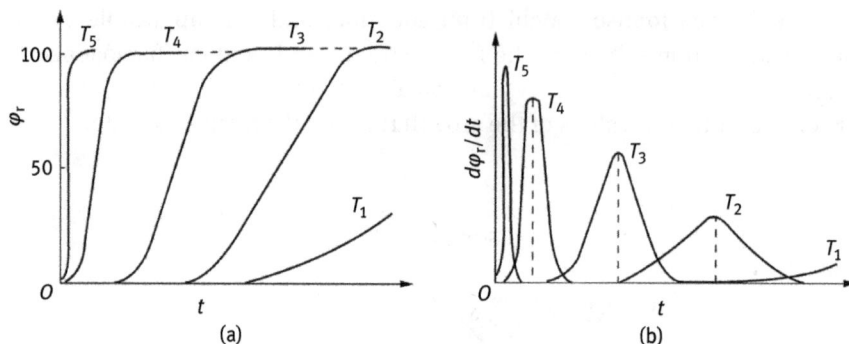

**Fig. 6.18:** Curves of transformation kinetic (a) and curves of transformation rate (b) at different temperatures.

Inserting the value of $t^A$ into J-M equation obtains crystallization fraction corresponding to the maximum crystallization rate $\varphi_r(\text{max.}) = 52.8\% \approx 50\%$. Therefore, the time $t$ corresponding to $\varphi_r(\text{max.}) = 50\%$ is usually taken as $t_{1/2}$ to reach the maximum crystallization rate, as shown in Fig. 6.18(b).

An interesting condition occurs when nucleation event occurs on specific sites (such as substrate or impurity surface), which saturate soon after the solidification begins (heterogeneous nucleation). Initially, nucleation event may occur randomly and freely at a relatively high rate. Once these sites are occupied, the nucleation event will cease. In addition, the growth direction of solid particle is also not certainly three dimensional. Especially, the volume fraction of uncrystallized melt decreases with growth. When these conditions are involved, another version of crystallization kinetics was proposed by Avrami as phenomenological equation:

$$\varphi_r = 1 - \exp(-kt^n) \tag{6.52}$$

similar to J-M equation in format but different in nature. For example, $n$ in eq. (6.52) has an integer value between 1 and 4 reflecting the crystallization mechanism. The value of 4 refers to the contributions from three dimensions of growth accompanying with a constant nucleation rate. For this case, Avrami equation regresses to event of homogeneous nucleation. If the nucleation sites is distributed nonrandomly, then the growth manner may be changed to one or two-dimensions. Site saturation can generally result in $n$ values of 1, 2, or 3 for different nucleation and growth manners, respectively.

## Crystal morphologies

In general, the distribution of temperature in the remnant melt is not homogeneous during solidification. The temperature gradient vertical to the solid–liquid interface,

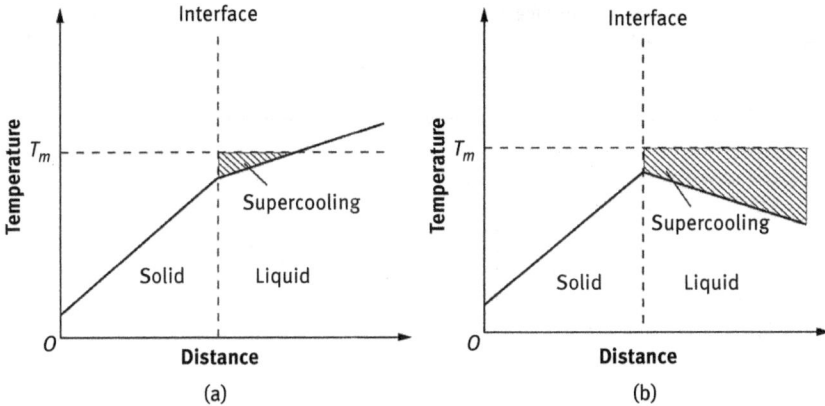

**Fig. 6.19:** The possible temperature gradient in the front of growth interface in liquid (a) positive gradient; (b) negative gradient.

together with the microstructure of interface, will affect growth manner of pure crystal, and finally to determine its morphology after solidification.

### a. Positive temperature gradient

When the remnant melt becomes hotter with the distance $z$ away from the liquid–solid interface, it is said that the melt is in the positive gradient state with $dT/dz > 0$, such as shown in Fig. 6.19(a). The latent heat released during crystallization cannot be transported into the melt inner. The opposite direction in the solid, therefore, acts as the tunnel for heat flow, including both from neighboring melt and latent heat. For this case, the movement velocity of interface is controlled by the efficiency of heat flow in the solid. Planar migration is more possible because the occasional protuberance that begins to grow on the surface enters itself into an environment surrounded by the hotter liquid or the smaller supercooling, the reduced migration velocity leads to the results that the remainder of interface catches up and the protuberance will be diminished. Macroscopically, the solid–liquid interface maintains planar morphology during the movement. When interfacial structure is considered, it may be true that:

(1) If interface is smooth with zigzag facets corresponding to specific crystal planes, such as those shown in Fig. 6.20(a). The direction of interface migration from solid to liquid is perpendicular to isothermal plane, but the faceted planes in Fig. 6.20(a) are generally at an angle to the $T_m$ isotherm.

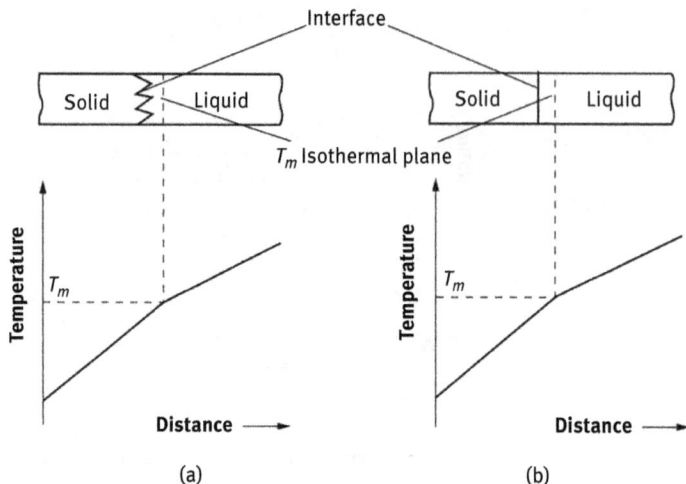

**Fig. 6.20:** Surface structures observed under the positive temperature gradient (a) smooth surface (steps feature) (b) rough surface (flat feature).

(2) If interface is rough in microscopic (atomic) scale, there is no facet feature. But in macroscopic scale the solid–liquid interface is flat plane, and the direction of its migration is perpendicular to isothermal plane directly, such as shown in Fig. 6.20(b).

### b. Negative temperature gradient

When the remnant melt becomes colder with the distance $z$ away from the liquid–solid interface, it is said that the melt is in the negative gradient state with $dT/dz < 0$, such as shown in Fig. 6.19(b). The latent heat released during crystallization can be transported into the inner of melt efficiently in addition to dissipation from the solid tunnel. For this case, the occasional protuberance that begins to grow on the interface touches with the colder or the larger supercooling melt, leading to an increase in its migration velocity in the inner liquid, and finally, the flat interface cannot be maintained. The protuberance can branch and then subbranching occurs repeatedly, resulting in the dendritic morphology of crystal (Fig. 6.21).

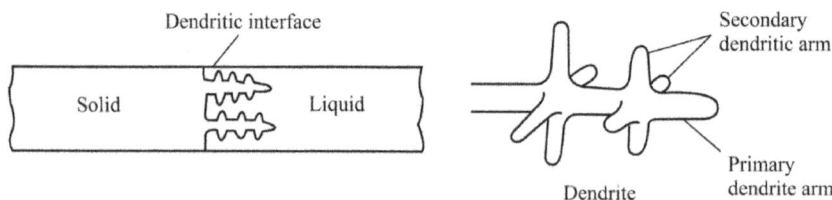

**Fig. 6.21:** Schematic illustration of dendritic growth of crystal.

It should be mentioned that the dendritic morphology is more obvious in crystals with rough interface (such as in metals). In metals, the preferred dendritic axis is related to crystalline lattice, and usually is

$$\text{fcc:} < 100 >$$

$$\text{bcc:} < 100 >$$

$$\text{hcp:} < 10\bar{1}0 >$$

When the solid–liquid interface appears as smooth type or the $\alpha$ (see eq. (6.30)) is relatively large, the crystal exhibits very weak inclination to form dendritic morphology even under the condition of negative gradient.

### 6.2.6 Structural modification and innovative technology

The principles discussed in previous sections are the basis to develop innovative processing methods to control or modify microstructure for desired physical and mechanical properties. A few examples including grain refinement, single crystal growth from the melt, and preparation of metallic glasses are described in this section.

### Grain refinement
Several techniques to prepare fine grained components have been developed based on balancing nucleation and growth during solidification. Among them, thermal methods, inoculation, and energy-induced methods are widely used in industry.

### a. Thermal method
This method is developed according to the fact that the high nucleation rate combined with low growth velocity will increase the number of crystal particles in solidified casting. In theory, this can be derived from the J-M equation that the nucleus number per unit volume solid, $P(t)$, formed in a period $(t)$, is determined by nucleation rate $N$ and growth velocity $v_g$ simultaneously:

$$P(t) = k\left(\frac{N}{v_g}\right)^{3/4} \tag{6.53}$$

where $k$ is a constant related to nucleus shape. As the

$$N \propto \exp\left(-\frac{1}{\Delta T^2}\right) \tag{6.54}$$

and

$$v_g \infty \Delta T \tag{6.55}$$

for continuous growth or

$$v_g \infty (\Delta T)^2 \tag{6.56}$$

for growth by screw dislocations. Since with the increase of supercooling the effect of nucleation rate $N$ on $P(t)$ is greater than that of growth velocity $v_g$ on $P(t)$, the refined grain can be obtained by large supercooling $\Delta T$.

### b. Grain refiners

Select a material with a higher melting point and can be wetted by the crystal, and throw it into the melt to increase sites for nucleation. This kind of materials is called as grain refiner (inoculant or modifier). To avoid dissolution in the liquid, the grain refiner usually comes from the insoluble compound or at least their surface is covered with insoluble materials. By now, it is still difficult to identify a good grain refiner just by simple rules, although it can be found that the low $\sigma_{aw}$ would be favorable to improve wetting capability (see eq. (6.20)). The half-empirical conclusion has been reached that the strong chemical bonding and the small lattice misfit ($\delta$) at refiner–crystal interface may act as the positive role in improving nucleation rate. Table 6.4 lists some experimental results on the nucleation of pure aluminum. The actual effect of different refiners fits well with respect to above prediction. However, some studies also showed that lattice misfit might not be so critical. In the case of pure gold, the actual effect of WC, ZrC, TiC, and TiN is stronger than that of $WO_2$, $ZrO_2$, and $TiO_2$ while their misfit is nearly same. Another example related to the nucleation of tin indicates that the metal refiner is more effective than nonmetal one, which has nothing to do with the factor of lattice misfit. Therefore, the effectiveness of grain refiner in engineering is still determined by actual test.

**Table 6.4:** Effect of different substances on the heterogeneous nucleation of pure aluminum.

| Compound | Lattice structure | δ among close-packed planes | Nucleation effect | Compound | Lattice structure | δ among close-packed planes | Nucleation effect |
|---|---|---|---|---|---|---|---|
| VC | Cubic | 0.014 | Strong | NbC | Cubic | 0.086 | Strong |
| TiC | Cubic | 0.060 | Strong | $W_2C$ | Hexagonal | 0.035 | Strong |
| $TiB_2$ | Hexagonal | 0.048 | Strong | $Cr_3C_2$ | Complex | – | Weak or none |
| $AlB_2$ | Hexagonal | 0.038 | Strong | $Mn_3C$ | Complex | – | Weak or none |
| ZrC | Cubic | 0.145 | Strong | $Fe_3C$ | Complex | – | Weak or none |

### c. Energy-induced method

The energy-induced method aims to employ the external energy to break coarsening grains during solidification. The available methods at present include mechanical vibration, magnetic–electric interactions, rotating magnetic fields, bubbling agitation, mold oscillation, and so on. The typical application of energy-induced method is to promote the fragmentation of the dendritic crystals, which can effectively increase the strength and toughness together of casting components.

### Single crystal growth from the melt

Single crystal growth is important for scientific study, such as the physical and mechanical properties of single crystal, and engineering application, such as single crystal silicon in the semiconductor industries or gas turbine blades in airplane. Thus, the preparation or production of single crystal is also an important example of microstructural control.

### a. Czochralski method

This method is named by its inventor, a Polish chemist. It is a typical method for single crystal preparation only through the growth process and there is no vessel around the growing crystal. Most of the silicon used for semiconductor applications and many artificial gemstones are grown by the Czochralski method. Figure 6.22(a) shows a schematic drawing of crystal growth in apparatus. The melt is slightly supercooled so as to inhibit nucleation completely. The crystal grows through atoms in melt attached the suspended "seed" in advance in coherent epitaxial manner. In the "seed" method, there is a drive mechanism that is used to adjust the velocity for pulling the crystal up and rotating seed rod so that uniform temperature and a small supercooling can be maintained. The crystal surface touching the melt thus stays approximately stationary as the crystal grows. In general, cylindrical crystal can be obtained in the process of continuous pulling and rotating. The desired orientation of single crystal can be realized by setting seed direction, which is an important advantage.

### b. Bridgeman method

This method is named by one of its inventors, an American physicist. As illustrated schematically in Fig. 6.22(b), the melt is contained in a tube surrounded by furnace initially. The tube is then forced to displace down slowly from the furnace. The nucleation event occurs when the tube bottom becomes cold enough. The bottom end of the tube is usually tapered to a point to limit nuclei number as much as possible. After formation of a single nucleus at bottom, it grows at the expense of remaining melt along the opposite displacive direction of tube. Therefore, different from Czochralski method, Bridgman method is unseeded and refers to both

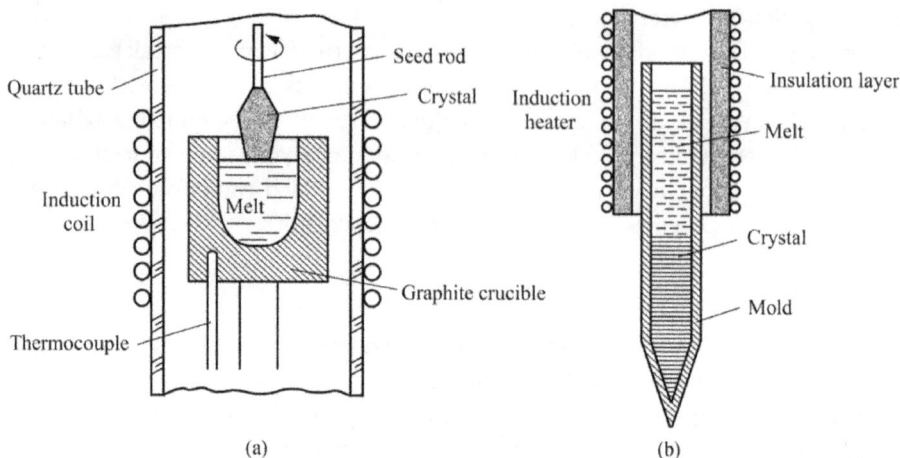

**Fig. 6.22:** Czochralski (a) and (b) Bridgeman single crystal preparation.

the nucleation and growth of crystal. In addition, the crystal orientation is randomly prepared by Bridgeman method, which is a disadvantage.

**Preparation of metallic glasses**

A glass (amorphous, noncrystalline) is formed when the crystallization in a supercooling melt is replaced by congealing continuously and uniformly. The formed solid keeps essentially the liquid structure but has the resistance to shear deformation.

The metallic systems are not easy to form glass due to the very low viscosity even when the melt is cooled down to the melting point $T_m$. The crystal is readily to nucleate in the melt with very low supercooling. Only under some special circumstances, the crystallization can be inhibited completely in metallic systems. Figure 6.23 shows the changes in volume when the crystallization or glass transition occurs in melt, respectively. Between melting temperature and glass transition temperature ($T_m \sim T_g$), where the supercooled melt has a low viscosity, nucleation and growth will occur. The crystallization can be inhibited only if the critical cooling rate is reached to eliminate time for nucleation. Also, the supercooled melt becomes very viscous at the lower temperature, the glassy state can be formed instead of crystallization even in the presence of some nuclei because the crystal growth is terminated by negligible atomic mobility.

Crystallization can occur only in the temperature range of $T_m \sim T_g$. Crystallization, when it occurs, is usually accompanied by abrupt changes in physical properties, such as the discontinuous change of volume shown in Fig. 6.23. When the melt is cooled down below $T_g$, The chance of crystallization is almost zero. The melt has to be congealed in a liquid configuration. As shown in the figure, the change in volume (also the changes in other properties not shown in the figure) of liquid/glass along with the

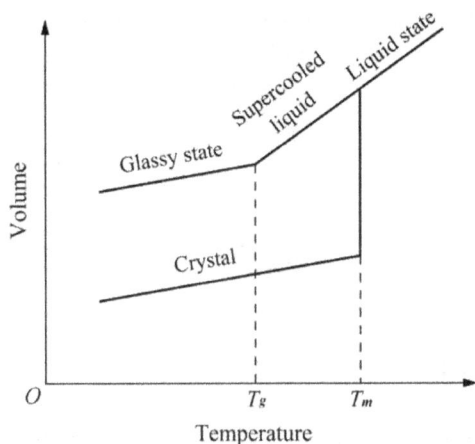

**Fig. 6.23:** Volume change of system caused by crystallization and glass transition, respectively.

temperature reflects the nature of $T_g$. Above $T_g$, the change in volume strongly relies on the temperature because it is dominated by the configurational evolution in liquid; while below $T_g$, the atomic motion ability is greatly weakened, so that it has to stay where it is. The configurational evolution in liquid is of course extinguished by quick cooling (quenching), and the heavily viscous melt becomes a rigid solid, glass finally. Because of the lack of configurational evolution, similar to crystal, its volume (also other properties) has a relatively weak temperature dependence.

It should be mentioned that $T_g$ is not a thermodynamically defined temperature. Its location changes along with the kinetic conditions completely. $T_g$ is usually lowered when the cooling rate is reduced, if the crystallization is not allowed to occur yet in the temperature range of $T_m \sim T_g$. Which means the liquid configurations can be kept at a larger supercooling, and thus a higher density and viscosity for the glass formation.

Based on the classic feature of glass formation above mentioned, it is easy to predict that the glass is preferred to form in the material with small $T_m \sim T_g$ interval, because the temperature can reach $T_g$ within a short time before the crystallization develops well. This phenomenon can be found in many oxides and polymers with small $T_m \sim T_g$ interval (e.g., $SiO_2$: $T_m = 1{,}993$ K, $T_g = 1600$ K, and $T_m - T_g = 393$ K); in the metallic systems, $T_m - T_g$ interval is very large, especially for those with high $T_m$ (e.g., Pd: $T_m = 1{,}825$ K, $T_g = 550$ K, and $T_m - T_g = 1{,}275$ K), it is difficult to obtain glass transition under the normal cooling rate.

In the early days of technology development, the bulk metallic glasses (with thickness over 1 millimeter at least) were difficult to be prepared through the rapid cooling methods. For example, the rapid cooling, on the order of $10^6$ K/s, can be realized by melt spinning. This cooling rate eliminates any probability of nucleation event. The amorphous ribbons are produced by sputtering melt onto a spinning substrate. With in-depth study, it has been found that the formation of amorphous structure in some alloys no longer requires such severe cooling condition. Their critical cooling rates are low enough to permit formation of metallic glasses with thicker layers. Furthermore, the amorphous steel with three times the strength of conventional steel can be prepared.

## 6.3 Vapor–solid phase transformation and growth of thin films

With the vapor deposition technology being widely used in the preparation of a variety of functional thin film materials, the vapor (gaseous)–solid phase transformation (transition) of materials also shows its importance. Vapor–solid and liquid–solid phase transformation have many similarities, but as for the control of evaporation and condensation, the microstructure and morphology of transformation products have their own characteristics. Around the vapor–solid phase transformation of vapor deposition, the relationship between saturated vapor pressure of materials and temperature, and two basic processes of vapor deposition: thermodynamic conditions of evaporation and condensation (sedimentary), nucleation and growth in condensation process are discussed in this section.

### 6.3.1 Vapor pressure

The pressure when the solid phase is in equilibrium with respect to the vapor phase is known as the saturation vapor pressure (or vapor pressure). When vapor phase is under the conditions of constant temperature, constant pressure, and closed container, its concentration will increase because of continuous evaporation process. Meanwhile, the gaseous phase will be condensed into solid because of the reversed condensation process. When two processes are carried out at the same rate, the vapor concentration will be maintained at a constant value, and this vapor pressure of dynamic equilibrium is called as saturated vapor pressure, which is represented by $P_e$, which corresponds to curve $aO$ in Fig. 6.1 with different temperatures or pressures

Vapor deposition includes two basic processes: the evaporation of materials on the high temperature evaporation source and the condensation of evaporated atoms on the low temperature substrate (slide undertaken evaporating gas atoms). For a given material, the vapor pressure will be changed with temperature, and this is necessary to understand the thermodynamic conditions of evaporation and condensation.

The relationship between vapor pressure of materials and temperature can be derived from the thermodynamic Clapeyron equation, that is, the vapor pressure equation. The molar volume of condensed phase (solid phase or liquid phase) is far less than that of gaseous phase, so in the Clapeyron equation, the volume change of two phase: $\Delta V \approx V_g$, and the gaseous phase is regarded as the ideal gas: $PV_g = RT$, the Clapeyron equation thus can be simplified as

$$\frac{1}{p}\frac{dp}{dT} = \frac{\Delta H}{RT^2} \tag{6.57}$$

Furthermore, it can be approximately considered that latent heat $\Delta H$ is independent to the change in temperature, so after integrals:

$$\ln p = -\frac{\Delta H}{RT} + A \tag{6.58}$$

or

$$\ln p = A - \frac{B}{T} \tag{6.59}$$

where $T$ is Kelvin temperature $(K)$, the unit of $p$ (i.e., $p_e$) is micron of mercury ($\mu$mHg, 1 $\mu$mHg = 0.133 Pa), $A$ and $B$ are constants related to the properties of evaporation material. Table 6.5 shows $A$ and $B$ values calculated by the above formulas. The relationship between vapor pressure and temperature of various evaporation materials is shown in Fig. 6.24. Their melting points are also presented in the same figure. The heating temperature of evaporation source (material) can alter vapor pressure, then directly affect the evaporation rate and manner. When the heating temperature is too low, the evaporation rate of materials is low, and then the growth rate of film is low as well. Conversely, the excessive heating temperature will lead to the excessive evaporation rate, the collision of gaseous atoms, and even the spattering randomly of evaporated atoms due to rapid expansion of gas in the container. Therefore, it is very important to determine the heating temperature of different evaporation materials. The temperature when the gaseous pressure reaches to several Pa is usually taken as the heating temperature. As shown in Fig. 6.24, besides Sr and Te, the heating temperature of most evaporation materials is higher than their melting points.

**Table 6.5:** The constants used for vapor pressure equation of some elementary substances.

| Metal | A | B | Metal | A | B |
|---|---|---|---|---|---|
| Li | 10.99 | $8.07 \times 10^3$ | In | 11.23 | $1.248 \times 10^4$ |
| Na | 1.72 | $5.49 \times 10^3$ | C | 15.73 | $4.0 \times 10^4$ |
| K | 10.28 | $4.48 \times 10^3$ | Co | 12.70 | $2.111 \times 10^4$ |
| Cs | 9.91 | $3.80 \times 10^3$ | Ni | 12.75 | $2.096 \times 10^4$ |
| Cu | 11.96 | $1.698 \times 10^4$ | Ru | 13.50 | $3.38 \times 10^4$ |
| Ag | 11.85 | $1.427 \times 10^4$ | Rh | 12.94 | $2.772 \times 10^4$ |
| Au | 11.89 | $1.758 \times 10^4$ | Pd | 11.78 | $1.971 \times 10^4$ |
| Be | 12.01 | $1.647 \times 10^4$ | Si | 12.72 | $2.13 \times 10^4$ |

**Table 6.5** (continued)

| Metal | A | B | Metal | A | B |
|-------|-------|-------------------------|-------|-------|-------------------------|
| Mg | 11.64 | $7.65 \times 10^3$ | Ti | 12.50 | $2.32 \times 10^4$ |
| Ca | 11.22 | $8.94 \times 10^3$ | Zr | 12.33 | $3.03 \times 10^4$ |
| Mo | 11.64 | $3.085 \times 10^4$ | Th | 12.52 | $2.84 \times 10^4$ |
| W | 12.40 | $4.068 \times 10^4$ | Ge | 11.71 | $1.803 \times 10^4$ |
| U | 11.59 | $2.331 \times 10^4$ | Sn | 10.88 | $1.487 \times 10^4$ |
| Mn | 12.14 | $1.374 \times 10^4$ | Pb | 10.77 | $9.71 \times 10^3$ |
| Fe | 12.44 | $1.997 \times 10^4$ | Sb | 11.15 | $8.63 \times 10^3$ |
| Sr | 10.71 | $7.83 \times 10^3$ | Bi | 11.18 | $9.53 \times 10^3$ |
| Ba | 10.70 | $8.76 \times 10^3$ | Cr | 12.94 | $2.0 \times 10^4$ |
| Zn | 11.63 | $6.54 \times 10^3$ | Os | 13.59 | $3.7 \times 10^4$ |
| Cd | 11.56 | $5.72 \times 10^3$ | Is | 13.07 | $3.123 \times 10^4$ |
| B | 13.07 | $2.962 \times 10^4$ | Pt | 12.53 | $2.728 \times 10^4$ |
| Al | 11.79 | $1.594 \times 10^4$ | V | 13.07 | $2.572 \times 10^4$ |
| La | 11.60 | $2.085 \times 10^4$ | Ta | 13.04 | $4.021 \times 10^4$ |
| Ga | 11.41 | $1.384 \times 10^4$ | | | |

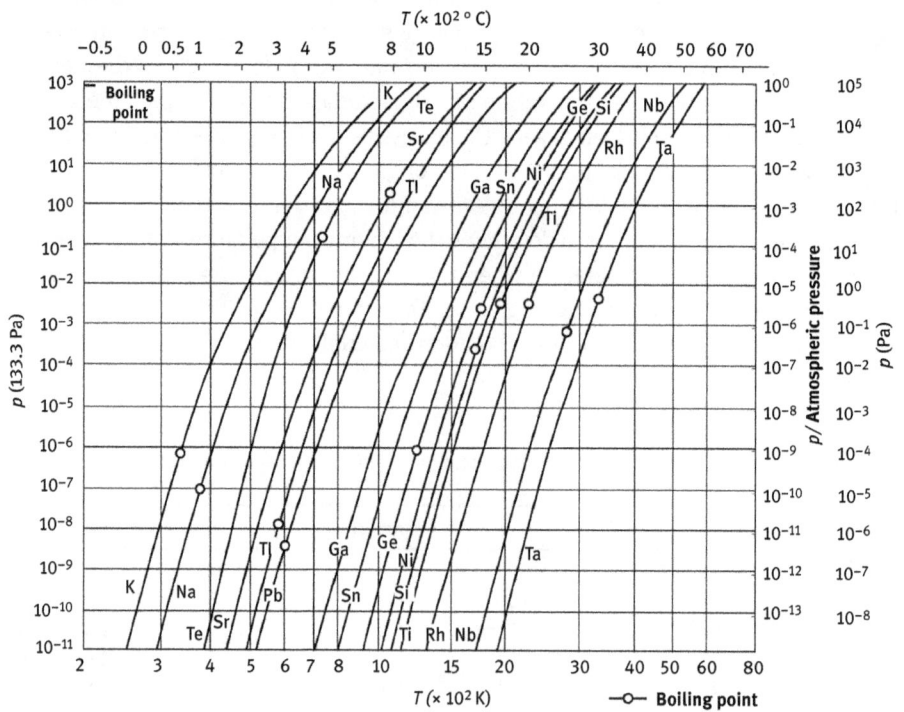

**Fig. 6.24:** The vapor pressure curves of some elementary substances.

### 6.3.2 Thermodynamic conditions of evaporation and condensation

When gaseous phase is approximately considered to be as ideal gas:

$$dG = -SdT + Vdp$$

When the temperature is constant $(dT = 0)$:

$$\Delta G = \int_{pe}^{p} Vdp$$

where $p$ is actual pressure.
    For ideal gas:

$$pV = nRT$$

So

$$\Delta G = \int_{pe}^{p} \frac{nRT}{p} dp$$

or

$$\Delta G = nRT \frac{p}{p_e} \tag{6.60}$$

According to eq. (6.60), the evaporation process can be carried out if $p < p_e$ accompanying with $\Delta G < 0$, and in this condition vapor phase is stale, as shown in Fig. 6.1. On the other hand, the condensation process can be carried out if $p > p_e$, and in this condition solid phase is stale, as shown in Fig. 6.1.When the material at evaporation source is heated to a very high temperature, the vapor pressure of the material is very high as well (see eq. (6.59) and curve $aO$ in Fig. 6.1), and the gaseous pressure in the vacuum container is much less than the vapor pressure of the material, meeting the requirement of evaporation. When the evaporated gaseous atoms reach to the cold substrate accompanying with the generation of low vapor pressure, the gaseous pressure in vacuum container is much higher than the vapor pressure of the material, meeting the requirement of condensation.

### 6.3.3 Mean free path of gaseous molecules

In order to meet the requirements of evaporation of solid materials, the background pressure in the container should be lower than the vapor pressure of the material. The background pressure must reduce the collision possibility between the gaseous atoms formed by the evaporation and the residual molecules in the background pressure (the collision will result in scattering so as to prevent gaseous atoms from the substrate surface), so the background pressure in the container must be low enough. In other words, it needs high vacuum degree, and this container is called vacuum cover.

Under the assumption of ideal gas, the average free path $L$ of the gaseous molecules is inversely proportional to the gaseous pressure $p$ according to statistical physics. At room temperature, it can also be approximately expressed as

$$L = \frac{6.5}{p} \tag{6.61}$$

where the unit of $L$ is mm and the unit of $p$ is Pa. When the pressure is 1 Pa, the mean free path $L = 6.5$ mm, and when the pressure is $10^{-3}$ Pa, $L = 6,500$ mm. During the movement of gas to substrate, the mean free path should be 10 times larger than the distance from evaporation source to substrate in order to make the collision probability between the evaporated atoms and the residual molecules in the background pressure is less than 10%. For normal evaporation coating equipment (as shown in Fig. 6.25), the distance from evaporation source to substrate is less than 650 mm, as a result, the background pressure in the vacuum cover is required to reach $10^{-2}$–$10^{-5}$ Pa, depending on the quality requirement of the thin film. It must be pointed out that the background pressure in the vacuum cover refers to the initial pressure before evaporation, also known as background vacuum. Even though the evaporation will cause the increase of pressure, it will not affect the above conclusion in essence.

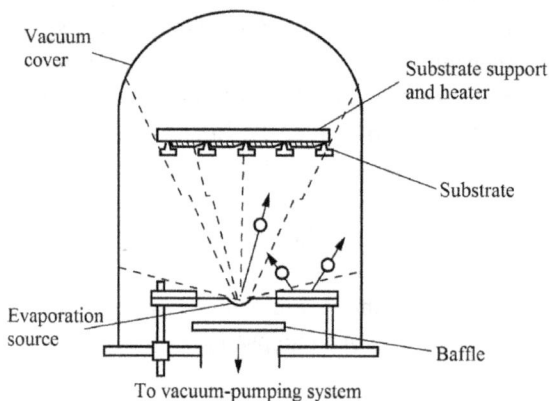

**Fig. 6.25:** Schematic diagram of vacuum evaporation coating equipment.

## 6.3.4 Nucleation

During the coating process, the heated evaporated atoms fly to the cold substrate (at room temperature). Due to sharp reduction of temperature of atoms after touching with the substrate, the local vapor pressure decreases rapidly so that the gaseous pressure in vacuum cover is much higher, leading to the occurrence of condensation event. When condensed area reaches to the critical size, other atoms can attached to the surface of grains automatically.

The critical grain size $r_c$ of condensed nucleus can be treated using the theory of solidification. Assuming the nucleus is spherical, $r_c$ is given by

$$r_c = \frac{2\sigma}{\Delta G_V} \tag{6.62}$$

where $\sigma$ is surface energy and $\Delta G_v$ is free energy per unit volume.

It is worth pointing out that the cooling rate of vapor deposition is very large, which generally is $10^7$–$10^{10}$ K/s, and the supercooling is much larger than that of solidification. Therefore, the critical grain size of deposition nuclei is very small. At the same time, the heat energy of the gaseous atoms dissipates rapidly on the large substrate, the grains are not easy to grow up. The most grains deposited at room temperature (i.e., the substrate is not heated) is very small, in the nanometer scale, or even amorphous state. Especially for alloys and compounds with high melting point, it is easier to obtain the amorphous clusters. Only when the substrate is heated, the grains can grow significantly. Figure 6.26 shows the schematic diagrams of deposited Ag at room temperature. It shows clearly the process of grain growth and formation of continuous thin films with the increase of evaporation time.

The nucleation rate of vapor deposition is similar to that of solidification, which is influenced by both the nucleation work and the atom diffusivity. Due to large supercooling, the nucleation rate of vapor deposition is mainly affected by the nucleation work, and it is easy to obtain the fine grains especially when the substrate is not heated.

## 6.3.5 Growth modes of thin films

There are three basic types of thin films growth: (1) three-dimensional growth (Volmer–Weber) model, (2) two-dimensional growth (Frank–van der Merwe) model, (3) layer nucleation growth (Stranski–Krastanov) model, as shown in Fig. 6.27.

In three-dimensional growth model, the growth process of thin films can be divided into four stages: nucleation stage, small island stage, network stage, and continuously film stage (as shown in Fig. 6.26). The specific processes are as follows. Through the migration on the substrate surface, the atoms attached can form atomic clusters, even stable crystal nuclei. By capturing the adsorbed atoms or directly accepting the incident atoms, each stable crystal nucleus grows up as a small

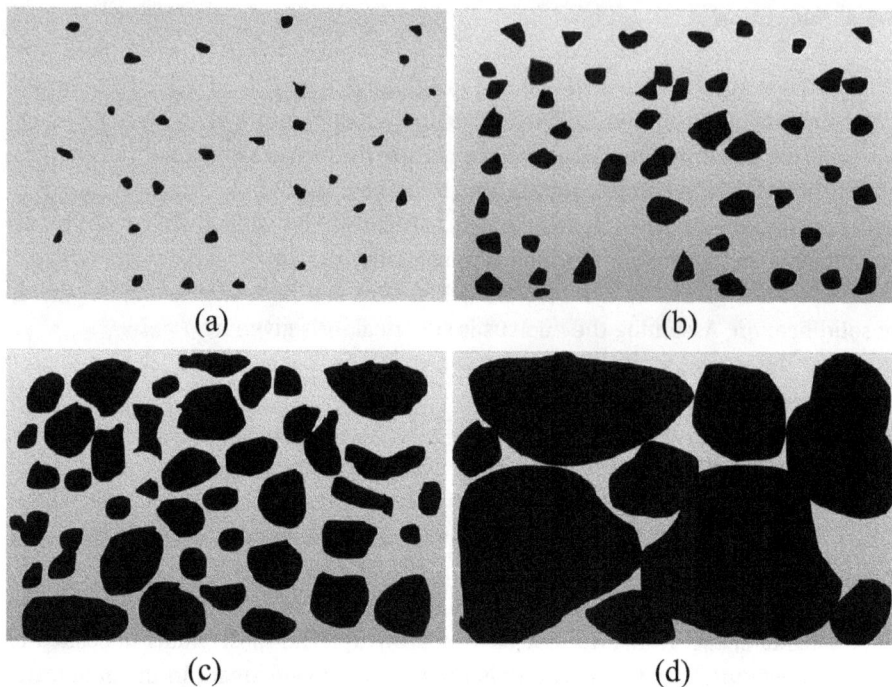

**Fig. 6.26:** Schematic diagrams of Ag films deposited on NaCl substrate.

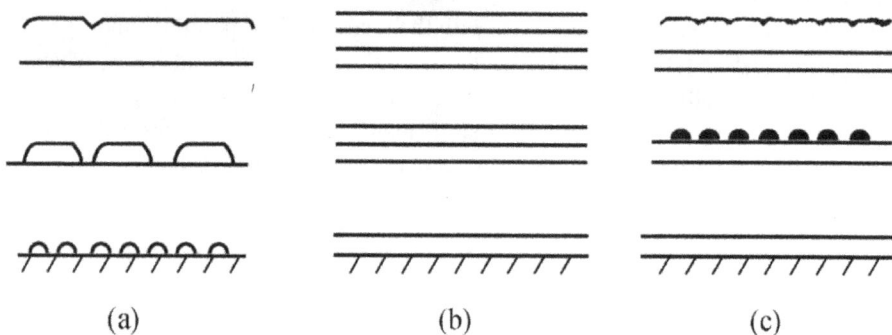

**Fig. 6.27:** Three types of thin films growth: (a) three-dimensional growth, (b) two-dimensional growth, (c) layer nucleation growth.

island in three-dimensional direction. At the same time, small islands meet each other and merge into large islands, and the large islands are connected further to form the network films. As for the channels in the network films, through the growth of network films or the formation of new small islands in channels, the channels are gradually filled out to form the continuous films.

In the model of two-dimensional growth, the substrate is single crystal. The adsorbed atoms can grow along with the crystal direction with coherent epitaxial manner. The coherent epitaxial growth can be divided into the same structure epitaxial growth and the different structure epitaxial growth. The same structure epitaxial growth means that the deposited films have same lattice type with respect to the substrate and they grow on the specific crystal planes (usually the low index close-packed plane). While the different structure epitaxial growth means that deposited films have different lattice type with respect to the substrate and they grow on the specific crystal planes.

The combination of above two kinds of growth models is layer nucleation growth, namely, one to two atomic layers are firstly formed on the substrate surface, and this two-dimensional structure is strongly influenced by the lattice of substrate, lattice mismatch leads to large lattice distortion, thereafter the atoms are adsorbed and grow up into the small islands by the model of three-dimensional growth, and ultimately the continuous films are formed.

In addition to vacuum evaporation coating, sputtering coating is a commonly used method of physical vapor deposition. In the process of sputtering, a small amount of inert gas (such as argon) needs to be inputted into the vacuum system. The sputtering materials is used as cathode (refers to as target) and the substrate is used as anode. Between them, the glow discharge by high voltage can produce ions ($Ar^+$). After accelerating in the electric field, $Ar^+$ ions impact target material (cathode), knocking the target atoms out, then they deposit on the substrate to form thin films. The sputtering is different from evaporation, the average energy of target particles (referred to as sputtering particles, which are mostly atoms and a small amount of ions) which are generated by impacting of inert gas ions is several electron volts. It is much higher than that of evaporation particles (the average kinetic energy of evaporated atom at 3,000 K is only 0.26 eV). By increasing the incident energy of inert gaseous ions, the sputtering efficiency can be evidently improved. To obtain higher sputtering efficiency, magnetron sputtering technology appeared in 1970s. By this method, the deposition rate of films was increased, gas content in the films was decreased, and the phenomenon of temperature rising in the substrate was improved.

### 6.3.6 Application examples (giant magnetoresistance of multilayer and granular films)

As is well known, the resistance of anisotropic magnetic metallic materials, such as Fe–Ni alloy, generally decrease in the magnetic field, and this phenomenon is called magnetoresistive effect, which is usually represented by $\Delta R/R$ ($R$ is

Resistance, $\Delta R/R = [R(H)\text{-}R(U)]/R(U)$, $R(H)$ and $R(U)$, respectively represent the resistance with and without the magnetic field $H$). In general, the change rate of magnetoresistance is about several percentages. In 1988, in the preparation of Fe/Cr/Fe multilayer films (ferromagnetic/nonmagnetic/ferromagnetic) by sputtering method, Fert et al. worked in Paris University of France observed that the change rate of magnetoresistance is negative isotropically, as high as 50%, which is one order in magnitude larger than that of normal magnetoresistance. This effect is called the giant magnetoresistance (GMR) effect. In 1994, IBM developed GMR read heads, increasing the recording density of disk by 17 times, which is 5 Gbit/in². Subsequently, there is another report that the density reaches 11 Gbit/in².

In 1992, Berkowtz and Xiao et al., respectively, found that the GMR effect also exists in the granular films of nonmagnetic metal Cu matrix embedded with nanoparticle Co. In 1996, another finding appeared that the GMR effect exists in the granular films of insulators ($SiO_2$) embedded with ferromagnetic particles Co(Ni). The preparation of granular films is easier than that of multilayer, and the magnetoresistance mechanism of granular films is more abundant than that of multilayer as well. However, the magnetoresistance of granular films is far less than that of multilayer (less than 10%), and it needs higher external magnetic field to generate saturation magnetoresistance. Therefore, the application of granular films is limited.

## 6.4 Crystallization characteristics of polymers

Like metals, ceramics, and low molecular weight organic compounds, the crystals of polymer have long-range order arrangement in three-dimensional direction. Therefore, the crystallization behavior of later is similar to that of former in many ways. However, because of the long chain structure in polymers, it is much more difficult than low molecule to arrange the spatial structure of the polymer chains into crystal lattice with high degree of regularity. Therefore, during the crystallization, the incompleteness, imperfection, temperature rise during melting, low crystallization rate, etc., are usually present. This section provides brief description on the similarities and differences in the crystallization between polymers and low molecules.

**Similarities**

(1) The grain size is affected by supercooling. When the temperature decreases to any temperature between the melting point ($T_m$) and glass transition temperature ($T_g$), the crystallization occurs. Crystallization needs supercooling, and nucleation rate increases with the increase of supercooling. In general, nucleation of polymer crystal forms spherulites. Under the optical microscope, spherulites grow in spherical

symmetric form. Moreover, it is composed by bundles of bifurcated lamellae. In industry, the spherulite size can be controlled by adjusting the cooling rate, as the smaller spherulite and the higher density of spherulites can be obtained by accelerating cooling and increasing supercooling. Figure 6.28 shows the polarized optical microscopy of spherulites obtained by a group of isotactic polybutene-1 melts at different cooling rates. The big spherulites are formed when the supercooling is small, so the crystal plates are thick and the density of internal defects is low, but the number of "joint chains" between crystal plates is small and the content of impurity or low molecule is high. In contrast, the small spherulites are formed when the supercooling is large, so the crystal plates of the spherulites are thin, and the density of internal defects is high, but the number of "joint chains" between crystal plates is big. The strength of crystallized polymers can be improved by the increase of "joint chains" number.

(a)            (b)            (c)

**Fig. 6.28:** The different spherulites sizes crystallized at different cooling rates for isotactic polybutene-1 melts: (a) quenched to room temperature, (b) 10 °C/min, (c) 1 °C/min.

(2) The crystallization process includes nucleation and growth. The nucleation can be divided into two types: homogeneous and heterogeneous. Homogeneous nucleation means that the nucleus is provided by the orderly arranged chains formed by thermal motion of chain segments in melts. While the heterogeneous nucleation relies on external impurities, the residual unmelted crystalline polymers, dispersed small particles, and the walls of containers as core, absorbing the polymer chains in melts and then arranging them orderly.

(3) The supercooling required for heterogeneous nucleation is less than that for homogeneous nucleation. Consequently, nucleating agents can effectively improve nucleation rate and accelerate crystallization rate of polymers. Nowadays, nucleating agents have been widely applied in the industry. Table 6.6 lists the effects of some nucleating agents on the crystallization rate and the spherulite size of

**Table 6.6:** The effect of nucleating agents on the crystallization rate and the spherulite size of Nylon-6.

| Nucleating agents | Content of nucleating agents (%) | Crystallization rate at 200 °C $t_{1/2}$ (min) | Sizes of spherulites ($\mu$m) | |
|---|---|---|---|---|
| | | | Crystallizing at 150 °C | Crystallizing at 5 °C |
| – | – | 20 | 50–60 | 15–20 |
| Nylon-66 | 0.2 | 10 | 10–15 | 5–10 |
| | 1 | | 4–5 | 4–5 |
| Polyethylene terephthalate | 0.2 | 6.5 | 10–15 | 5–10 |
| | 1 | | 4–5 | 4–5 |
| Lead phosphate | 0.05 | 5.5 | 10–15 | 8–10 |
| | 0.1 | | 4–5 | 4–5 |

Nylon-6. As shown in Table 6.6, when the amount of various nucleating agents is up to 1%, the crystallization rate can be increased by two to three times, and in this case the size of spherulites is independent of the crystallization temperature (i.e., supercooling), being of great significance in the industries. Adjusting cooling rate is widely used in industry because it is a simple but effective method to control the spherulite size. But for thick-walled products, because the polymer is a poor conductor, a large temperature gradient from the surface to center is generated, and the cooling rate of each part is not identical. This affects crystallization quality because spherulite size is different at various positions. If the nucleating agents are used, the different supercoolings at different positions can be ignored so that the size uniformity of spherulite can be improved. Figure 6.26 shows a photograph which was taken during the crystallization of polymer melts mixed with several carbon fibers. From Fig. 6.29, it is obvious that the density of spherulites along the carbon fiber is much higher than that along other parts, and this evidence directly confirms the role of nucleating agents.

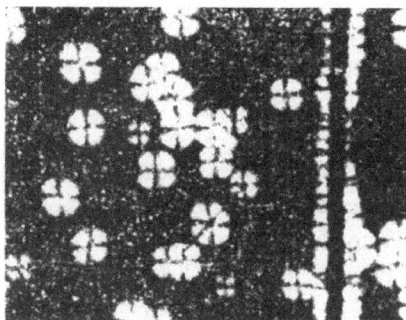

**Fig. 6.29:** The effects of carbon fibers on the nucleation of polymer crystal.

(4) The crystallization kinetics isothermally can be described by Avrami equation. When the crystallization of polymer melt occurs during cooling, the volume of the polymers shrinks continuously, and the volume shrinkage during the crystallization can be measured by dilatometer. If $V_0$, $V_t$, and $V\infty$ denote the volume per unit mass of melt at initial time, $t$ time and terminal time during the crystallization respectively, the Avrami equation is given by

$$\varphi_u = \frac{V_t - V_\infty}{V_0 - V_\infty} = e^{-kt^n} \tag{6.63}$$

where $\varphi_u$ is volume fraction of melt, $k$ is crystallization rate constant, $n$ is Avrami index related to nucleation mechanism and growth manner (see Table 6.7).

Table 6.7: The Avrami index of different nucleation mechanisms and crystal growth manners.

| Nucleation modes | Homogeneous nucleation | Heterogeneous nucleation |
|---|---|---|
| Three-dimensional growth (bulk (spherical) crystal) | $n = 3 + 1 = 4$ | $n = 3 + 0 = 3$ |
| Two-dimensional growth (lamellar crystal) | $n = 2 + 1 = 3$ | $n = 2 + 0 = 2$ |
| One-dimensional growth (needle-like crystal) | $n = 1 + 1 = 2$ | $n = 1 + 0 = 1$ |

After the logarithm of both side of the eq. (6.63), it is

$$\lg(-\ln \varphi_u) = \lg k + n \lg t \tag{6.64}$$

Plotting $\lg(-\ln \varphi_u)$ against $\lg t$ obtains the straight lines in the Fig. 6.27. $n$ and $k$ can be calculated by using their slopes and intercepts, respectively.
When $\varphi_u = 1/2$,

$$k = \frac{\ln 2}{t_{1/2}^n} \tag{6.65}$$

where $t_{1/2}$ is half-crystallization period, denoting the time it takes to correspond to half of the crystallization amount ($\varphi_u = 1/2$). In addition, eq. (6.65) exhibits the significance of crystallization rate constant, that is, the $1/t_{1/2}$ can be used to judge crystallization rate.

Avrami equation had been used for many crystallization of polymers, with various degrees of success. However, there are many deviations as well, for example, $n$ is not equal to integer; or experimental data from later stage of crystallization deviates from the straight line, indicating that crystallization processes of polymers are actually much more complicate than Avrami's model. Figure 6.30 shows the comparison of $\lg(-\ln \varphi_u)$ against $\lg t$ at different temperatures for Nylon 1010.

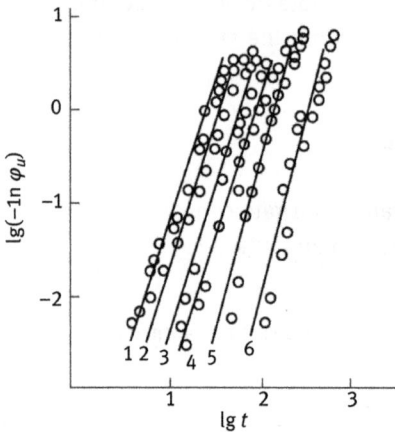

**Fig. 6.30:** The comparison of crystallization kinetics at different temperature for nylon 1010

1, 189.5 °C; 2, 190.3 °C; 3, 191.5 °C; 4, 193.4 °C; 5, 195.5 °C; 6, 197.8 °C.

### Differences

The crystallization of polymers exhibits the feature of incompleteness. Among the polymers, the crystallization capability of polyethylene is highest. Even for this, only 95% melts can transform to crystal. For other polymers, generally only 50% melts can transform to crystal. The incompleteness of crystallization and the crystallization capability of polymers result from the structure of macromolecular chain. Structural factors affecting the crystallization capability are as follows.

(1) Symmetry of chains. The higher the structure symmetry of the polymer chains is, the easier is the crystallization of the polymers. For example, in Fig. 6.31(a), the main chains of polyethylene are all carbon atoms that are bonded with hydrogen atoms. The symmetrical degree is high enough to promote crystallization in any harsh condition (such as quenched in liquid nitrogen). However, when the symmetry is destroyed by chlorination (Fig. 6.31(b)), the crystallization capability of polyethylene weakens rapidly, and even disappears completely.

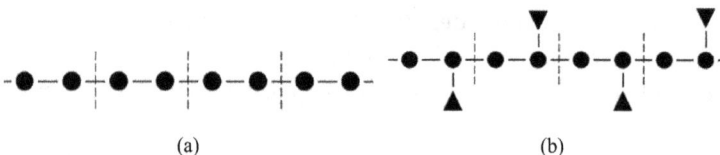

(a)                                          (b)

**Fig. 6.31:** Schematic illustration of structural symmetry between polyethylene (a) and polyvinyl chloride (b). Each triangle in (b) means a hydrogen atom is replaced by chlorine atom, reducing the symmetry of main chain.

(2) Regularity of chains. As for the polymers with irregular main chains completely and without the symmetry center, generally its crystallization capability is negligible. For example, in Fig. 6.32, polystyrene prepared by the free radical polymerization and polymethylmethacrylate are the amorphous polymers without any capability to crystallize (Fig. 6.32(a)). With the method of stereo specific polymerization, the asymmetric center of the main chains can obtain the regular configuration, such as syndiotactic polymer (Fig. 6.32(b)) and isotactic polymer (Fig. 6.32(c), and the polymer can obtain various degrees of crystallization capability.

‒▲‒▲‒▼‒▲‒▲‒▲‒▼‒▼‒  ‒▲‒▼‒▲‒▼‒▲‒▼‒▲‒▼‒  ‒▲‒▲‒▲‒▲‒▲‒▲‒▲‒▲‒

      (a)                (b)                (c)

**Fig. 6.32:** Schematic illustration of structural symmetry of polystyrene prepared by different manners. From (a) to (c), the crystallization capability increases due to the improvement of structural symmetry.

(3) Copolymerization effect. The polymers formed by two or more than two different monomer molecules are called the copolymer. Corresponding to different copolymerization, their crystallization capability is different as shown in Fig. 6.33. The random copolymerization is usually destructive to the symmetry and regularity of the chains, so the crystallization capability can be weakened or even disappeared. However, if two kinds of homopolymer (homopolymer is a polymer produced by one kind of monomer), which is the unit of copolymers, have the same type of crystalline structure, these copolymers also have crystallization capability. Each block of copolymers basically maintains the relative independence, the blocks which can be crystallized will form their own crystalline regions, such as, polyester–polybutadiene–polyester block copolymer, the polyester segments have strong crystallization capability.

(4) Flexibility of chains. Flexibility of chains is a necessary condition for diffusing chains to the crystalline surface and arranging their segments regularly. Therefore, the structural factors that can reduce the flexibility of chain will weaken the crystallization capability of the polymers. For example, the main chain of polyethylene is very flexible, if the flexibility is reduced by polymerization of benzene ring in polyethylene glycol terephthalate, the crystallization capability will weaken significantly. For another instance, the branching will destroy the symmetry and regularity of chains, the cross-linking will greatly restrict the activity of chains, and these will weakens the crystallization capability of polymers.

    Another difference between the polymer and the low molecule lies in the temperature rise phenomenon (melting along with the rising of temperature) which

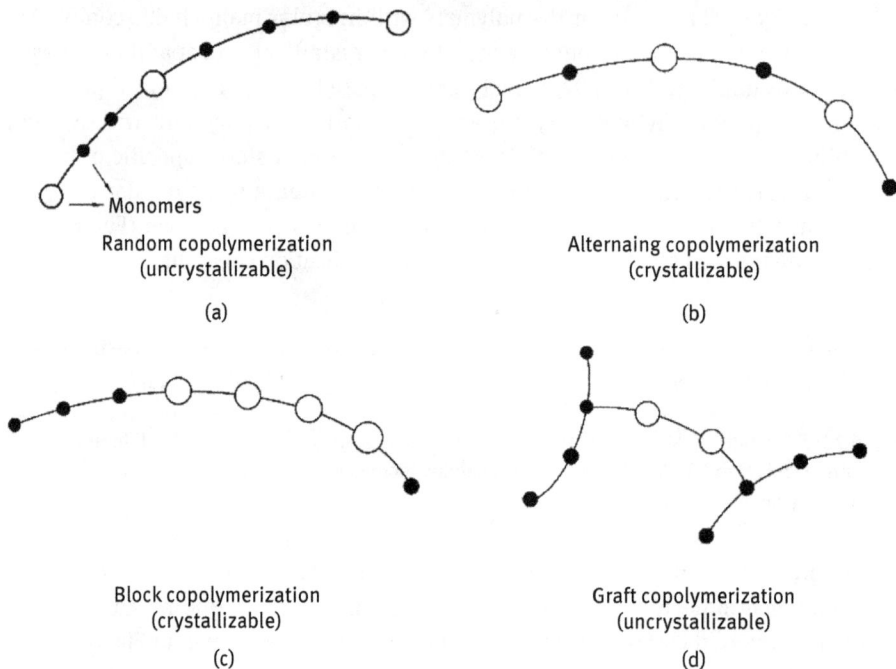

**Fig. 6.33:** Schematic illustration of crystallization capability of copolymers prepared by different manners.

usually occurs during the melting process. Figure 6.34(a) and (b) illustrates the curve of mass per unit volume against temperature during the melting of crystalline polymer and low molecule, respectively. As shown in these figures, there is no essential difference between polymer and low molecule during melting, but only differences in degree, such as the sudden change of thermodynamic functions (e.g., mass per unit volume and specific heat capacity). Unlike low molecule, whose thermodynamic functions changes suddenly within a narrow temperature range of 0.2 °C, the change of thermodynamic function of polymers occurs in a wide range of temperature, and this range of temperature is called melting range. In the melting range, the polymer crystals melt accompanying with the increase in temperature. This is different from the low molecule that crystallizes at a constant temperature under the equilibrium state of liquid and solid phases. The phenomenon of temperature rise along with melting originates from the slow crystallization of polymer because it is difficult to diffuse chains to proper position and arrange segments regularly to form relatively perfect crystals under the normal cooling rate. Under the normal heating rate, the relatively imperfect parts (with thin crystal plates and many defects) will melt at a relatively low temperature, but the relatively perfect parts have to melt at a relatively high temperature. As a result, a wide range of melting temperature appears. If the heating rate is decreased, for

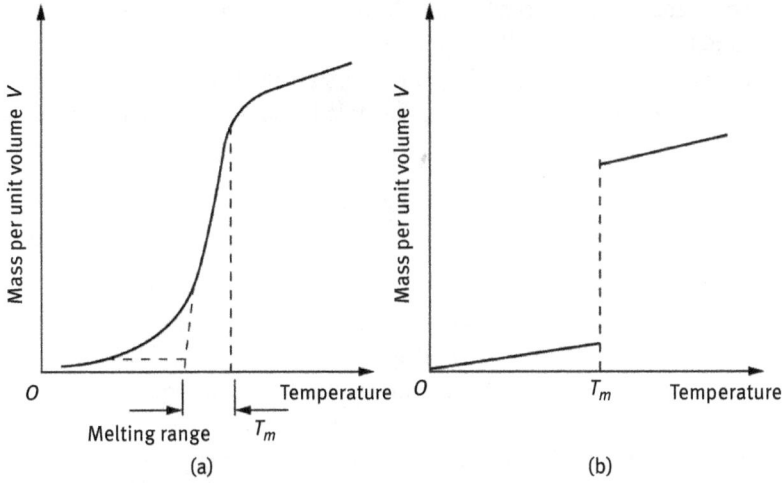

**Fig. 6.34:** The comparison of the volume per unit mass (V)–temperature (T) curves of polymer crystals (a) and low molecule crystals (b) during melting.

example, increasing 1 °C and then maintaining it for 24 h, the melting process of polymer crystals will be similar to that in low molecule, that is, the sudden change of thermodynamic functions within relatively narrow temperature range of 3–4 °C (as shown in Fig. 6.35), according to the measurement of volume per unit mass until there is no further change in the volume (for metals, the heating rate can reached at 0.5–0.15 °C per minute). At the end of melting, the curve exhibits an obvious turning point, which determine the melting point of the polymer crystal. The reason why the melting range is narrowed under the low cooling rate can be

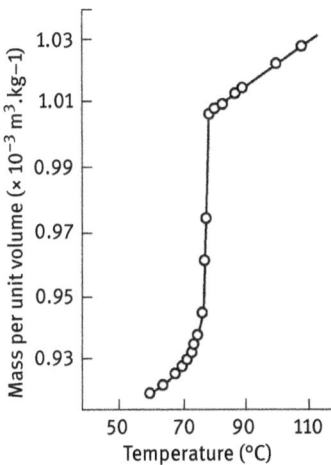

**Fig. 6.35:** The volume per unit mass–temperature curve of polydecamethylene adipate.

explained as: when the imperfect crystals are destroyed at a relatively low temperature, there is enough time to recrystallization to form more perfect and stable crystals.
At a higher temperature, all the relatively perfect crystals melt in a narrow range of
temperature. According to thermodynamics, the relationship between the melting
points and the thickness of crystal plates can be expressed as

$$T_{m,l} = T_{m,\infty}\left(1 - \frac{2\sigma_c}{l\Delta H}\right) \tag{6.66}$$

where $l$ is the thickness of crystal plate, $T_{m,l}$ and $T_{m,\infty}$ are the melting points when
the crystal plates are $l$ and $\infty$, $\Delta H$ is the melting enthalpy per unit volume of polymer
crystal, $\sigma_c$ is specific surface energy. Obviously, the smaller $l$ is, the lower $T_{m,l}$ is.
When $l = \infty$, the melting point reaches the limiting value $T_{m,\infty}$, the equilibrium melting point. Thus, it is true that the change degree of the melting range $(T_{m,\infty} - T_{m,l})$ is
related to the thickness of crystal plates. Table 6.8 lists the melting points of polyethylene with different thickness of crystal plates. Generally, its $T_{m,\infty}$ is 145 °C.

**Table 6.8:** The thickness values melting points of polyethylene crystal plates.

| $l$ (nm) | 28.2 | 29.2 | 30.9 | 32.3 | 33.9 | 34.5 | 35.1 | 36.5 | 39.8 | 44.3 | 48.3 |
|---|---|---|---|---|---|---|---|---|---|---|---|
| $T_m$ (°C) | 131.5 | 131.9 | 132.2 | 132.7 | 134.1 | 133.7 | 134.4 | 134.3 | 135.5 | 136.5 | 136.7 |

# Chapter 7
# Binary phase diagrams and solidification of alloys

In the practical industry, materials widely used are multicomponent materials composed of at least two components, rather than single-component materials. The addition of multicomponent makes the solidification process and phase transformation more complex, which provides the opportunity for the variability and its choice of material properties. Among the multicomponent systems, the binary system is the most basic and well-studied system at present. The binary phase diagram is a powerful tool to study the relationships between temperature, compositions, and phases under equilibrium thermodynamic conditions. It has been widely used in metallic, ceramic, and polymer materials. As the viscosity of metal alloy melts is small and easy to flow, it can often be directly solidified into the required parts, or poured into an ingot which can be made into products through subsequent hot or cold processing. Because of the high viscosity of ceramic melt and its poor mobility, ceramic products are less made via direct solidification from melts, but usually made by powder sintering. Polymer alloys can be prepared by physical (mechanical) or chemical blending, from the molten (liquid) state of direct molding or extrusion.

This chapter briefly describes the representation and determination method of binary phase diagram, reviews the basic points of thermodynamics of phase diagram, focuses on the characteristics of different types of phase diagram, and analyzes microstructure evolution and the microstructures/defects of alloy castings. Finally, a brief introduction of polymer alloy will be given.

## 7.1 Representation and determination method of phase diagram

The binary system is one that has more than one component than the one-component system. Considering the composition, temperature, and pressure at the same time, the binary phase diagram must be three-dimensional phase diagram. In view of the complexity of the three-dimensional stereogram and the study of system normally under 1 atm, the binary phase diagram only considers the thermodynamic equilibrium state of the system under the two variables of composition and temperature. The abscissa of the binary phase diagram indicates the composition, and the ordinate indicates the temperature. If the system consists of two components of $A$ and $B$, one end of the abscissa represents component $A$, and the other end represents component $B$, then the point corresponding to any

https://doi.org/10.1515/9783110495379-007

composition of different ratios between two components in the system can be found on the abscissa.

There are two ways to represent the composition on the binary phase diagram: mass fraction ($w$) and mole fraction ($x$). If $A$ and $B$ are elementary substances, the conversion between them are as follows:

$$\left. \begin{array}{l} w_A = \dfrac{A_{rA}x_A}{A_{rA}x_A + A_{rB}x_B} \\[3mm] w_B = \dfrac{A_{rB}x_B}{A_{rA}x_A + A_{rB}x_B} \end{array} \right\} \qquad (7.1)$$

$$\left. \begin{array}{l} x_A = \dfrac{w_A/A_{rA}}{w_A/A_{rA} + w_B/A_{rB}} \\[3mm] x_B = \dfrac{w_B/A_{rB}}{w_A/A_{rA} + w_B/A_{rB}} \end{array} \right\} \qquad (7.2)$$

where $w_A$ and $w_B$ are the mass fractions of $A$ and $B$, respectively; $A_{rA}$ and $A_{rB}$ are the relative atomic masses of $A$ and $B$, respectively; $x_A$ and $x_B$ are the mole fractions of $A$ and $B$, respectively; and $w_A+w_B=1$ (or 100%), $x_A+x_B=1$ (or 100%).

When the components $A$ and $B$ in the binary phase diagram are stable compounds, the relative atomic mass $A_{rA}$ (or $A_{rB}$) of component $A$ (or $B$) in eq. (7.2) is replaced by the relative molecular mass $M_{rA}$ (or $M_{rB}$) of the compound $A$ (or $B$) and the atomic mass fraction of $A$ (or $B$) in eq. (7.2) is replaced by the corresponding molecular mass fraction. Then we can obtain the mole fraction expression of the compound, which is more commonly used in the binary phase diagrams of ceramics and polymer.

Unless specified, the compositions of binary phase diagrams in this textbook are shown in the mass fraction.

The binary phase diagram is drawn from the critical points of the various materials, and the critical point represents the phase transition point where the state of the material structure changes essentially. There are two methods to determine the critical points: dynamic method and static method. The former includes thermal analysis method, thermal expansion method (dilatometry), and electrical resistance method, while the latter includes metallographic method and X-ray structure analysis. Accurate determination of the phase diagram must be used in conjunction with a variety of methods. As an example, the process to draw the binary Cu–Ni phase diagram by measuring the critical points using the thermal analysis method will be described as follows.

First, a series of Cu–Ni alloys with different Ni contents were prepared, and their cooling curves were measured from liquid to room temperature, and the critical points were obtained. Figure 7.1(a) shows the cooling curves of pure Cu, Cu–Ni alloys with 30%, 50%, and 70% of $w$(Ni) and pure Ni. The cooling curves of pure Cu and Ni are similar that both have a horizontal plateau, which indicates that the solidification occurs at a constant temperature of 1,083 and 1,452 °C, respectively. The other three binary alloy curves do not show the horizontal plateau, but have the second transition point. The turning point (the critical point) of the higher temperature indicates the start temperature of solidification, and the lower turning point corresponds to the finish temperature of solidification. This indicates that the solidification of the three alloys is different from the pure metals and occurs in a certain temperature range. Marking the temperature and composition corresponding to the critical points on the ordinate and the abscissa of the binary phase diagram, each critical point corresponds to a point on the binary phase diagram. Then the Cu–Ni binary phase diagram can be obtained by connecting the start temperature points and the finish temperature points of solidification, respectively, as shown in Fig. 7.1(b). The phase boundary line connecting the start temperature of solidification is called the liquidus line and that connecting the finish temperature of solidification is called the solidus line. In order to accurately determine the critical points of phase transformation, the thermal analysis method must be operated at a very slow cooling rate to achieve the thermodynamic equilibrium conditions, generally controlled at 0.5–0.15 °C per minute.

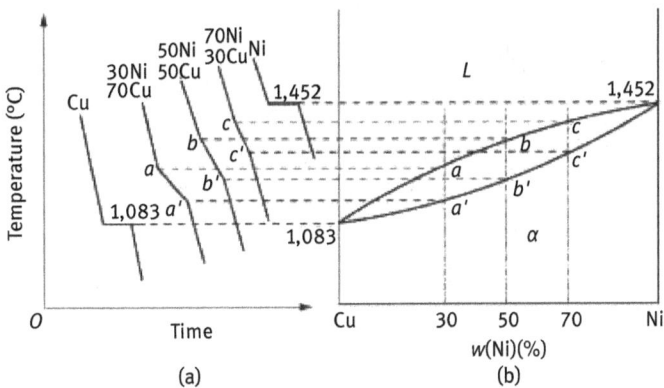

**Fig. 7.1:** Cu–Ni phase diagram established by the thermal analysis method. (a) Cooling curves and (b) phase diagram.

The regions separated by boundary lines in the phase diagram are called phase regions, indicating the type and number of equilibrium phases present in this range. In the binary phase diagram, there are single-phase and two-phase regions. According to the phase rule, in the single-phase region, $f = 2-1+1 = 2$, indicating that the alloy in this phase range can be independent of temperature and composition to maintain the original state. If in the two-phase region, $f = 1$, showing that there is only one independent variable between the temperature and compositions. This means the compositions must change with any change in temperature in this phase region, and cannot change independent of temperature and vice versa. If there are three phases coexisting in the alloy, then $f = 0$, which indicates that the compositions and temperature of the three equilibrium phases are fixed and this is an invariant reaction. Hence, the phase diagram is expressed as a horizontal line which is called the three-phase equilibrium line, such as that in the binary $Al_2O_3$–$ZrO_2$ phase diagram in the ceramic material system (see Fig. 7.2). The $Al_2O_3$ and $ZrO_2$ solid phases simultaneously crystallize out of the liquid phase with $w(ZrO_2) = 42.6\%$ at 1,710 °C and the three phases coexist at this temperature. From the phase rule, we can see that at most three phases can coexist in the binary system.

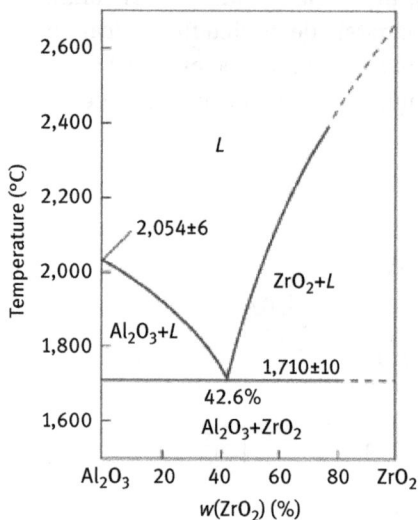

Fig. 7.2: Phase diagram of $Al_2O_3$–$ZrO_2$.

## 7.2 Essentials of phase diagram thermodynamics

Phase diagram is usually drawn after a lot of experimental data. But for various reasons, it may be difficult to measure some phase region or bring errors during the measurement of the phase diagram. For this reason, we need to use thermodynamics knowledge to calculate the phase diagram, which has made great progress with the development of the computer. In this section, we will only learn to use the fundamental principle of thermodynamics to analyze the phase diagram.

### 7.2.1 Free energy–composition curve of solid solution

The quasichemical model of solid solution can be used to calculate the free energy of solid solution. This model only considers the bond energy between the nearest neighbor atoms, so the mixing enthalpy $\Delta H_m$ is approximated. Assuming that the solvent atoms of the solid solution and the solute atoms have the same radius and crystal structures and they are infinitely miscible, the volume of the mixture before and after mixing is constant and the volume change $\Delta V_m = 0$ after mixing. Additionally, the quasichemical model only considers the mixed entropy produced by the different arrangement of the two components, irrespective of the temperature-induced vibration entropy. So the Gibbs free energy of the solid solution can be obtained as

$$G = \underbrace{x_A\mu_A^\circ + x_B\mu_B^\circ}_{G^\circ} + \underbrace{\Omega x_A x_B}_{\Delta H_m} + \underbrace{RT(x_A\ln x_A + x_B\ln x_B)}_{-T\Delta S_m} \tag{7.3}$$

where $x_A$ and $x_B$ are the mole fractions of $A$ and $B$, respectively; $\mu_A^\circ$ and $\mu_B^\circ$ are the molar free energies of $A$ and $B$ at temperature $T$ (K), respectively; $R$ is the gas constant and $\Omega$ is interaction parameter which is expressed by

$$\Omega = N_A z \left( e_{AB} - \frac{e_{AA} + e_{BB}}{2} \right) \tag{7.4}$$

where $N_A$ is the Avogadro constant, $z$ is coordination number, and $e_{AA}$, $e_{BB}$, and $e_{AB}$ are the binding energies of component pairs of $A$–$A$, $B$–$B$, and $A$–$B$, respectively. By considering zero energy to be the state where the atoms are separated to infinity, $e_{AA}$, $e_{BB}$, and $e_{AB}$ are negative quantities, and become increasingly more negative as the bonds become stronger.

In eq. (7.3), the first term, $G^\circ$, corresponds to the Gibbs energy of a mechanical mixture of the constituents of the phase; the second term, $\Delta H_m$, is the so-called excess Gibbs energy term; and the third term, $-T\Delta S_m$, corresponds to the entropy of mixing for an ideal solution. The free energy $G$ of solid solution is the sum of $G^\circ$, $\Delta H_m$ and $-T\Delta S_m$, which are functions of the composition (mole fraction $x$). Therefore, the free energy–composition curve of the solid solution can be drawn at any given temperature for three different cases of $\Omega$, as shown in Fig. 7.3.

If $\Omega < 0$, the curve is U-shaped with only a minimum value within the whole range of the composition and its curvature $d^2G/dx^2$ is positive as shown in Fig. 7.3(a).

If $\Omega = 0$, the curve is also U-shaped as shown in Fig. 7.3(b).

If $\Omega > 0$, however, the free energy–composition curve has two minimum values of $E$ and $F$ as shown in Fig. 7.3(c). The curve is $\cap$-shaped with $d^2G/dx^2 < 0$ among the composition range between the saddle points $q$ and $r$ ($d^2G/dx^2 = 0$). In the composition range between $E$ and $F$, the system is decomposed into two solid solutions with different compositions, that is, solid solution has a certain miscibility gap which will be analyzed in section 7.3.4.

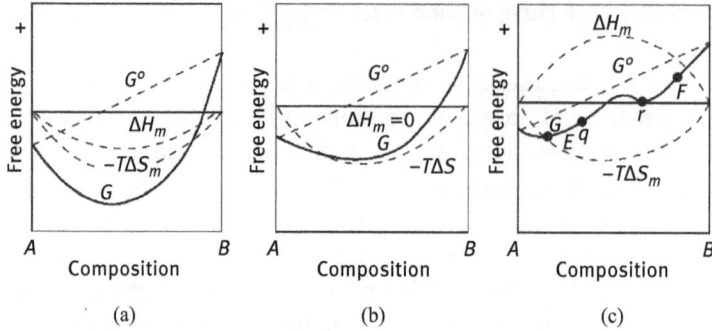

**Fig. 7.3:** Free energy–composition curve of solid solution: (a) $\Omega < 0$, (b) $\Omega = 0$, and (c) $\Omega > 0$.

Different interaction parameters lead to different types of free energy–composition curves, the physical meaning of which is as follows:

When $\Omega < 0$, that is, $e_{AB} < (e_{AA} + e_{BB})/2$ from eq. (7.4), the energy of $A$–$B$ bond is below the average energy of $A$–$A$ and $B$–$B$, so the components of $A$ and $B$ in the solid solution attract each other to form short-range ordering distribution, with an extreme case forming a long-range order, and in this case $\Delta H_m < 0$.

When $\Omega = 0$, that is, $e_{AB} = (e_{AA} + e_{BB})/2$, the solid solution is called ideal solid solution as the energy of $A$–$B$ bond is equal to the average energy of $A$–$A$ and $B$–$B$, and the atomic site occupation of components is completely random, and in this case $\Delta H_m = 0$.

When $\Omega > 0$, that is, $e_{AB} > (e_{AA} + e_{BB})/2$, the energy of $A$–$B$ bond is above the average energy of $A$–$A$ and $B$–$B$, which means that the binding of $A$–$B$ is unstable and $A$ and $B$ components tend to accumulate to form a segregated state, and in this case $\Delta H_m > 0$.

## 7.2.2 The principle of common tangent of multiphase equilibrium

For the tangent line to the Gibbs free energy–composition curve at each point, the ends of the tangent line intersect with the two component ordinates, where the intercept of the A component ordinate represents the chemical potential ($\mu_A$) of A component and the intercept of the B component ordinate represents the chemical potential ($\mu_B$) of B component in the solid solution at this point. In the binary system, the condition for thermodynamic equilibrium of two phases (e.g., solid phase $\alpha$ and solid phase $\beta$) is expressed by $\mu_A^\alpha = \mu_A^\beta$ and $\mu_B^\alpha = \mu_B^\beta$, that is, each component must have the same chemical potential in the two phases. Thus, when two phases exist in equilibrium, the compositions of two phases are determined by the common tangent of the two-phase free energy–composition curves as shown in Fig. 7.4. The slope of the tangent line can be expressed by

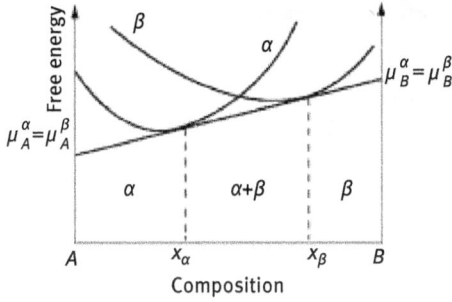

**Fig. 7.4:** The free energy–composition curves of two-phase equilibrium.

$$\begin{rcases} \dfrac{dG_\alpha}{dx} = \dfrac{\mu_B^\alpha - \mu_A^\alpha}{\overline{AB}} = \mu_B^\alpha - \mu_A^\alpha \\[2mm] \dfrac{dG_\beta}{dx} = \dfrac{\mu_B^\beta - \mu_A^\beta}{\overline{AB}} = \mu_B^\beta - \mu_A^\beta \end{rcases} \tag{7.5}$$

where $\overline{AB} = 1$. For the binary system, there can be three-phase equilibrium at a certain temperature, such as the three-phase equilibrium of $\alpha$, $\beta$, and $\gamma$. In this case, the thermodynamic conditions are $\mu_A^\alpha = \mu_A^\beta = \mu_A^\gamma$ and $\mu_B^\alpha = \mu_B^\beta = \mu_B^\gamma$. Therefore, the three points at the common tangent line represent the compositions of $\alpha$, $\beta$, and $y$ at equilibrium. The intercept on the coordinate of $A$ and $B$ components are the chemical potentials of $A$ and $B$, respectively, under this condition as shown in Fig. 7.5.

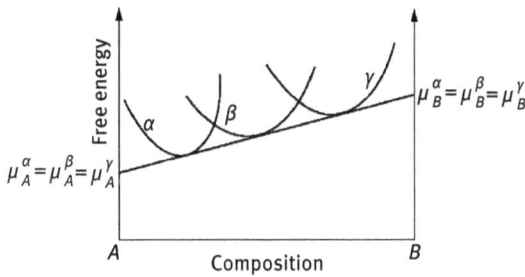

**Fig. 7.5:** The free energy–composition curves of three-phase equilibrium in binary system.

## 7.2.3 Free energy and lever rule of mixture

Assuming $\alpha$ and $\beta$ phases formed by components $A$ and $B$, the mole amounts of two phases are $n_1$ and $n_2$ and their molar Gibbs free energy are $G_{m1}$ and $G_{m2}$, respectively. If the mole fractions of component $B$ in $\alpha$ and $\beta$ phases are $x_1$ and $x_2$, respectively, the molar fraction of component $B$ in the mixture can be expressed by

$$X = \frac{n_1 x_1 + n_2 x_2}{n_1 + n_2}$$

The molar Gibbs free energy of the mixture can be expressed by

$$G_m = \frac{n_1 G_{m1} + n_2 G_{m2}}{n_1 + n_2}$$

Then

$$\frac{G_m - G_{m1}}{x - x_1} = \frac{G_{m2} - G_m}{x_2 - x} \tag{7.6}$$

The above equation indicates that the molar Gibbs free energy $G_m$ of the mixture should be on the same line as the molar Gibbs free energies $G_{m1}$ of phase $\alpha$ and $G_{m2}$ of phase $\beta$; moreover, $x$ should lie between $x_1$ and $x_2$. This straight line is the common tangent line when $\alpha$ and $\beta$ phases are at equilibrium as shown in Fig. 7.6.

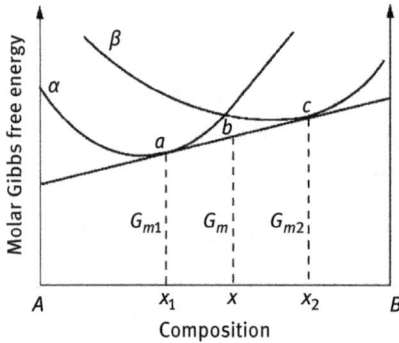

Fig. 7.6: The free energy of mixture.

When the composition $x \leq x_1$, the molar Gibbs free energy of the $\alpha$ solid solution is lower than that of the $\beta$ solid solution. Thus, the solid $\alpha$ phase is the stable phase and the system is in the single-phase $\alpha$ state. Similarly, the system is in the single-phase $\beta$ state when $x \geq x_2$, since the molar Gibbs free energy of the $\beta$ solid solution is lower than that of the $\alpha$ solid solution. But when $x_1 < x < x_2$, the Gibbs free energy of the mixture is lower than that of the $\alpha$ phase or $\beta$ phase, so the energy of the system is a minimum when the two phases coexist (forms a mixture). When the two equilibrium phases coexist, the compositions of the multiphase are fixed corresponding to those of the tangent points of $x_1$ and $x_2$. Then we can obtain

$$\left. \begin{array}{l} \frac{n_1}{n_1 + n_2} = \frac{x_2 - x}{x_2 - x_1} \\[2mm] \frac{n_2}{n_1 + n_2} = \frac{x - x_1}{x_2 - x_1} \end{array} \right\} \tag{7.7}$$

The relative amounts of two phases $\alpha$ and $\beta$ can be calculated by eq. (7.7), which is called the lever rule. When two phases coexist, the relative amount of $\alpha$ phase is $\frac{x_2 - x}{x_2 - x_1}$ and $\beta$ phase is $\frac{x - x_1}{x_2 - x_1}$, both of which are changed with the composition $x$.

## 7.2.4 Phase diagram deduced from the free energy–composition curve

The compositions of equilibrium phases of a system at a certain temperature can be obtained according to the principle of common tangent line. Therefore, the binary system phase diagram can be drawn from the free energy–composition curves at different temperatures of the system. Figure 7.7 shows the complete miscible phase diagram of the two components A and B obtained from the free energy–composition curves of liquid (L) and solid phase (S) at temperatures $T_1$, $T_2$, $T_3$, $T_4$, and $T_5$. Figure 7.8 shows the phase diagrams of the eutectic phases of the $A$, $B$ components obtained from the free energy–composition curves of the L, $\alpha$, and $\beta$ phases at the above five different temperatures. Figures 7.9–7.11 show the peritectic phase diagram, the miscibility gap phase diagram and the phase diagram containing compound obtained from the free energy–composition curves, respectively.

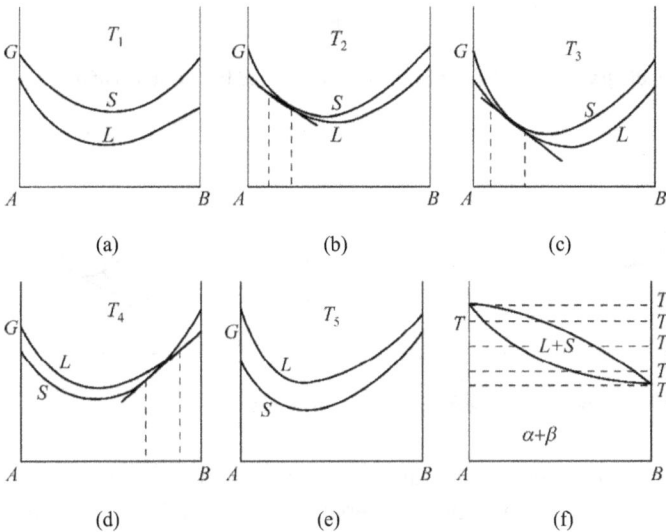

Fig. 7.7: The completely miscible phase diagram obtained from a series of the free energy curves.

## 7.2.5 Geometric rules of binary phase diagram

According to the basic principle of thermodynamics, some geometrical rules of phase diagram can be derived, which can help us to understand the framework of phase diagram and judge the possible error of phase diagram.

(1) All the lines in the phase diagram represent the temperature at which the phase transition takes place and the composition of the equilibrium phases, so the phase boundary is a reflection of phase equilibrium and the equilibrium phase compositions must change with temperature along the boundary line.

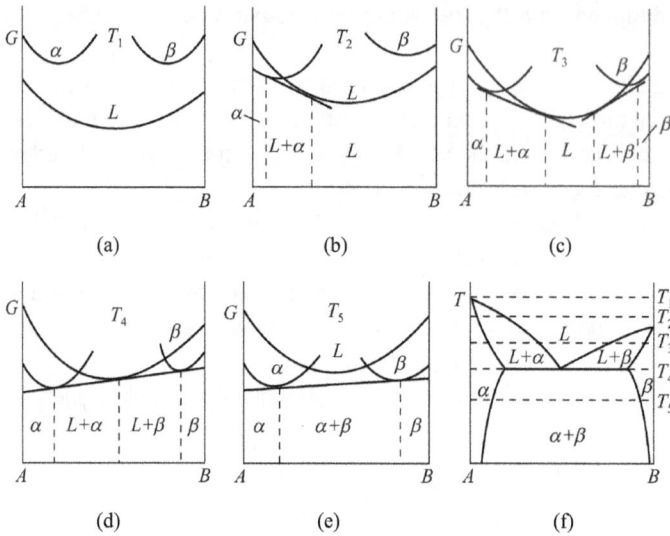

**Fig. 7.8:** The eutectic phase diagram formed by two components obtained from a series of free energy curves.

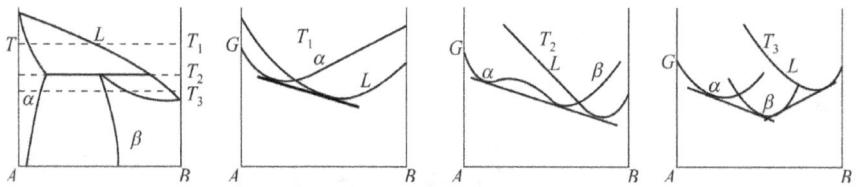

**Fig. 7.9:** The relation between the peritectic phase diagram and free energy.

(2) There must be a two-phase region separating two single-phase regions instead of one line. Two two-phase regions must be separated by a single phase region or a three-phase horizontal line. That is, in the binary phase diagram, the phase number difference of adjacent phase regions must be 1 (except for the case of point contact), and this rule is called the phase contact rule.

(3) In binary phase diagram, the three-phase equilibrium must be a horizontal line, which represents the constant temperature reaction. There are three com- position points on this horizontal line corresponding to the equilibrium three phases, where two points should be at both ends of the horizontal line and the other between the end points. The upper and lower regions of the horizontal line are connected with three two-phase regions.

(4) When the boundary between a two-phase region and a single-phase region in- tersects with a three-phase isotherm, the extension line of the boundary line should enter another two-phase region, rather than a single-phase region.

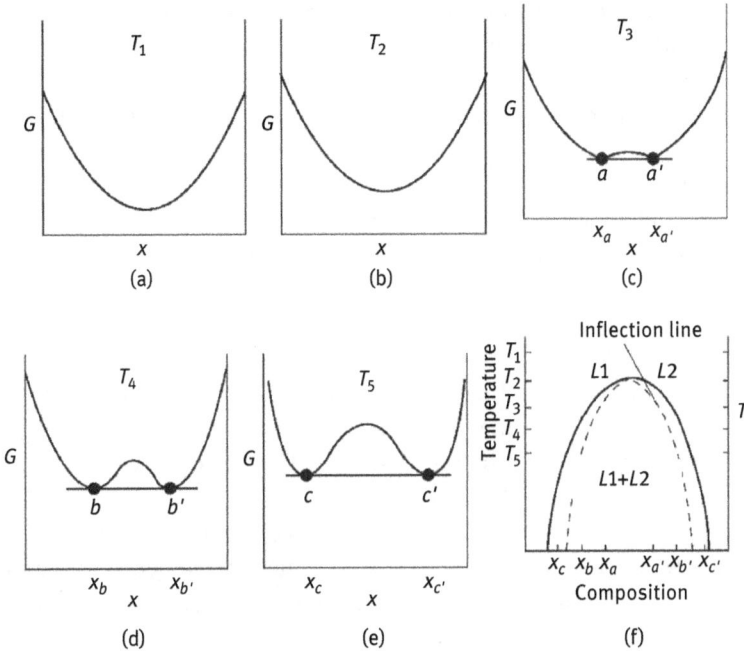

**Fig. 7.10:** The relation between the miscibility gap phase diagram and free energy ((a)–(e) are the series of Gibbs free energy curves at different temperatures).
(a) $T = T_1$, (b) $T = T_2$, (c) $T = T_3$, (d) $T = T_4$, (e) $T = T_5$, (f) phase diagram.

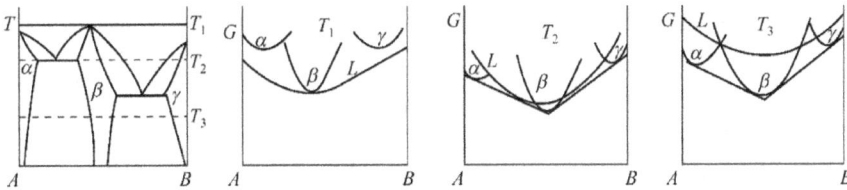

**Fig. 7.11:** The relation between phase diagram containing compound and free energy.

### 7.2.6 Calculation of phase diagram

The experimental determination of phase diagrams is very time-consuming and sometimes difficult, which becomes even more complicated as the number of components increase. For this concern, the CALPHAD method (CALculation of PHAse Diagrams, also expanded to refer to Computer Coupling of Phase Diagrams and Thermochemistry) has been developed since the early 1970s and enormous progress has been made over the past 30 years, which significantly reduces the effort required to determine equilibrium conditions in a multicomponent system.

In the CALPHAD method, all available experimental and theoretical information on phase equilibria, thermochemical, and thermophysical properties should be collected, such as crystallography, atomic bonding, order–disorder transitions, and magnetic properties. The thermodynamic properties of each phase are then represented by a variety of mathematical models to describe the temperature, pressure, and concentration dependencies of the free-energy functions of the various phases. For the calculation of phase equilibria in a multicomponent system, it is necessary to minimize the total Gibbs energy of all the phases that take part in this equilibrium, which involves the solution of a series of nonlinear equations on thermodynamic equilibrium conditions.

Additionally, the CALPHAD method can also be extended to model kinetic processes in multicomponent systems. In a similar way as the Gibbs free energy, the atomic mobilities and chemical driving force can be calculated, which provides the basis for calculating critical kinetic properties such as diffusion coefficients.

There are a variety of commercial software products available for the calculation of phase diagrams, such as Thermo-Calc/DICTRA, FactSage, Pandat, JMatPro, and MTDATA. These software packages have graphical user-friendly interfaces, powerful functions, and systematic thermodynamic and kinetic databases for alloys and other materials, which make phase diagram information more easily accessible for the nonexpert user. They are widely used in research and industrial development of various alloys and processing technologies, where the calculated phase diagrams save large amount of time and resources by reducing the experimental effort and costs.

## 7.3 Binary phase diagram

In this section, three types of basic phase diagrams, that is, isomorphous, eutectic, and peritectic diagram, will be mainly analyzed by in-depth discussion on solidification process and final structures of the binary systems, in order to obtain a systematic understanding of the relationship between composition and structure under equilibrium or nonequilibrium solidification conditions. Besides, we will analyze the binary phase diagram with miscibility gap and spinodal decomposition. Finally, other types of binary phase diagrams are introduced, and the methods for analysis of binary phase diagrams are summarized.

### 7.3.1 Isomorphous phase diagram and solidification of solid solution

#### 1. Isomorphous phase diagram
The transition from a liquid phase to a single-phase solid solution is called isomorphous phase transformation. Most of binary phase diagrams contain the isomorphous

transformation at least part. Some binary alloys, such as Cu–Ni, Au–Ag, Au–Pt, and some binary ceramics like NiO–CoO, CoO–MgO, and NiO–MgO, only have isomorphous transformation. In these systems, the two elements or components are completely soluble each other in both the liquid state and solid state and only a single type of crystal structure exists in whole composition range of two components. Therefore, they are called isomorphous systems. For the complete solid solubility of two elements each other, they must obey the following conditions formulated by William Hume-Rothery and known as the Hume-Rothery solid solubility rules:

(1) The crystal structure of each element of the solid solution must be the same.
(2) The difference of the atomic sizes of two elements must be less than 15%.
(3) There should be no appreciable difference in the electronegativities of the two elements, so that they cannot form compounds each other.

Basically, these rules are also suitable for ionic compound-based solid solutions, with just a modification that the atomic radius is replaced by the ionic radius. For instance, unlimited solubility between NiO and MgO is attributed to the same NaCl-style crystal structure, very close ionic radius of $Ni^{2+}$ and $Mg^{2+}$ (0.069 and 0.066 nm, respectively) and the same valence. On the contrary, CaO and MgO can only form a solid solution with limited solubility, because of a larger ionic radius of $Ca^{2+}$ of 0.099 nm, although both have the same crystal structure and valence. Binary isomorphous phase diagram for the Cu–Ni and NiO–MgO systems is shown in Figs. 7.12 and 7.13, respectively.

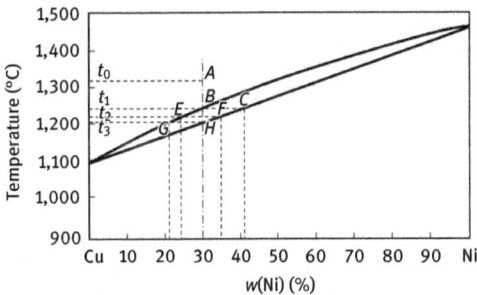

**Fig. 7.12:** Cu–Ni phase diagram.

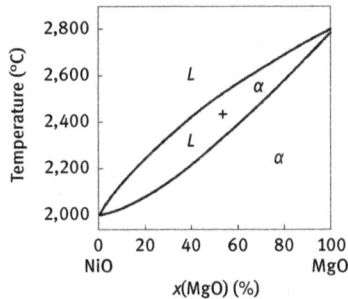

**Fig. 7.13:** NiO–MgO phase diagram.

There can be other types of isomorphous phase diagram. For instance, Au–Cu and Fe–Co have a minimum point while Pb–Ti has a maximum point on the phase diagram, as shown in Fig. 7.14(a) and (b), respectively. For the alloy corresponding to the minimum or maximum point, the number of variables for determination of the state of the system should be subtracted by one, because the liquid and solid phases have the same composition. Thus, the degree of freedom is $f = C - P + 1 = 1 - 2 + 1 = 0$, namely, isothermal transformation.

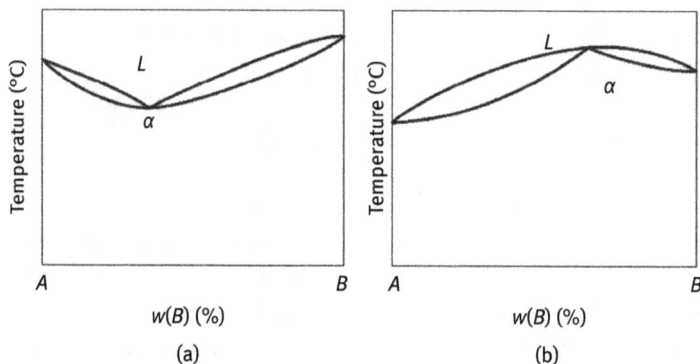

**Fig. 7.14:** Phase diagram with (a) a minimum point and (b) a maximum point.

## 2. Equilibrium solidification of solid solution

Equilibrium solidification is known as a freezing process in which each solidifying step can reach equilibrium, that is, components have enough time for completely mutual diffusion to gain the composition of equilibrium phases. A Cu–Ni alloy, with w(Ni) of 30% (Fig. 7.12), will be taken as an example to describe the equilibrium solidification process.

When the liquid alloy is cooled from point A of a higher temperature to point B ($t_1 = 1,245$ °C), which intersects with the liquidus line, it will start to crystallize. The composition of the solid phase can be determined from the intersection point C of the tie (horizontal) line BC and the solidus line, that is, approximately 41% Ni, which indicates that the liquid phase of composition B and the solid phase of composition C are at equilibrium at this temperature. Since the formation of nucleus needs undercooling to overcome energy barrier of nucleation, the liquid alloy must be cooled to a temperature lower than $t_1$ to initiate nucleation and growth, and the composition of crystallized solid solution is close to that of C at this moment. With the drop of temperature, the composition of the solid phase varies along the solidus line while that of the liquid phase varies along the liquidus line. When cooling to $t_2$ (1,220 °C), the composition of liquid is E, about 24% Ni, and the composition of solid is F, about 36% Ni, obtained by intersecting the tie line EF with the liquidus and solidus lines. Meanwhile, the relative amounts of the solid and liquid are both 50% based on the lever rule. When cooling to $t_3$ (1,210 °C), the composition of solid solution becomes the initial alloy composition (30% Ni), which is at equilibrium with a last little bit of liquid phase (composition $G$). Once the temperature is slightly lower than $t_3$, the last liquid also freezes into solid. After completely solidified, a homogeneous single-phase solid solution is obtained. The structural evolution of the alloy during whole solidification process is shown in Fig. 7.15.

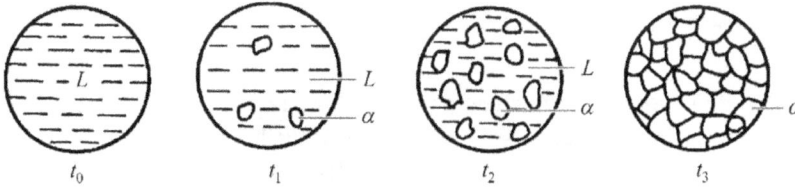

**Fig. 7.15:** Schematic representation showing structural evolution of a Cu–Ni solid solution during equilibrium solidification.

Like pure metals, the solidification of a solid solution also includes two stages: nucleation and growth, but the secondary component in the alloy makes the solidification process more complicated. For instance, the crystallized solid has a different composition from liquid alloy, thus the fluctuations in both constituent and thermal energy are required in the nucleation stage. In addition, the alloy freezes in a temperature interval during which the compositions of both liquid and solid phases vary with the decreasing temperature. Therefore, the solidification process is necessarily dependent on atom diffusion of two components. It should be emphasized that at each temperature the equilibrium solidification essentially experiences three processes: diffusion process within the liquid phase; then the growth of the solid phase; finally diffusion process within the solid phase. The concept will be better understood by a following example.

Consider the equilibrium solidification of the above alloy when cooling from $t_1$ to $t_2$, as shown in Fig. 7.16, where $L$ and $S$ indicate the liquid and solid phases, respectively; $w_L$ and $w_s$ represent their compositions at the phase interface which are determined by the corresponding compositions of the liquidus and solidus lines at $t_2$; and $z$ indicates the distance from the initially solidified front. As the liquid alloy partially crystallizes into solid phase, the Ni content in the liquid phase surrounding the solid phase decreases to $w_L$ while the Ni content in liquid phase far away from the solid phase maintains its initial content. Thus, diffusion occurs in the liquid due to the concentration gradient. During diffusion process, the solid continues to grow until the composition throughout the liquid reaches $w(\text{Ni}) = 24\%$. Similarly, concentration gradient also exists in solid and leads to diffusion as well. But since the diffusion of atoms in solid is much slower than that in liquid, it will take much more time for the diffusion of atoms in the solid to achieve a uniform composition of $w_s$.

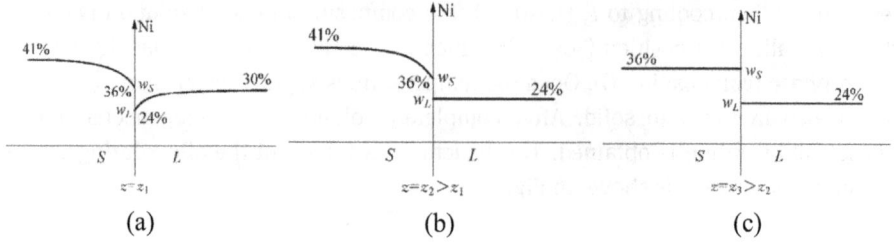

**Fig. 7.16:** Three processes in equilibrium solidification.

During solidification, each nucleus grows into a grain and the composition within each grain is uniform due to sufficient atomic diffusion at each temperature. Therefore, the microstructure of solid solution under equilibrium solidification is same as pure metal, and the compositions among different grains and within them are identical except at grain boundary.

### 3. Nonequilibrium solidification of solid solution

The crystallization of solid solution depends on diffusion of components. Conditions of equilibrium solidification are realized only for extremely slow cooling rates, where there is sufficient time for complete diffusion. In all practical solidification situations, however, cooling rates are much too rapid to ensure sufficient diffusion at each temperature. As a result, the solidification process always deviates from the equilibrium conditions, and it is termed nonequilibrium solidification.

During nonequilibrium solidification, the compositions of the liquid and solid phase will deviate from the liquidus and solidus line, respectively. The deviation of solid composition is more serious than that of liquid due to a much slower diffusion rate in solid, which governs the nonequilibrium solidification process. Figure 7.17 demonstrates the composition variations of the liquid and solid phases during nonequilibrium solidification. At the temperature $t_1$, the alloy I first crystallizes into a solid with composition $\alpha$. Inasmuch as the Cu content of the solid is far below the initial composition, the adjacent liquid will certainly increase its Cu content to $L_1$. Upon cooling to $t_2$, equilibrium composition of the solid should be $\alpha_2$, and the liquid should be $L_2$ by tie line. However, due to a fast cooling rate, diffusion is not sufficient, especially in the solid. The composition within the solid is thus lower than $\alpha_2$ or even close to as $\alpha_1$, leading to a nonuniform composition. At this moment, the average composition of the solid $\alpha_2'$ should lie in a range from $\alpha_1$ to $\alpha_2$ while that of the overall liquid $L_2'$ lies between $L_1$ and $L_2$. Upon cooling to $t_3$, equilibrium composition of the solid and the liquid should be $\alpha_3$ and $L_3$, respectively. But actually, the solid has an average composition $\alpha_3'$ of $\alpha_1$, $\alpha_2$ and $\alpha_3$, while the liquid has an average composition $L_3'$ of $L_1$, $L_2$, and $L_3$, owing to the same reason of insufficient diffusion to achieve equilibrium compositions. Solidification terminates until cooling to the temperature $t_4$, and the average solid composition then shifts from $\alpha'_3$ to $\alpha'_4$, that is, the initial alloy composition. If we link the average composition points of the solid and liquid phases at each temperature, dash lines $\alpha_1\alpha'_2\alpha'_3\alpha'_4$ and $L_1L'_2L'_3L'_4$ can be obtained as shown in Fig. 7.17(a), which are termed the average composition lines of the solid and liquid phases, respectively. Figure 7.17(b) illustrates the composition and the microstructure evolution of both liquid and solid phases.

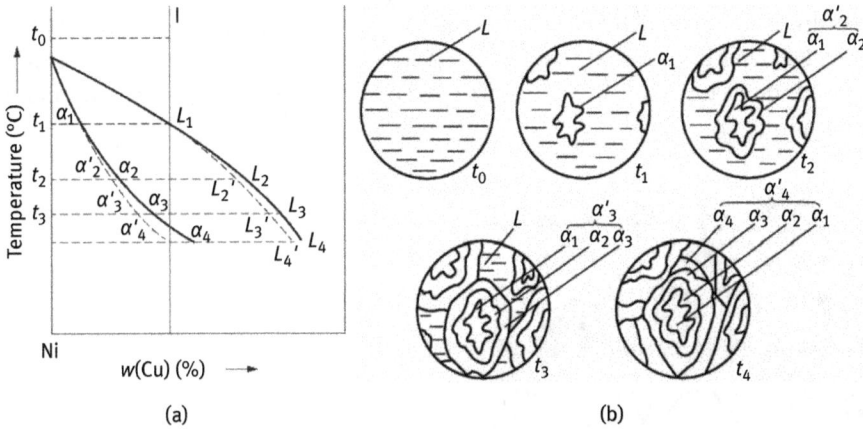

**Fig. 7.17:** Schematic representation showing the composition and microstructure evolution of both liquid and solid during nonequilibrium solidification.

On basis of analysis on nonequilibrium solidification mentioned above, the following conclusions can be drawn:

(1) The average composition lines of the solid and liquid phases are different from the solidus and liquidus lines, which depend on the cooling rate. The higher the cooling rate is, the greater the average composition lines of the solid and liquid phases deviate from the solidus and liquidus lines becomes. Conversely, the lower the cooling rate is, the smaller they deviate from the solidus and liquidus lines, manifesting the cooling rate is much closer to equilibrium cooling rate.

(2) The center part of each grain, which freezes first, is always rich in the high-melting element (e.g., Ni for this Cu–Ni system), while the outer part is rich in the low-melting element (Cu). The concentration of the high-melting element (Ni) decreases with position from the center to the grain boundary in each grain, and this is termed a cored structure.

(3) Nonequilibrium solidification always leads to a lower final solidification temperature than that of equilibrium solidification.

Usually, solid solution crystallizes in a pattern of dendrite growth. Nonequilibrium solidification causes a different composition between early crystallized dendrite trunks and lately crystallized interdendritic regions, which is termed dendritic segregation. Because one dendritic grain is developed from one nucleus, dendritic segregation belongs to intragranular segregation. Figure 7.18 shows as-cast structure of a Cu–Ni alloy, where the dendrite morphology is revealed with different contrasts after erosion attributed to different compositions between trunks and interdendritic regions. With the help of the electronic probe microscope analyzer (EPMA), the trunks are found to be rich in Ni (bright contrast as invulnerable to erosion) and the regions between branches are rich in Cu (dark contrast as vulnerable to erosion). The dendritic segregation is often found in solid solution during nonequilibrium solidification.

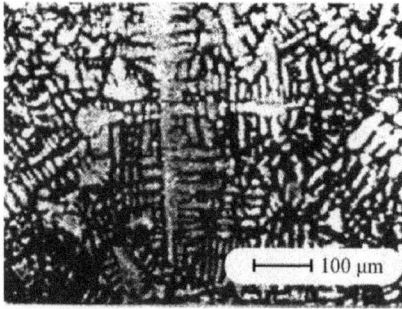

**Fig. 7.18:** As-cast structure of a Cu–Ni alloy (dendritic crystals).

The *cored* structure mentioned above usually does not possess the optimal properties due to composition segregation. As a casting component or ingot having a cored structure is reheated, the intergranular and interdendritic regions will melt first even at a temperature below the equilibrium solidus temperature of the alloy because they are richer in the low-melting component. The thin liquid film that appears in these regions separates branches and the gains, leading to a sudden loss in mechanical integrity. To eliminate cored structure, a homogenization annealing can be carried out at a temperature below the solidus point for the particular alloy composition. During this process, atomic diffusion occurs to produce grains with homogeneous compositions. Figure 7.19 shows microstructure of a Cu–Ni alloy after diffusion annealing, where dendrite morphology has disappeared. Further EPMA analysis also verifies the elimination of dendritic segregation.

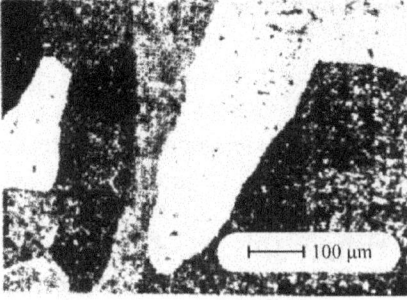

**Fig. 7.19:** Microstructure of a Cu–Ni alloy after diffusion annealing.

## 7.3.2 Eutectic phase diagram and solidification of alloys

### 1. Eutectic phase diagram

Two components which constitute a eutectic phase diagram are infinitely soluble each other in the liquid state but partially soluble or even mutually insoluble in the solid state. The mixing of the two components lowers the melting point of alloys with regard to each single component. As a result, the liquidus lines are concave from the pure components at two ends of the diagram to the middle, which intersect at a *eutectic point* corresponding to the *eutectic temperature*, the lowest temperature at which the liquid phase can exist. At this temperature, two solid phases are simultaneously crystallized through eutectic solidification, and the mixture of the two phases is called *eutectic structure or eutectic.*

Figure 7.20 shows a typical binary eutectic phase diagram of Pb–Sn alloy. Many other alloys, such as Al–Si, Pb–Sb, Pb–Sn, and Ag–Cu, have a similar type of phase diagram. Eutectic alloys are very important in casting industry because of some unique properties: (1) lower melting points than the pure components to

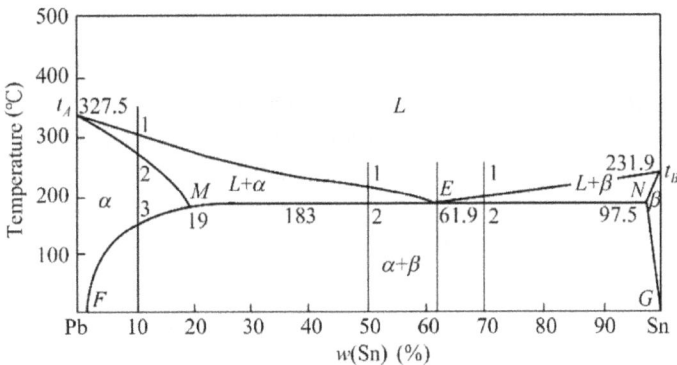

**Fig. 7.20:** Pb–Sn binary phase diagram.

simplify melting and casting operation; (2) better fluidity of eutectic alloys than solid solution alloys, without formation of dendritic crystals hindering liquid flow, and hence improved castability; (3) isothermal transformation (without temperature range of solidification) with reduced casting defects, such as segregation and shrinkage; (4) various types of microstructural morphology obtained from eutectic solidification, especially regularly arranged lamellar or rod-like eutectic structure, to form in situ composites with excellent properties.

As shown in Fig. 7.20, the melting point of Pb ($t_A$) is 327.5 °C and that of Sn ($t_B$) is 231.9 °C. Two liquidus lines intersect at point $E$ of eutectic temperature 183 °C. $\alpha$ is a Pb-based solid solution with dissolved Sn and $\beta$ is a Sn-based solid solution with dissolved Pb. They are called *terminal solid solutions* since they appear at the ends of the diagram. Liquidus lines $t_A E$ and $t_B E$ represent the start temperatures of crystallization of $\alpha$ and $\beta$ phases, while solidus lines $t_A M$ and $t_B N$ represent their finish temperatures, respectively. Horizontal line *MEN* indicates the eutectic temperature of coexisting three phases and their compositions and is called *eutectic line*. The eutectic line demonstrates a liquid phase $L_E$ with composition $E$ will simultaneously transform into a solid phase $\alpha_M$ with composition $M$ and the other solid phase $\beta_N$ with composition $N$. The mixture of two phases ($\alpha_M + \beta_N$) is called eutectic structure, and the eutectic reaction is written as follows:

$$L_E \rightarrow \alpha_M + \beta_N$$

Often, the horizontal solidus line *MN* is called the *eutectic isotherm*. According to the phase rule, the degree of freedom is zero when three phases coexist in a binary system. Therefore, the eutectic reaction is called an *invariant reaction*, point $E$ is called an invariant point, which is designated by composition $w(E)$ and temperature $T_E$. Lines *MF* and *NG* are the *solvus* lines of $\alpha$ and $\beta$ solid solution, respectively, showing that the solubility of $\alpha$ and $\beta$ decreases with decreasing temperature.

In Fig. 7.20, phase equilibrium lines separate the phase diagram into three single-phase regions ($L$, $\alpha$, and $\beta$) and three two-phase regions ($L + \alpha$, $L + \beta$, and $\alpha + \beta$). $L$ phase region locates just above the eutectic line while $\alpha$ and $\beta$ phase regions are at both sides of the eutectic line.

### 2. Equilibrium solidification of eutectic alloys and microstructures

Consider Pb–Sn alloys, the equilibrium solidification and microstructures for typical alloys of different compositions will be discussed as follows.

### a. $w(Sn) < 19\%$

As shown in Fig. 7.20, when a liquid Pb–Sn alloy with $w(Sn) = 10\%$ is slowly cooled to temperature $t_1$ (marked "1" in the figure), $\alpha$ solid solution begins to crystallize from liquid. With decreasing temperature, the amount of the primary $\alpha$ solid solution increases whilst that of liquid decreases, and compositions of the solid and liquid vary along the solidus line $t_A E$ and liquidus line $t_A M$, respectively. On cooling to $t_2$, solidification completes and the liquid all transforms into the singe-phase $\alpha$ solid solution. This process is the same as equilibrium solidification in the isomorphous phase diagram. The $\alpha$ solid solution will have no change in the temperature range between $t_2$ and $t_3$. However, once cooling below $t_3$, Sn will be in a saturated sate in the $\alpha$ solid solution. Therefore, excess Sn will precipitate in the form of $\beta$ solid solution from the $\alpha$ solid solution, which is called *secondary* $\beta$ solid solution (marked by $\beta_{\mathrm{II}}$) in order to differentiate the primary $\beta$ solid solution directly formed from liquid. The secondary $\beta$ solid solution preferably precipitates along grain boundaries or at defects inside grains. With further drop of temperature, the amount of $\beta_{\mathrm{II}}$ increases gradually whilst equilibrium compositions of $\alpha$ and $\beta_{\mathrm{II}}$ phases vary along the solvus lines $MF$ and $NG$, respectively. As previously pointed out, relative amounts of phases in two-phase regions, such as $L$ and $\alpha$ phases in $(L + \alpha)$ two-phase region, and $\alpha$ and $\beta$ phases in $(\alpha + \beta)$ two-phase region, can be calculated by the lever rule.

Figure 7.21 is a schematic representation of solidification process for a Pb–10% Sn alloy. All the alloys with compositions ranging between $M$ and $F$ experience similar equilibrium solidification process mentioned above, with same equilibrium microstructure of $\alpha + \beta_{\mathrm{II}}$ at ambient temperature but different relative amounts. For alloys in the composition range between $N$ and $G$, the equilibrium solidification processes are basically similar but the equilibrium microstructure is $\beta + \alpha_{\mathrm{II}}$ after solidification.

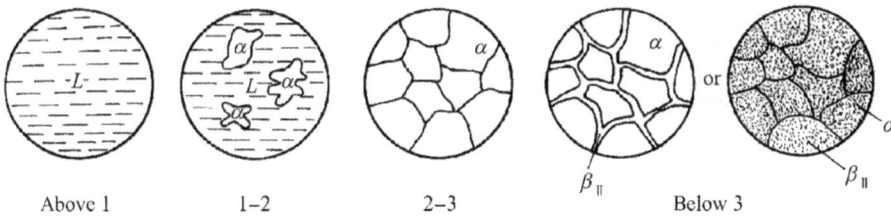

Above 1     1–2     2–3     Below 3

**Fig. 7.21:** Schematic view of equilibrium solidification for a Pb–10% Sn alloy.

### b. Eutectic alloy
The composition of eutectic alloy is $w(\mathrm{Sn}) = 61.9\%$ (as shown in Fig. 7.20). Upon the slow cooling to 183 °C, the liquid phase $L_E$ simultaneously transforms into $\alpha$ and $\beta$ solid solutions at a constant temperature until whole solidification completes. The relative amount of $\alpha$ and $\beta$ phases can be determined by the lever rule, that is, constructing a tie line below the eutectic line in $(\alpha + \beta)$ two-phase region, the length of which can be approximately taken as $MN$, then

$$w(\alpha_M) = \frac{EN}{MN} \times 100\% = \frac{97.5 - 61.9}{97.5 - 19} \times 100\% = 45.4\%$$

$$w(\beta_N) = \frac{ME}{MN} \times 100\% = \frac{61.9 - 19}{97.5 - 19} \times 100\% = 54.6\%$$

During cooling to room temperature after the eutectic reaction for the alloy, there is a decrease in solid solubility of solute in the $\alpha$ and $\beta$ solid solutions along $MF$ and $NG$ solvus lines, respectively. Secondary $\beta_{II}$ and $\alpha_{II}$ precipitate from $\alpha$ and $\beta$, respectively. It is difficult to distinguish the secondary phases from the corresponding phases in the eutectic mixture under an optical microscope, because the secondary phases generally combine with each of same phases in the eutectic. Figure 7.22 shows alternatively distributed lamellar microstructure of the eutectic alloy at ambient temperature, where $\alpha$ phase is shown in black and $\beta$ phase in white (etched in a 4% nitric acid alcohol solution). The equilibrium solidification process of the eutectic alloy is shown in Fig. 7.23.

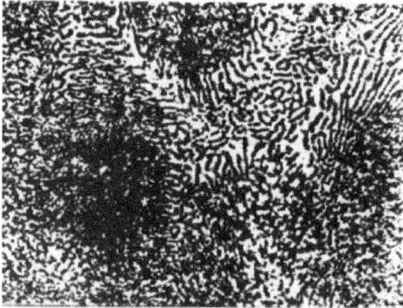

**Fig. 7.22:** Eutectic structure of a Pb–Sn eutectic alloy (250 ×).

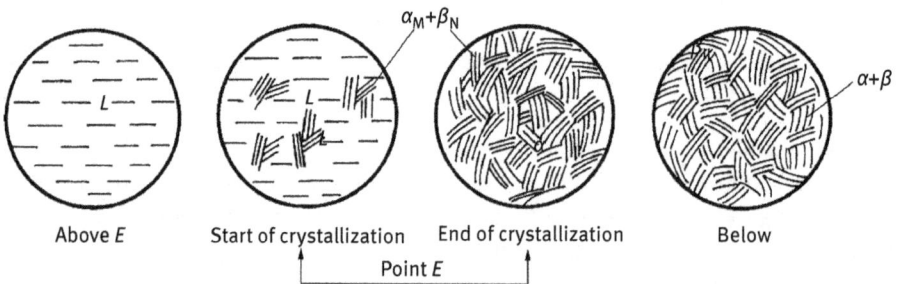

**Fig. 7.23:** Schematic view of equilibrium solidification of a Pb–Sn eutectic alloy.

## c. Hypoeutectic alloy

In Fig. 7.20, alloys with compositions in the range between $M$ and $E$ are called *hypoeutectic alloys* because the compositions are lower than the eutectic composition and only a part of liquid phase can transform into eutectics. Let us take a Pb–50% Sn alloy as an example to analyze its equilibrium solidification process (Fig. 7.24).

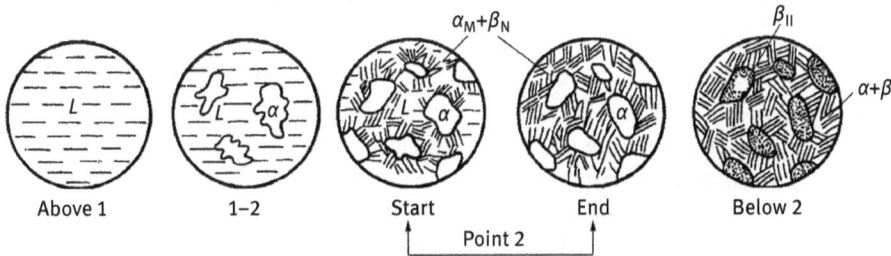

**Fig. 7.24:** Schematic view of equilibrium solidification for a Pb–Sn hypoeutectic alloy.

When slowly cooling to a temperature between $t_1$ and $t_2$, primary $\alpha$ phase continually precipitates from liquid by isomorphous transformation. Compositions of $\alpha$ phase change along the solidus line $t_A M$ while those of the liquid phase vary along the liquidus line $t_A E$. Upon cooling to $t_2$, the residual liquid reaches composition point $E$, and eutectic transformation occurs to form eutectics while the primary $\alpha$ phase do not take part in the eutectic reaction. Once the eutectic transformation is completed, the equilibrium structure consists of the primary $\alpha$ solid solution and eutectics $(\alpha + \beta)$, which can be abbreviated as $\alpha + (\alpha + \beta)$. The primary $\alpha$ phase (also called proeutectic $\alpha$ phase) usually exhibits dendritic morphology and the eutectics $(\alpha + \beta)$ exhibit alternatively lamellar morphology, moreover, the primary phase is dispersed between the colonies of the eutectic microconstituent. Relative amounts of the two structures, also called the *microconstituents*, can be calculated by the lever rule, that is, constructing a tie line in the two-phase region $(L + \alpha)$ above the eutectic line, and the length of tie line is close to $ME$, then the mass fractions are obtained as follows:

$$w(\alpha + \beta) = w(L) = \frac{50 - 19}{61.9 - 19} \approx 72\%$$

$$w(\alpha) = \frac{61.9 - 50}{61.9 - 19} \approx 28\%$$

Above results show that there are 28% of the primary $\alpha$ phase and 72% of the $(\alpha + \beta)$ eutectics after eutectic transformation of the Pb–50% Sn alloy. However, these two microconstituents are both composed by $\alpha$ and $\beta$ phases, also termed constituent phases. The relative amounts of $\alpha$ and $\beta$ phases can be calculated by the lever rule, that is, constructing a tie line in the two-phase region $(\alpha + \beta)$ below the eutectic line, and the length of tie line is close to $MN$, so

$$w(\alpha) = \frac{97.5 - 50}{97.5 - 19} \approx 60.5\%$$

$$w(\beta) = \frac{50 - 19}{97.5 - 19} \approx 39.5\%$$

Note that in above calculations, $\alpha$ phase includes primary $\alpha$ phase and $\alpha$ phase in eutectics. Hypoeutectic alloys of different compositions get similar microconstituents of $\alpha + (\alpha + \beta)$ after eutectic reaction, but the relative amounts of primary and eutectic microconstituent are different. The closer to eutectic composition $E$ point the composition of a hypoeutectic alloy is, the more eutectics will be obtained. Conversely, the closer to composition point $M$ of $\alpha$ phase, the more primary $\alpha$ phase will be got. The key point of the above analysis is to construct corresponding tie lines to determine the relative amounts of microconstituents and phases based on the lever rule. The microconstituents not only reflect the difference in the structure of phases, such as $\alpha$ phase and $\beta$ phase, but in phase morphologies, such as primary $\alpha$ phase and $\alpha$ phase in eutectic structure.

When the alloy continues to cool below temperature $t_2$, $\beta_{\mathrm{II}}$ phase will precipitate from both primary $\alpha$ and eutectic $\alpha$ phase, and $\alpha_{\mathrm{II}}$ phase will precipitate from eutectic $\beta$ phase due to decreasing solubility of the solid solutions. The final structure at ambient temperature should be $\alpha_{\mathrm{primary}} + (\alpha + \beta) + \alpha_{\mathrm{II}} + \beta_{\mathrm{II}}$. But since the amounts of the $\alpha_{\mathrm{II}}$ and $\beta_{\mathrm{II}}$ phase are very small, only $\beta_{\mathrm{II}}$ could be found in the primary $\alpha$ phase in the observation of optical microscope and the characteristic of the eutectic microconstituent remains unchanged. Thus, the final structure can be written as $\alpha_{\mathrm{primary}} + (\alpha + \beta) + \beta_{\mathrm{II}}$, or even $\alpha_{\mathrm{primary}} + (\alpha + \beta)$.

Figure 7.25 shows microstructure of a hypoeutectic Pb–Sn alloy after etched by a 4% nitric acid alcohol solution at ambient temperature. Dark black dendrites are primary $\alpha$ solid solution, inside which the white dots are $\beta_{\mathrm{II}}$ phase, and small black and white phases are $(\alpha + \beta)$ eutectic structure.

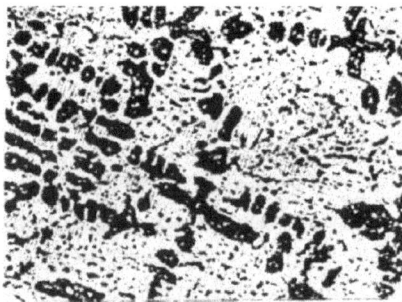

**Fig. 7.25:** Microstructure of a hypoeutectic Pb–Sn alloy (200 ×).

## d. Hypereutectic alloy

Alloys with compositions between $E$ and $N$ are called *hypoeutectic alloys*. The equilibrium solidification process and the equilibrium structure are similar as hypoeutectic alloys except that the primary phase is dendritic $\beta$ solid solution instead of $\alpha$, and the final structure at ambient temperature is $\beta_{primary} +(\alpha + \beta)$, as show in Fig. 7.26.

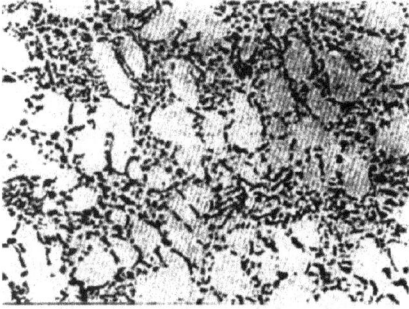

Fig. 7.26: Microstructure of a hypereutectic Pb–Sn alloy: primary $\beta$ phases in ellipse shapes are dispersed in black Pb–Sn eutectics (200×).

Based on the above analysis, although alloys of compositions in a range from $F$ to $G$ have different microstructures, all of them are composed of two basic phases, $\alpha$ and $\beta$. Hence, microstructures of two-phase alloys are determined by the different morphologies, amounts, sizes, and distributions of constituent phases, resulting in different properties.

## 3. Nonequilibrium solidification of eutectic alloys

### a. Pseudoeutectic

Under equilibrium solidification condition, only eutectic alloys can get complete eutectic structures, but under nonequilibrium solidification condition, some hypoeutectic or hypereutectic alloy may also obtain complete eutectic structures. Such eutectic structures obtained in the noneutectic alloys are termed *pseudoeutectic*.

Alloys with eutectic transformation (such as alloy I in Fig. 7.27) can get eutectic structures if the melts are undercooled to a shadow region embraced by extended lines of the two liquidus (as shown in Fig. 7.27). The shadow region is called pseudoeutectic region, outside which the alloys exhibit both eutectic and dendrite structures. The pseudoeutectic region is enlarged with increasing undercooling in certain range.

Configurations of pseudoeutectic regions are different for different alloys in the phase diagram. It is usually symmetrically distributed as shown in Fig. 7.27 when the melting points of the two components are close to each other. Otherwise, it usually

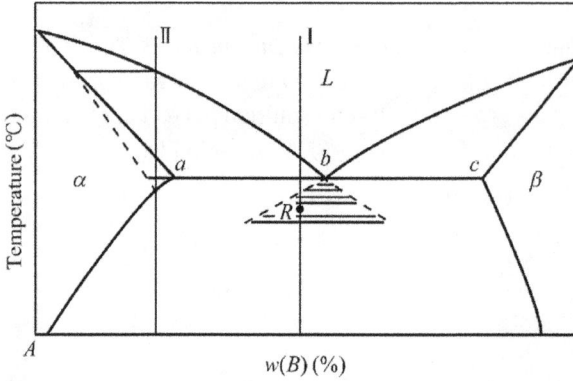

**Fig. 7.27:** Nonequilibrium solidification of eutectic alloy system.

deviates to the side of the component with a higher melting point, such as in Al–Si alloy as shown in Fig. 7.28. The reason is generally that the nucleation and growth of the two phases in eutectics require diffusion of two components, because they have different compositions from the liquid alloy. Since the constituent phase based on the low melting-point component is much closer to the liquid alloy in compositions, it is easier to reach the phase composition through diffusion, and hence gets a faster growth rate. As a result, in order to meet the requirement of formation of the two constituent phases on diffusion, the position of the pseudoeutectic region must deviate to the side of the high melting-point component.

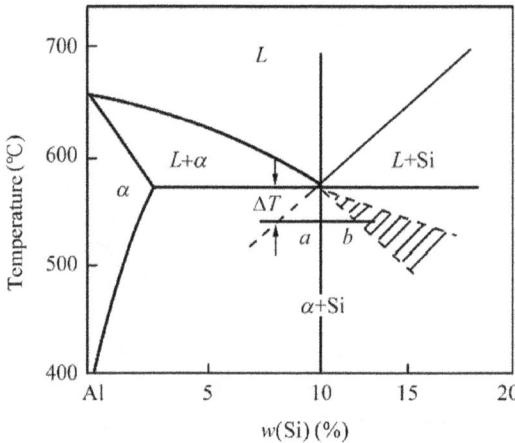

**Fig. 7.28:** Pseudoeutectic region of Al–Si alloys.

It is extremely important to learn about the location and size of the pseudoeutectic region in order to properly interpret the formation of nonequilibrium microstructure of alloys. The accurate configuration of the pseudoeutectic region in the phase diagram must be determined by experiments. But qualitative information about the distribution of pseudoeutectic region in the phase diagram is helpful to explain some abnormal phenomena that cannot be explained by equilibrium phase diagrams. For example, in Al–Si alloys, hypoeutectic rather than eutectic structure will be obtained for the alloy of the eutectic composition under rapid cooling conditions while eutectic or hypoeutectic structure may be gained for hypereutectic alloys. These abnormal phenomena are reasonably explained by considering the configuration of the pseudoeutectic region as shown in Fig. 7.28.

### b. Nonequilibrium eutectic structure

Some alloys, such as alloys to the left side of point $a$ or right side of pint $c$ in Fig. 7.27, may have a small amount of nonequilibrium eutectics under fast cooling rather than the single-phase solid solution under equilibrium conditions. For alloy $\text{II}$, the solid solution shows the cored structure with dendrite segregation, and the average compositions deviate from the solidus line. The dash line in Fig. 7.27 indicates the average composition line of solid under fast cooling. There still remains a small amount of liquid that have not crystallized when the alloy cools to the solidus line. Upon further cooling to the eutectic temperature, the remained liquid reaches the eutectic composition and then eutectic transformation takes place, resulting in nonequilibrium eutectic structures distributed at grain boundaries and interdendritic regions of $\alpha$ phase which are the last locations to solidify. Nonequilibrium eutectic structures seriously affect material properties and should be eliminated. They are unstable thermodynamically and can be eliminated by diffusion annealing slightly below the eutectic temperature to obtain a uniform structure of single-phase $\alpha$ solid solution. Because the amount of nonequilibrium eutectics is very small, the eutectic $\alpha$ phase usually grows attached to the primary $\alpha$ phase, and the other eutectic $\beta$ phase is pushed to the grain boundaries solidified last. This results in the vanishing of characteristics of the constituent two phases in eutectic structure, accordingly, and such separated two eutectic phases are termed *divorced eutectic*. For example, in as-cast Al-4% Cu alloy, the nonequilibrium eutectic $\alpha$ phase may grow attached to the primary $\alpha$ phase, leaving the other eutectic $CuAl_2$ phase distributed at grain boundaries or interdendritic regions to form divorce eutectic as shown in Fig. 7.29.

It should be noted that divorce eutectic can be got not only by nonequilibrium solidification but also under equilibrium solidification conditions. For instance, hypo- or hypereutectic alloys close to the solubility limits, such as alloys adjacent to the right side of point $a$ or left side of point $c$ in Fig. 7.27, have a great amount of primary phases and few eutectics, thus it is also possible to form divorce eutectics.

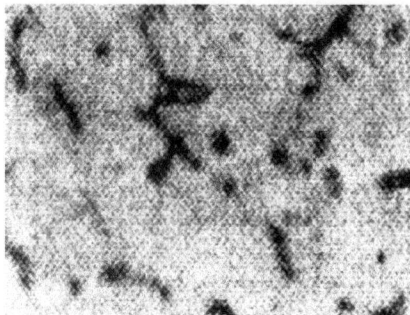

**Fig. 7.29:** Divorce eutectic structure in Al–4% Cu alloy (300×).

### 7.3.3 Peritectic phase diagram and solidification of alloys

#### 1. Peritectic phase diagram

The two components of a binary peritectic alloy system are unlimitedly mutually soluble in the liquid state while have limited solubility in the solid state, and their melting points are often quite different. In the *peritectic reaction*, a liquid phase reacts with a solid phase to form a new and different solid phase at a constant temperature. Such alloy systems include Fe–C, Cu–Zn, Ag–Sn, and Ag–Pt.

Figure 7.30 shows a typical Pt–Ag peritectic phase diagram, where $ACB$ is the liquidus line, $AD$ and $PB$ are the solidus lines, $DE$ is the solvus line of Ag dissolved in Pt-based $\alpha$ solid solution, and $PF$ is the solvus line of Pt dissolved in Ag-based $\beta$ solid solution. The horizontal line $DPC$ is the peritectic line, and alloys of compositions within $DC$ will experience the peritectic reaction at the corresponding temperature:

$$L_c + \alpha_D \rightarrow \beta_p$$

The peritectic reaction is an isothermal transformation, and $P$ is termed *peritectic point*.

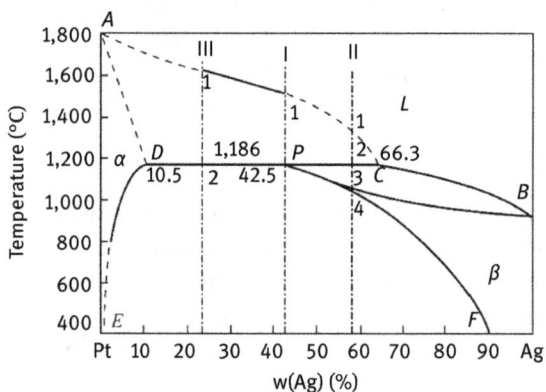

**Fig. 7.30:** The Pt–Ag phase diagram.

## 2. Solidification of peritectic alloys and the equilibrium structures

### a. Pt–42.4% Ag alloy (alloy I)

As shown in Fig. 7.30, when the alloy is slowly cooled from the liquid state to temperature 1 which intersects the liquidus line, primary $\alpha$ phase begins to precipitate. As the alloy continues to cool, the amount of $\alpha$ phase gradually increases while the liquid phase continues to decrease. The compositions of $\alpha$ phase and the liquid phase vary along the solidus line AD and the liquidus line AC, respectively. Their compositions at different temperatures can be determined by a tie lie in $(L + \alpha)$ region. On cooling to the peritectic temperature of 1,186 °C, the composition of the primary $\alpha$ phase reaches point D and the composition of liquid phase reaches point C. The relative amounts of the two phases prior to the peritectic reaction can be obtained by the lever rule:

$$w(L) = \frac{DP}{DC} \times 100\% = \frac{42.5 - 10.5}{66.3 - 10.5} \times 100\% = 57.3\%$$

$$w(\alpha) = \frac{PC}{DC} \times 100\% = \frac{66.3 - 42.5}{66.3 - 10.5} \times 100\% = 42.7\%$$

where $w(L)$ and $w(\alpha)$ are mass fractions of the liquid and solid, respectively. After the peritectic reaction, the liquid and the solid are completely transformed into $\beta$ solid solution.

With further decreasing temperature, the solubility of Pt in the $\beta$ phase gradually decreases along line PF and $\alpha_{II}$ phase continues to precipitate from the $\beta$ solid solution. As a result, the final equilibrium structure at ambient temperature is $\beta + \alpha_{II}$. The whole solidification process is illustrated in Fig. 7.31.

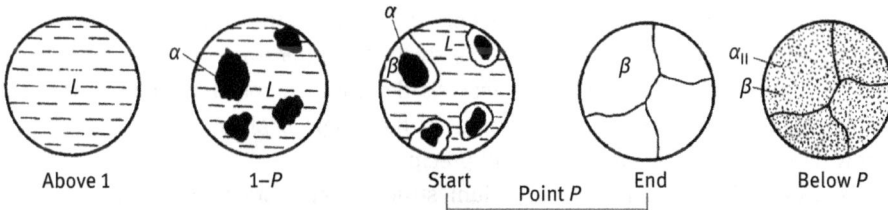

Fig. 7.31: Schematic view of equilibrium solidification of alloy I.

In most cases, the $\beta$ phase formed via the peritectic reaction prefers to nucleate by absorption on surfaces of the primary $\alpha$ phase to reduce the nucleation energy barrier and grows by consuming the liquid and $\alpha$ phase. The $\alpha$ phase no more contacts with the liquid phase $L$ once embraced by the newly formed $\beta$ phase. As shown in Fig. 7.30, Ag is more rich in the liquid phase than in the $\beta$ phase while the content of Ag in the $\beta$ phase is higher than in the $\alpha$ phase. Therefore, Ag atoms need to diffuse from the liquid to the $\alpha$ phase through the $\beta$ phase, while Pt atoms diffuse from the $\alpha$ phase to the liquid in the opposite direction, as illustrated in Fig. 7.32. As a result, the $\beta$ phase grows toward the liquid and $\alpha$ phase simultaneously until it completely swallows the liquid and $\alpha$ phase. Just because the $\beta$ phase grows by embracing the primary $\alpha$ phase and isolating the $\alpha$ phase from the liquid phase, it is called peritectic reaction.

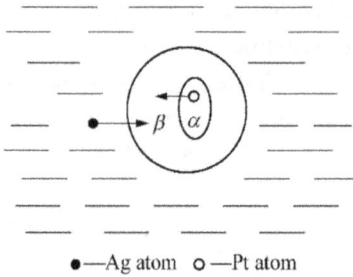

●—Ag atom  ○—Pt atom

**Fig. 7.32:** Schematic representation of atomic migration during the peritectic reaction.

In a few cases, such as a high $\alpha$–$\beta$ interface energy or a large undercooling, the $\beta$ phase may nucleate directly in the liquid phase $L$, rather than relying on the primary $\alpha$ phase, and consistently contact with $L$ and $\alpha$ during growth so as to produce the $\beta$ phase by direct reaction between $L$ and $\alpha$. Apparently, this form of peritectic reaction is faster than the above-mentioned one.

### b. Pt–Ag alloy with 42.4% < w(Ag) < 66.3% (alloy II)

Freezing process of alloy II before peritectic transformation is similar as alloy I. Due to the larger amount of liquid than that needed for the peritectic transformation, after peritectic reaction the remaining liquid continues to crystallize the $\beta$ phase via isomorphous transformation. Compositions of the liquid and the $\beta$ phase vary along the liquidus line CB and solidus line PB, respectively. Until freezing completes at temperature 3, the composition of $\beta$ phase reaches the alloy composition. Then, between temperature 3 and 4, there is no change of the single $\beta$ phase. However, cooling below temperature 4, the $\alpha_{II}$ phase will precipitate from the $\beta$ phase with decreasing temperature. Thus, the final equilibrium structure at room temperature is $\beta + \alpha$. Figure 7.33 schematically shows the equilibrium solidification process of alloy II.

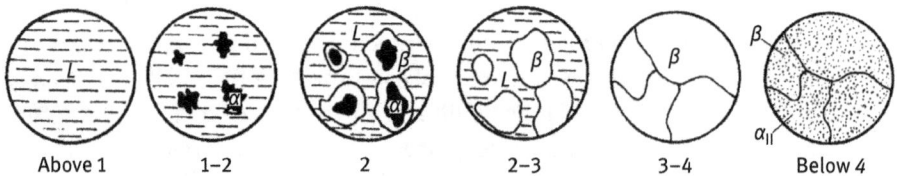

Above 1          1–2          2          2–3          3–4          Below 4

**Fig. 7.33:** Schematic view of equilibrium solidification of alloy II.

### c. Pt–Ag alloy with 10.5% < w(Ag) < 42.4% (alloy III)

Freezing process of alloy III before peritectic transformation is similar as the above cases. Because the relative amount of the $\alpha$ phase is larger than that needed for the peritectic reaction before the peritectic transformation, there is excess $\alpha$ phase besides the newly formed $\beta$ phase after the peritectic transformation. Below the peritectic temperature, the $\alpha_{II}$ phase will precipitate from the $\beta$ phase and the $\beta_{II}$ phase will precipitate from the $\alpha$ phase, resulting in the final equilibrium structure at room temperature of $\alpha + \beta + \alpha_{II} + \beta_{II}$. Figure 7.34 schematically shows the equilibrium solidification process of alloy III.

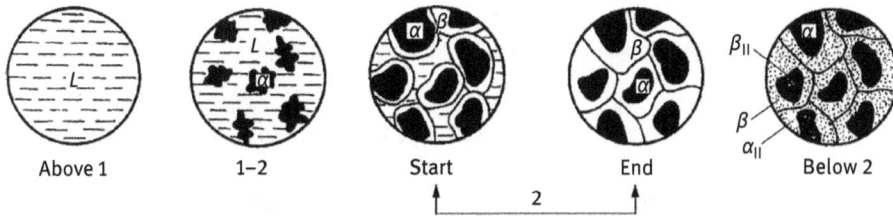

**Fig. 7.34:** Schematic view of equilibrium solidification of alloy III.

### 3. Nonequilibrium solidification of peritectic alloys

As previously analyzed, the product of peritectic transformation, $\beta$ phase, embraces the primary $\alpha$ phase, isolating the $\alpha$ phase from the liquid so as to hinder direct atom interdiffusion between them. Instead, the diffusion must proceed through the $\beta$ phase, thus the rate of peritectic transformation is usually extremely slow. It is apparent that the key factor that governs the rate of peritectic transformation is the diffusion rate inside the newly formed $\beta$ phase.

In practice, the cooling rate is usually high, the atomic diffusion in solid that the peritectic reaction is relying on is usually insufficient, resulting in an incomplete peritectic reaction; namely, there coexist the liquid phase and the $\alpha$ phase below the peritectic temperature, where the liquid phase may directly transform to $\beta$ phase or participate in other reactions and the $\alpha$ phase remains in the center of the $\beta$ phase, forming a nonequilibrium structure of peritectic reaction. For example, when a Sn–35% Cu alloy cools to 415 °C, the peritectic transformation of $L + \varepsilon \rightarrow \eta$ takes place as shown in Fig. 7.35(a). Subsequently, the remaining liquid phase $L$ experiences a eutectic transformation upon cooling to 227 °C, resulting in a final equilibrium structure of $\eta + (\eta + \text{Sn})$. However, the actual nonequilibrium structure (shown in Fig. 7.35(b)) retains considerable amount of the primary $\varepsilon$ phase (gray) which is surrounded by the $\eta$ phase (white) and the outside is the black eutectic structure.

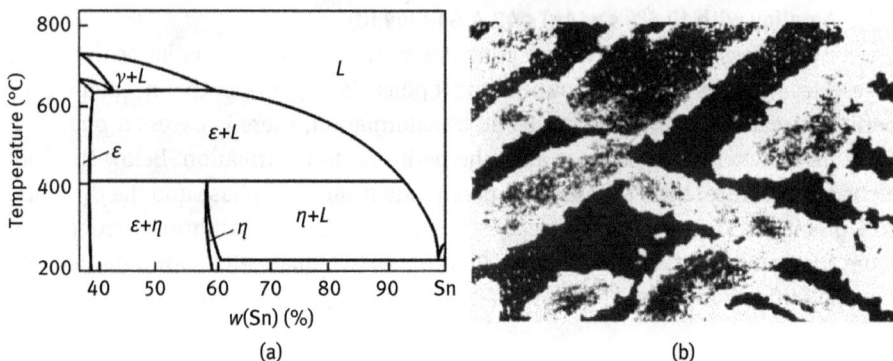

Fig. 7.35: A portion of phase diagram for a Cu–Sn alloy (a) and the nonequilibrium structure (b).

In addition, some alloys without peritectic reaction process originally, such as alloy I shown in Fig. 7.36, will experience the peritectic reaction between the primary $\alpha$ phase and the remaining liquid due to dendritic segregation of solidified $\alpha$ phase under rapid cooling condition. As a result, some phases unexpected in equilibrium state will appear.

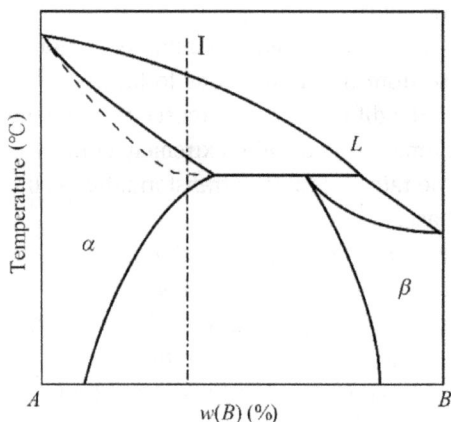

Fig. 7.36: Schematic representation of peritectic reaction due to rapid cooling.

It should be noted that the incompleteness of the peritectic reaction mentioned above is mainly related to the growth pattern of the $\alpha$ phase embraced by the newly formed $\beta$ phase. Therefore, for some alloys (such as Al–Mn), if the peritectic phase solely nucleates and grows in the liquid phase, the peritectic transformation could be completed very quickly. The incomplete peritectic reaction easily occurs in alloys of low peritectic transformation temperatures or diffusion rates.

The nonequilibrium structure produced by peritectic transformation can also be eliminated by annealing diffusion at some temperature below solidus line similar to the nonequilibrium eutectic structure.

### 7.3.4 Miscibility gap phase diagram and spinodal decomposition

Many binary phase diagrams contain miscibility gap, such as Cu–Pb, Cu–Ni, Au–Ni, Cu–Mn alloys and NiO–CoO, $SiO_2$–$Al_2O_3$ ceramics. The miscibility gap shown in Fig. 7.10(f) exhibits mutual immiscibility of two liquid phases. Miscibility gap can also appear in a single-phase solution region, indicating that the single phase will decompose into two phases with a same structure but different compositions. In a word, miscibility gap can be written as $L \rightarrow L_1 + L_2$ or $\alpha \rightarrow \alpha_1 + \alpha_2$, and the latter has two kinds of transformation modes from single phase into the two phases: one is the classic nucleation and growth mode which needs to overcome energy barrier; and the other is unstable decomposition without the nucleation stage, well known as *spinodal decomposition*.

In the section of phase diagram thermodynamics, the physical significance of the interaction parameter $\Omega$ has been explained. When $\Omega > 0$, components $A$ and $B$ prefer to aggregate, respectively, forming a segregation state. There are two minimal values and two spinodal points on the free energy–composition curve when $\Omega > 0$, accompanying with $d^2G/dx^2 < 0$ for compositions in the range between the two spinodal points. The miscibility gap curve (such as the solid line in Fig. 7.10(f)) can be drawn based on the two minimal values on the free energy–composition curves at different temperatures, and the dashed line could be determined by the compositions corresponding to the spinodal points (usually nonappeared in phase diagrams).

Based on the free energy–composition curves, there are thermodynamically unstable regions between the two minimal values and any solid solution phase falling in this composition region will decompose into two phases with compositions corresponding to the two minimal values, respectively. Nevertheless, solid solutions with compositions located in the miscibility gap inside and outside the trace line of spinodal points exhibit different modes of decomposition: the former spontaneously decomposes into two phases of different compositions, while the latter has to overcome energy barrier to nucleate first and then grow up.

The reason why spinodal decomposition occurs inside the trace line of spinodal points in the miscibility gap can be attributed to the difference of free energy before and after the spinodal decomposition $\Delta G$. Assuming the composition of parent phase is $x$, and those of two decomposed phase are $x + \Delta x$ and $x - \Delta x$, respectively. Then $\Delta G = G_{\alpha_1 + \alpha_2} - G_\alpha$.

Taking the second-order Taylor series expansion of the above expression, it can be obtained:

$$\Delta G = \frac{1}{2}[G(\chi + \Delta\chi) + G(\chi + \Delta\chi)] - G(\chi)$$

$$\approx \frac{1}{2}\left[G(\chi) + \frac{dG}{d\chi}\Delta\chi + \frac{d^2G}{d\chi^2}\frac{(\Delta\chi)^2}{2} + G(\chi) + \frac{dG}{d\chi}(-\Delta\chi) + \frac{d^2G}{d\chi^2}\frac{(-\Delta\chi)^2}{2}\right] - G(\chi)$$

$$= \frac{1}{2}\frac{d^2G}{d\chi^2}(\Delta\chi)^2$$

In the above equation, $(\Delta x)^2$ is constantly positive. Therefore, $\Delta G$ is positive when $d^2G/dx^2 > 0$, that is, any minor composition fluctuation could lead to the increase of the free energy of the system, which manifests that the parent phase must overcome an energy barrier to decompose into two phases with different compositions outside the trace line of spinodal points in the miscibility gap. Conversely, since $d^2G/dx^2 < 0$ inside the trace line of inflection points in the miscibility gap, that is, $\Delta G < 0$, any minor composition fluctuation of $\Delta x$ leads to the decrease of the free energy of the system. Subsequently, the parent phase should experience spinodal decomposition without thermodynamic barrier and form new phases by increasing composition fluctuation via uphill diffusion.

### 7.3.5 Binary phase diagrams of other forms

#### 1. Binary phase diagram containing compounds
In some binary systems, one or several compounds appear in the middle of phase diagram, also known as intermediate phases. The compounds can be classified into stable compounds and unstable compounds according to their stability. A stable compound has a definite melting point and can melt into a liquid phase with the same composition of solid phase, while an unstable compound cannot melt into a liquid phase with the same composition of solid phase, instead, it would decompose into two phases as heated to a certain temperature. Here are some examples to illustrate the characteristics of the two types of compounds.

#### a. Phase diagrams containing stable compounds
The zero solubility of a compound in the phase diagram emerges as a vertical line which can be seen as an independent component and splits the whole phase diagram into two separate parts for analysis. Figure 7.37 shows a Mg–Si phase diagram where a stable $Mg_2Si$ compound forms at $w(Si) = 36.6\%$. The compound has a definite melting point (1,087 °C) and the content of Si does not change after melting. Therefore, the stable compound $Mg_2Si$ can be seen as an independent component,

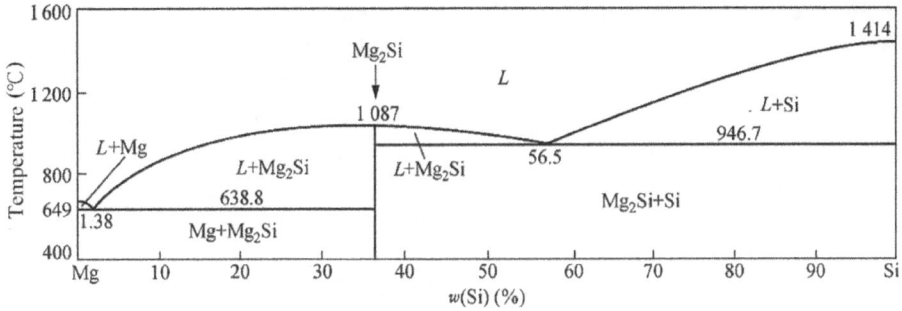

**Fig. 7.37:** Mg–Si phase diagram.

and the Mg–Si phase diagram can be analyzed by dividing into two independent binary phase diagrams, Mg–Mg$_2$Si and Mg$_2$Si–Si. If the compound formed possesses solubility of the component, namely forming a compound-based solid solution, then the compound has a composition range in the phase diagram. For example, in the Cd–Sb phase diagram shown in Fig. 7.38, $\beta$ phase has a certain composition range, and the phase diagram can be divided into two independent phase diagrams by a vertical line through the composition point corresponding to the melting point of the compound (456 °C) (indicated by the dash line). There are many binary systems which can form stable compounds, such as alloy series Cu–Mg, Fe–P, Mn–Si, Ag–Sr, and ceramic series Na$_2$SiO$_3$–SiO$_2$, BeO–Al$_2$O$_3$, SiO$_2$–MgO, and so on.

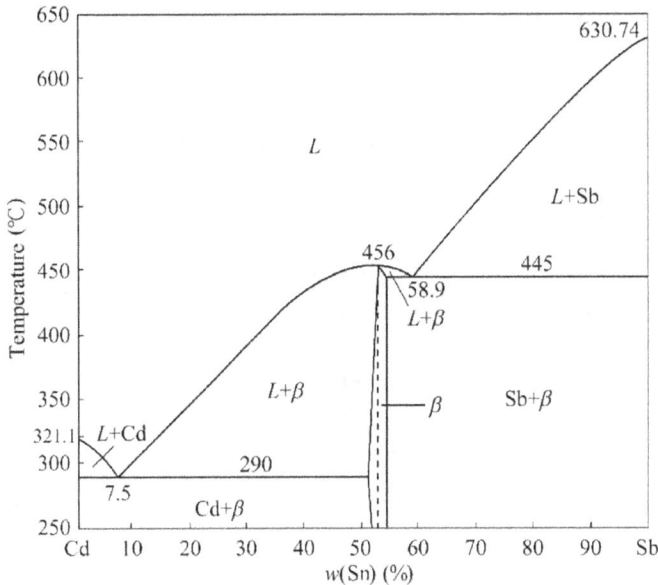

**Fig. 7.38:** Cd–Sb phase diagram.

### b. Phase diagrams containing unstable compounds

Figure 7.39 shows a K–Na phase diagram containing an unstable compound $KNa_2$. For a K–Na alloy with composition $w(Na) = 54.4\%$, the unstable compound $KNa_2$ will decompose into Na crystal and a liquid phase of a different composition as heated to 6.9 °C. Actually, it forms via a peritectic transformation: $L + Na \rightarrow KNa_2$. An unstable compound may also possess a certain of solubility, and in this case it appears as a phase region in the phase diagram. It is noted that an unstable compound cannot be considered as an independent component, and thus we cannot split the whole phase diagram into two parts for analysis, wherever it exhibits in a vertical line or in a phase region with a specific solubility. Other examples containing unstable compounds in their binary phase diagrams include Al–Mn, Be–Ce, Mn–P alloys, and $SiO_2$–MgO, $ZrO_2$–CaO, BaO–$TiO_2$ ceramics.

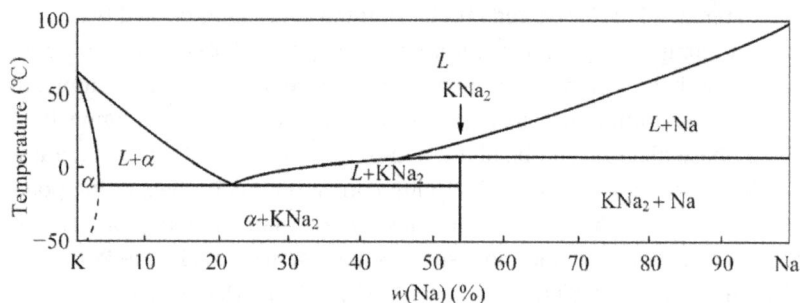

**Fig. 7.39:** K–Na phase diagram.

### 2. Phase diagram containing monotectic transformation

Monotectic transformation is an isothermal transformation that a liquid phase decomposes into a solid phase and another liquid phase. Figure 7.40 shows a Cu–Pb binary phase diagram, where monotectic transformation occurs at 955 °C:

$$L_{36} \rightarrow Cu + L_{87}$$

In Fig. 7.40, the 955 °C isotherm is called the monotectic line, and the composition point $w(Pb) = 36\%$ is called the monotectic point. The 326 °C isotherm is the eutectic line and the composition point $w(Pb) = 99.94\%$ is the eutectic point, which is not shown here because it is very close to that of pure Pb with melting point of 327.5 °C. Other binary phase diagrams containing monotectic transformation include Cu–S, Mn–Pb, and Cu–O.

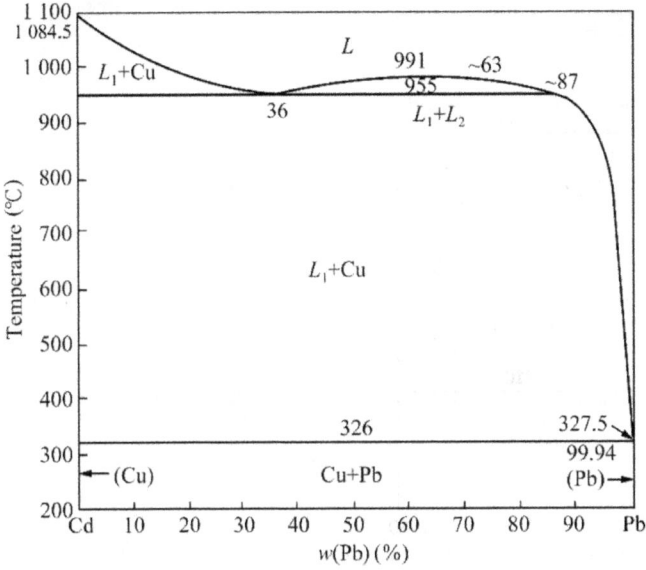

**Fig. 7.40:** Cu–Pb phase diagram.

## 3. Phase diagram containing syntectic transformation

There are very few binary alloy series containing syntectic transformation, such as Na–Zn and K–Zn. Syntectic transformation is that two liquid phases of different compositions react to form a solid phase, which occurs at the temperature $asb$ as schematically shown in Fig. 7.41:

$$L_{1a} + L_{2b} \rightarrow \beta_s$$

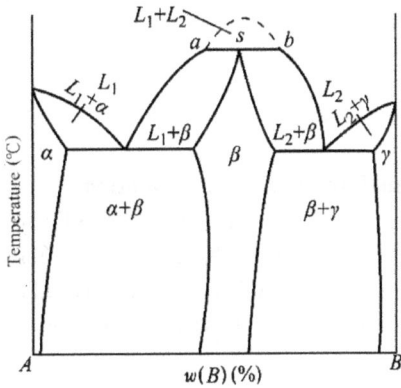

**Fig. 7.41:** A phase diagram containing syntectic transformation.

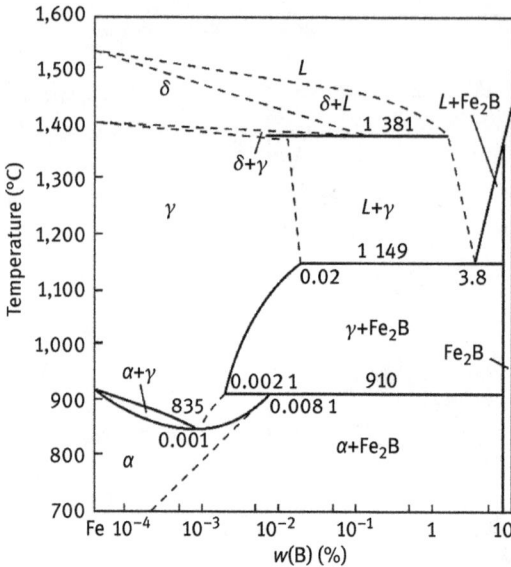

**Fig. 7.42:** Fe–B phase diagram.

### 4. Phase diagram containing metatectic transformation

Figure 7.42 shows a binary Fe–B phase diagram containing metatectic transformation. A Fe–B alloy with trace B experiences a solid-phase isothermal decomposition into a liquid phase and another solid phase, namely

$$\delta \rightarrow \gamma + L$$

which is termed metatectic transformation or catatectic transformation. It can be regarded as the inverse peritectic transformation. Only a few of alloy series contain metatectic transformation, such as Fe–S and Cu–Sb.

### 5. Phase diagram containing solid-state transformation

#### a. Phase diagram containing polymorphism transformation of solid solution

When components possess allotropic transformation in a system, the formed solid solution usually has polymorphism transformation or congruent phase transformation. Figure 7.43 shows a Fe-Ti binary phase diagram in which both Fe and Ti have allotropic transformation. A polymorphism transformation of $\beta$ phase (bcc) $\rightarrow$ $\alpha$ phase

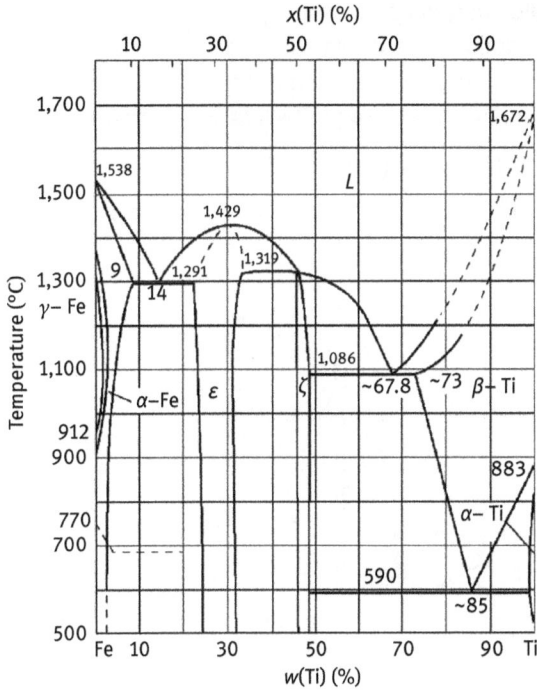

**Fig. 7.43:** Fe–Ti alloy phase diagram.

(hcp) emerges at the Ti side and another polymorphism transformation of $\alpha$ (or $\delta$) $\to \gamma$ $\to \alpha$ occurs at the Fe side.

### b. Phase diagram containing eutectoid transformation

Eutectoid transformation is that a solid phase isothermally transforms into the other two solid phases, the form of which is similar with eutectic transformation. Figure 7.44 shows a Cu–Sn phase diagram, where $y$ is $Cu_3Sn$, $\delta$ is $Cu_{31}Sn_8$, $\varepsilon$ is $Cu_3Sn$, $\zeta$ is $Cu_{20}Sn_6$, $\eta$ and $\eta'$ are $Cu_6Sn_5$. They have a little solubility. In the phase diagram, there are four eutectoid transformations:

$$IV : \beta \to \alpha + \gamma$$

$$V : \gamma \to \alpha + \delta$$

$$VI : \delta \to \alpha + \varepsilon$$

$$VII : \zeta \to \delta + \varepsilon$$

## c. Phase diagram containing peritectoid transformation

Peritectoid transformation is that a solid phase reacts with the other solid phase to isothermally transform into a third solid phase, the form of which is similar with peritectic transformation. In Fig. 7.44, there are two peritectoid transformations:

$$VIII : \gamma + \varepsilon \rightarrow \zeta$$

$$IX : \gamma + \zeta \rightarrow \delta$$

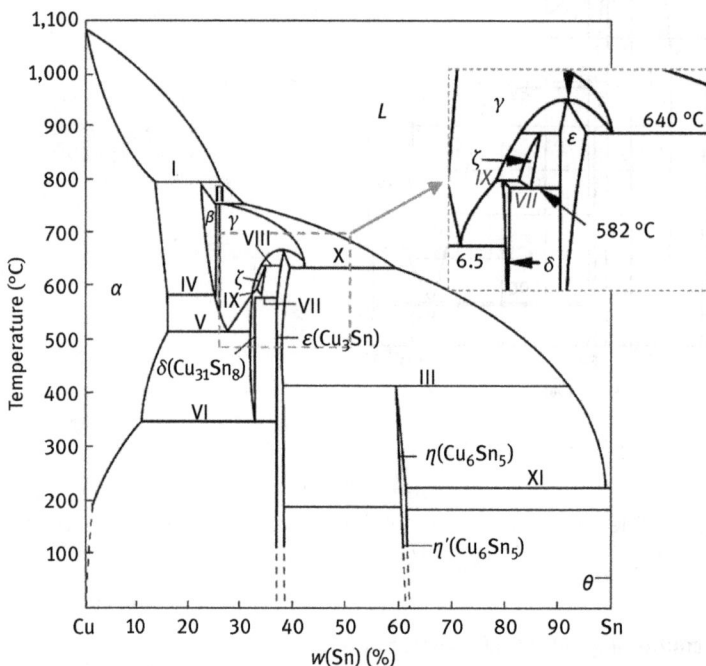

**Fig. 7.44:** Cu–Sn phase diagram.

## d. Phase diagram containing precipitation process

Usually, a second phase precipitates from a solid solution due to the reduced solubility with decreasing temperature. For example, in the Cu–Sn phase diagram as shown in Fig. 7.44, the $\alpha$ solid solution has the maximum solubility of $w(Sn) =$ 11.0% at 350 °C. The solubility continuously decreases with the drop of temperature and becomes nearly zero at room temperature. As a result, the $\varepsilon$ phases (Cu$_3$Sn) precipitates from the $\alpha$ solid solution continuously upon cooling below 350 °C, which is termed precipitation process (reaction).

## e. Phase diagram containing order–disorder transformation

Some alloys within a specific composition and temperature range will experience the order–disorder transformation. If a disordered solid solution transforms to an ordered solid solution as a first-order phase transformation, the two single-phase regions should be separated by a two-phase region in the phase diagram. For example, as shown in Fig. 7.45, the alloy with $w(Au) = 50.8\%$ is a disordered solid solution at temperature above 390 °C and will transform into an ordered solid solution of $\alpha'$ (AuCu$_3$) once below 390 °C in the Cu–Au phase diagram. In addition, $\alpha_1''$ (AuCu I), $\alpha_2''$ (AuCu II), and $\alpha'''$ (Au$_3$Cu) are also ordered solid solutions. If the transformation of an ordered solid solution into a disordered one is a second-order phase transformation, there is no two-phase region between the two solid solutions but a dashed or fine straight line. For instance, in the Cu–Sn phase diagram shown in Fig. 7.44, a fine straight line is used to represent the $\eta \rightarrow \eta'$ order–disorder transformation. It should be noted that the so-called first-order phase transformation is that the new phase and the old one have an identical chemical potential but different first-order partial derivatives of chemical potential, while for the second-order phase transformation, the chemical potential and its first-order derivatives are identical but the second-order derivatives for the two phases are different. It can be proved that there is only one single line between the two single-phase regions for a second-order phase transformation in binary systems, namely, the two equilibrium phases have the same composition at any equilibrium temperature and concentration.

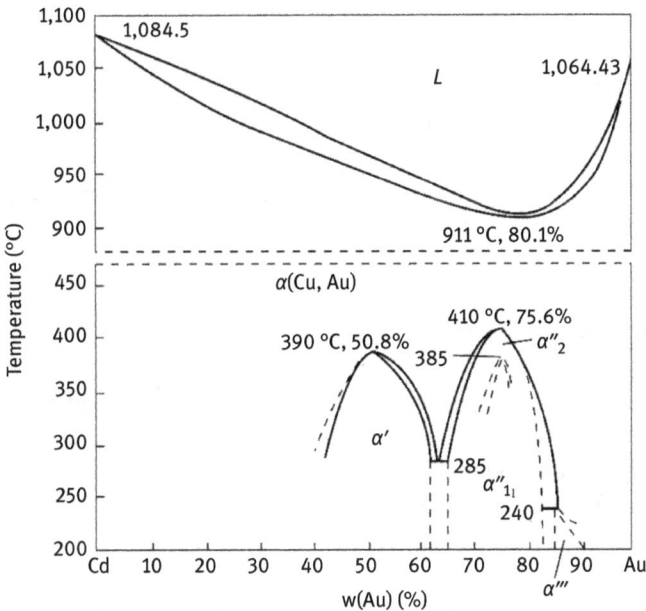

Fig. 7.45: Cu–Au phase diagram.

**f. Phase diagram containing intermediate phase transformed from solid solution**

Intermediate phases formed in some alloys are not obtained via direct reactions between two components as previously described, but formed from a solid solution. Figure 7.46 shows the Fe–Cr binary phase diagram. When the $\alpha$ solid solution with $w(Cr) = 46\%$ experiences a transformation of $\alpha \rightarrow \sigma$ at 821 °C, the obtained $\sigma$ phase is an intermetallic compound FeCr-based solid solution.

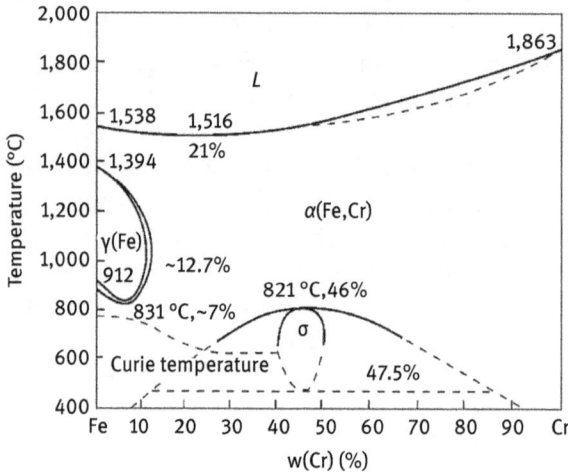

**Fig. 7.46:** Fe–Cr phase diagram.

**g. Phase diagram containing magnetic transition**

Magnetic transition belongs to the second-order phase transformation. Solid solutions or pure components are paramagnetic at high temperatures and ferromagnetic below $T_c$ temperature. $T_c$ is called *Curie temperature*, which is represented by a dash line on phase diagrams as shown in Fig. 7.46.

**7.3.6 Analysis of complicated binary phase diagram**

Complicated binary phase diagrams are combined by basic ones previously mentioned and can be simplified according to their characteristics and transformation rules. Generally, the analyzing method is as follows:

(1)  First, we find out whether any stable compound exists in the phase diagram, and if so, divide the diagram into several regions by these compounds to conduct analysis.

(2)  Distinguish each phase region on basis of the contact rule of phase region.

(3) Find out the horizontal line of three-phase coexistence and analyze types of the isothermal transformations. Table 7.1 lists all the isothermal transformations for three phases in binary systems to help the analysis. These isothermal transformations are divided into two large types: eutectic type (four reactions) and peritectic type (three reactions). Since one phase transforms into two phases with endpoint compositions in eutectic type during cooling, this single phase locates above the horizontal line and contacts it. On the contrary, two phases with endpoint compositions transform into one phase in peritectic type during cooling, so this single phase locates below the horizontal line and contacts it.

**Table 7.1:** Types of isothermal transformations in binary systems.

| Type of isothermal transformations | | Reactions | Graphic patterns |
|---|---|---|---|
| Eutectic type | Eutectic | $L \rightarrow \alpha + \beta$ | |
| | Eutectoid | $\gamma \rightarrow \alpha + \beta$ | |
| | Monotectic | $L_1 \rightarrow L_2 + \alpha$ | |
| | Metatectic | $\delta \rightarrow L + \gamma$ | |
| Peritectic type | Peritectic | $L + \beta \rightarrow \alpha$ | |
| | Peritectoid | $\gamma + \beta \rightarrow \alpha$ | |
| | Syntectic | $L_1 + L_2 \rightarrow \alpha$ | |

(4) Apply phase diagrams to analyze phase transformations and structure evolution with temperature variation for a given alloy. In the single-phase region, phase composition is the same with the original alloy, while in the two-phase region, the compositions of two phases vary along with the phase boundaries at different temperatures. At a specific temperature, draw the tie line, which intersects with two phase boundaries at the both ends, and the relative amounts of the two phases can be obtained by the lever rule. When three phases coexist, their compositions and transformation temperature are fixed based on the

phase rule, and the relative amounts of microconstituents or two phases can be calculated out after isothermal transformations according to lever rule.

(5) When phase diagrams are used to analyze practical cases, keep in mind that phase diagrams just give phases and the relative amounts at equilibrium conditions, and cannot show the morphologies, sizes, and distributions. Moreover, phase diagrams represent only equilibrium conditions which will rarely happen in practical production of alloys and ceramics. Hence, phases and structures under nonequilibrium conditions should be paid special attention; especially for ceramics, which possess higher liquid viscosities than alloy melts and slower diffusion of the components, resulting in a preferable formation of amorphous or metastable phases after solidification.

(6) Errors or even mistakes may exist when building phase diagrams for some reasons, which can be identified by the phase rule. Alloys in practical production are different from those in the phase diagrams due to the purify level of raw materials, which also affects the accuracy of the analysis results.

### 7.3.7 Alloy properties deduced from phase diagrams

Alloy performance largely depends on characteristics of elements and the properties and relative amounts of the constituent phases, which can be indicated by phase diagrams. With help of these characteristics and parameters mentioned earlier, the performance (such as mechanical and physical properties) and processing properties (such as castability, formability and heat treatment properties) of alloys can be deduced, which is beneficial to actual application.

#### 1. The performance of alloys deduced from phase diagrams
Figure 7.47 shows relationships between the mechanical/physical properties and several categories of basic binary alloy phase diagrams. It is manifested that the properties of alloys constituted by two mechanically mixed phases are average values of the properties of the two phases, namely properties exhibit a linear relationship with compositions. The properties of solid solutions vary in a curve with alloy compositions. When a stable compound (intermetallic phase) forms, a singularity will appear on the curve. In addition, the dispersion of phases has large impact on some properties since their dispersion is sensitive to structure. For instance, alloys with the compositions close to eutectic points could improve the strength and hardness because their constituted phases are usually fine and dispersed, as shown by the dash lines in Fig. 7.47.

#### 2. Processing properties deduced from phase diagrams
Figure 7.48 shows relationships between casting properties of alloys and phase diagrams. Eutectic alloys are prone to produce centered shrinkage pores and a densified

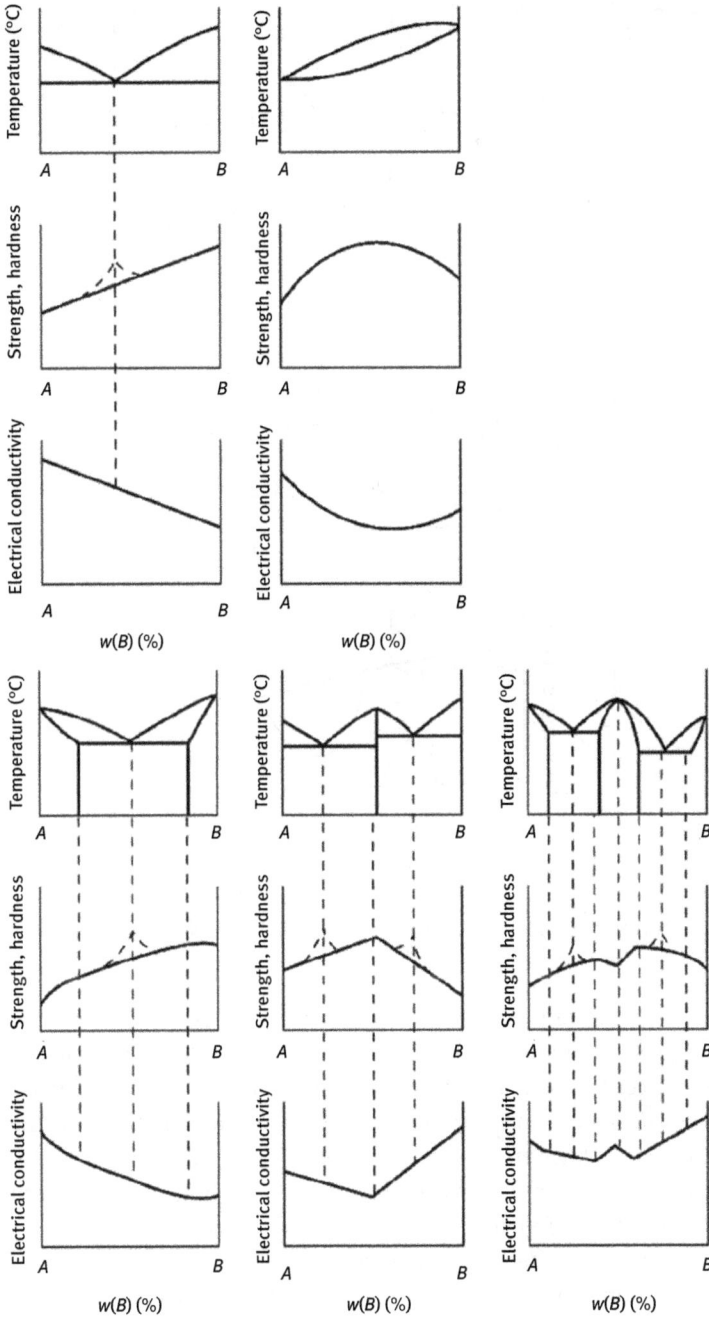

**Fig. 7.47:** Relationships between phase diagram and alloy hardness, strength and conductivity.

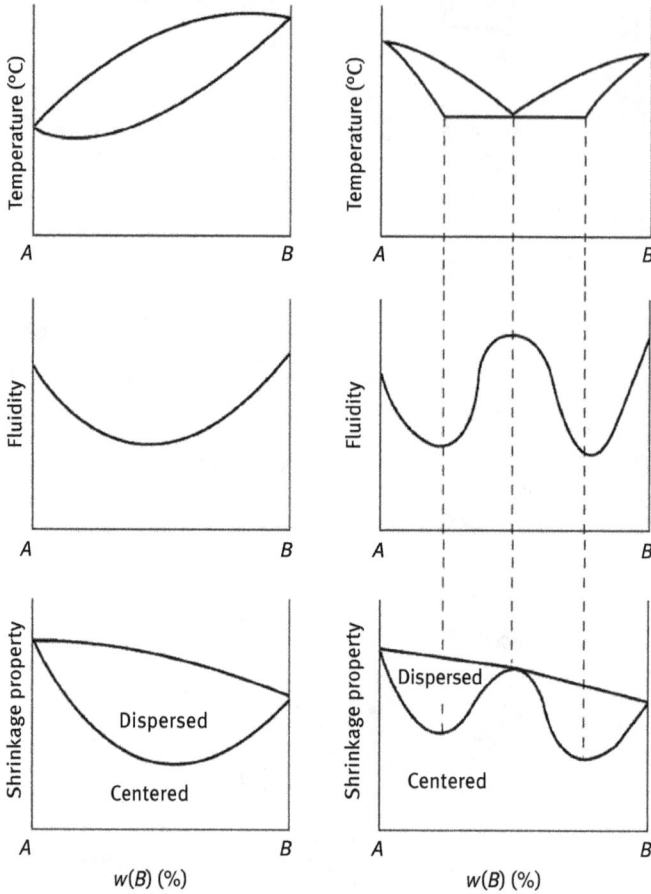

**Fig. 7.48:** Relationships between the fluidity, shrinkage property and phase diagrams.

structure after solidification due to the low melting points, the isothermal eutectic transformation, and excellent fluidity. Hence, alloys of compositions close to eutectic points are optimum choices for casting. Solid solution alloys, however, possess worse fluidity than the eutectic alloys and pure metals, and the intervals between the liquidus lines and the solidus lines are usually large, namely they have wide crystallization temperature ranges. As a result, coarse dendrites are likely to appear, which significantly retard the fluidity of melts, leading to more dispersed shrinkage pores and relative low density as well as serious segregation.

Alloys with excellent formability are usually single-phase solid solutions owing to their low strength, good ductility, and uniform deformation. However, two-phase mixtures are prone to nonuniform deformation due to different strengths of the two phases and cause crack at the interface between the two phases in case of large deformation. The presence of brittle intermediate phases is particularly harmful to the deformation.

Therefore, alloys for press working are always the single-phase solid solutions or alloys close to the single-phase solid solutions with small amount of secondary phase.

Phase diagrams can be used to estimate the possibility of heat treatment for alloys. For alloys without solid-state phase transformations on the phase diagrams, only diffusion annealing can be used to eliminate dendritic segregation, rather than other types of heat treatment. For alloys containing allotropic transformations, recrystallization annealing and normalizing treatment can be used to refine grains. For alloys with varied solubility, aging treatment can be used to strengthen by the precipitation of secondary phase. For some alloys containing eutectoid transformations, such as carbon steels in Fe–C alloys, they are at first heated to form $y$ phase (austenite), then quickly cooled (quenching) to room temperature, in which austenite transforms to metastable phase: bainite or martensite through nonequilibrium transformations, replacing equilibrium transformation: eutectoid transformation. Such a quenching process can evidently raise the strength of carbon steels.

### 7.3.8 Examples of binary phase diagrams

$SiO_2$–$Al_2O_3$ system and Fe–C system are the most important binary systems of ceramics and alloys, respectively. Their phase diagrams will be analyzed as examples in the following section.

#### 1. Structure and properties of $SiO_2$–$Al_2O_3$ system

Figure 7.49 shows a binary phase diagram of $Al_2O_3$–$SiO_2$, which is very important to the researches of ceramics and refractory materials. There are tens of different phase diagrams since the first $Al_2O_3$–$SiO_2$ phase diagram published in 1909. The focus of debate is whether the intermediate phase mullite is a stable or unstable compound, and whether its composition is constant or not. The former is that mullite is an unstable phase because it has no fixed melting point, and it is product of peritectic transformation, so it cannot be considered as a component in phase diagram. The latter question also has been solved that the composition of mullite is not fixed, the mass fraction of $Al_2O_3$ in which fluctuates in a range from 72% to 78%, equivalent to a stoichiometric molecular formula between $3Al_2O_3$–$SiO_2$ and $2Al_2O_3$–$SiO_2$. Therefore, there is a solid-solution range in the phase diagram for mullite.

The phase diagram of $SiO_2$–$Al_2O_3$ system includes three compounds with complex structures. The component $\alpha$-$Al_2O_3$ (known as corundum) belongs to the rhombohedral lattice; $SiO_2$ has various types of lattice (as shown in Table 7.2) for different polymorphic forms; and the intermediate phase mullite has the monoclinic lattice. Silica and alumina are not mutually soluble in one another, which is evidenced by the absence of terminal solid solutions at both extremities of the phase diagram.

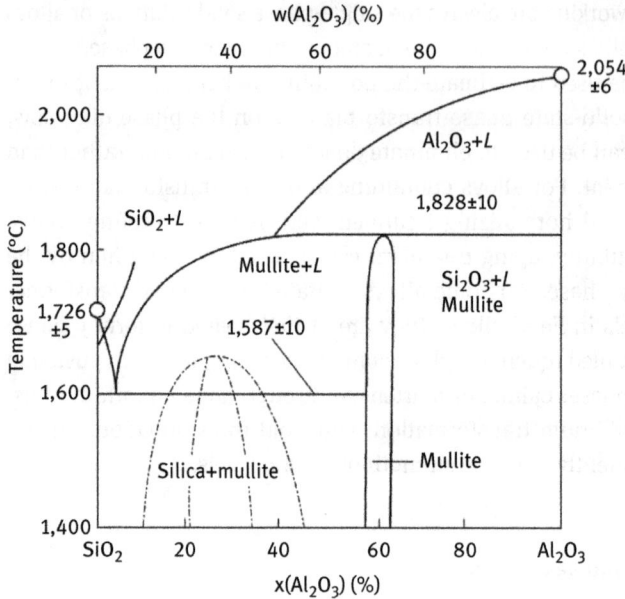

**Fig. 7.49:** Phase diagram of $SiO_2$–$Al_2O_3$ system.

**Table 7.2:** The polymorphic forms of $SiO_2$.

| Stable form | Lattice type | Temperature range /°C |
| --- | --- | --- |
| α quartz | Hexagonal lattice | (Room temperature) to 573 |
| β quartz | Hexagonal lattice | 573–870 |
| $β_2$ tridymite | Rhombohedral lattice | 870–1,470 |
| β cristobalite | Tetragonal lattice | 1,470–1,713 |
| Silicate | Amorphous | Above 1,713 |

The phase diagram contains two isothermal transformations: one is a eutectic transformation at 1,587 °C: $L{\rightarrow}SiO_2$ + mullite; the other is a peritectic transformation: L + $Al_2O_3{\rightarrow}$mullite. A miscibility gap emerges at the side of $SiO_2$-riched region on the diagram, and in miscibility gap the two phases will automatically separate by spinodal decomposition or the nucleation and growth mode. Such a separation of two metastable phases usually occurs in most of $SiO_2$ contained systems.

### a. Ceramic with $w(Al_2O_3)$ < 10% (hypoeutectic)
The molten $SiO_2$–$Al_2O_3$ ceramic with w ($Al_2O_3$) less than 10% begins to crystallize into $SiO_2$ (β cristobalite) through isomorphous transformation when cooling to the liquidus temperature. Upon further cooling, the amount of $SiO_2$ increases gradually accompanying with the increase of $Al_2O_3$ concentration in the liquid phase. As

cooling to 1,587 °C, the composition of the liquid phase reaches the eutectic point, $w(Al_2O_3) = 10\%$, and eutectic transformation occurs, resulting in eutectics of mechanical mixture of $SiO_2$ and mullite. After the transformation, the microstructure is composed by the primary cristobalite and the eutectics. On further cooling, the primary $SiO_2$ and eutectic $SiO_2$ phases will experience polymorphic transformation to high-temperature $\beta_2$ tridymite at 1,470 °C by a reconstructive transformation, then to high-temperature $\beta$ quartz at 867 °C by a reconstructive transformation, and finally to low-temperature $\alpha$ quartz at 573 °C through a displacive transformation. Because the reconstructive transformation is extremely slow, high-temperature $\beta$ cristobalite may transform into low-temperature cristobalite at 200–270 °C by a displacive transformation, or may transform into high-temperature tridymite by a reconstructive transformation (marked on some phase diagrams), then into intermediate-temperature tridymite at 160 °C by a displacive transformation, and finally transform into low-temperature tridymite at 105 °C by a displacive transformation. Whether $SiO_2$ is low-temperature cristobalite, low-temperature tridymite, or low-temperature quartz at room temperature is dependent on the cooling rate and if there is added solvent to promote the reconstructive transformation. During cooling process after the eutectic reaction, there is no precipitation phenomenon between $SiO_2$ and mullite because they have almost no solubility each other.

## b. Ceramic with $w(Al_2O_3) = 10\%$ (eutectic)

The molten $SiO_2$–$Al_2O_3$ ceramic with $w(Al_2O_3) = 10\%$ experiences eutectic reaction at 1,587 °C, resulting in formation of eutectics: $Al_2O_3$ + mullite. The relative amount of two phases in eutectics can be calculated by the lever rule:

$$w(SiO_2) = \frac{72-10}{72-0} \times 100\% = 86\%$$

$$w(mullite) = \frac{10-0}{72-0} \times 100\% = 14\%$$

After the reaction, $SiO_2$ will transform into one of three kinds of low-temperature quartz from the high-temperature cristobalite depended on different cooling rates.

## c. Ceramic with $10\% < w(Al_2O_3) < 55\%$ (hypereutectic)

The molten ceramics with compositions in this range starts to crystallize into mullite through isomorphous transformation when cooling to the liquidus temperature. Upon further cooling, the amount of mullite increases while the content of $Al_2O_3$ in liquid phase decreases and the liquid composition varies along the liquidus line. When the temperature is lowered to 1,587 °C, the liquid composition reaches the eutectic point and eutectic transformation occurs. The structure after eutectic reaction is constituted by mullite and eutectics. In this composition range, the maximum relative amount of primary mullite is

$$w(\text{mullite}_{max}) = \frac{55 - 10}{72 - 10} \times 100\% = 72.5\%$$

Similarly, eutectic $SiO_2$ will experience polymorphic transformation after the eutectic reaction.

### d. Ceramic with 55% < $w(Al_2O_3)$ < 72%

The molten ceramics with compositions in this range first crystallize into $Al_2O_3$ through isomorphous transformation when cooling to the liquidus temperature. Upon further cooling, the amount of $Al_2O_3$ increases while that of the liquid phase decreases. When cooling to 1,828 °C, a peritectic reaction: $L + Al_2O_3 \rightarrow$ mullite, occurs. At the end of the peritectic reaction, the primary $Al_2O_3$ is consumed completely but there is still some retained liquid. The liquid continues crystallize into mullite by isomorphous transformation, and the resulting mullite combines with the mullite from peritectic reaction. Accordingly, the liquid composition varies along the liquidus line, and finally eutectic transformation occurs at 1,587 °C when $w(Al_2O_3) = 10\%$, resulting in eutectics. The resultant structure after eutectic reaction is constituted by mullite and eutectics.

### e. Ceramic with 72% < $w(Al_2O_3)$ < 78%

The molten ceramics with compositions in this range crystallizes into $Al_2O_3$ when cooling to the liquidus temperature and experiences peritectic transformation until cooling to 1,828 °C. Taking the peritectic composition as $w(Al_2O_3) = 75\%$, the relative amounts of the liquid phase and $Al_2O_3$ required by peritectic reaction are, respectively:

$$w(\text{liquid}) = \frac{100 - 75}{100 - 55} \times 100\% = 55.6\%$$

$$w(Al_2O_3) = 100\% - 55.6\% = 44.4\%$$

It enters into a single-phase region of mullite at the end of peritectic reaction. The structure remains a single-phase mullite as cooling to room temperature.

### f. Ceramic with $w(Al_2O_3)$ > 78%

The molten ceramics with compositions in this range crystallizes into $Al_2O_3$ when cooling to the liquidus temperature and experiences a peritectic transformation until cooling to 1,828 °C. At the end of the peritectic reaction, the liquid phase is consumed completely while there exist some primary $Al_2O_3$ phase. Thus, the structure is constituted by the primary $Al_2O_3$ phase and mullite from peritectic transformation. It remains unchanged as cooling to room temperature because both mullite and $Al_2O_3$ have no change in the solubility.

In the binary system of $SiO_2$–$Al_2O_3$, different amounts of $Al_2O_3$ correspond to several kinds of widely used refractory products. For instance, the mass fraction of $Al_2O_3$ is 0.2–1.0% for silica brick, 35–50% for clay brick, and 60–90% for high-alumina brick.

Silica bricks are usually used as top bricks of an open-hearth furnace, the temperature inside which is usually 1,625–1,650 °C. As known from the phase diagram, part of the bricks is in liquid state in that temperature range, which is harmful to brick lifetime. For this reason, high siliceous bricks of a low alumina content that can be made from special materials or processes are used to improve the service temperature and prolong the lifetime.

Clay bricks are constituted by equilibrium mullite and silica phase at temperatures below 1,587 °C. The relative amounts of both phases vary with the $Al_2O_3$ content in the bricks, and the properties of bricks vary accordingly. Since plenty of liquid phase will appear at temperature above 1,600 °C, the clay bricks are not suitable for use in this case.

Refractories with $w(Al_2O_3) > 10\%$ will increase high temperature resistance with increasing the $Al_2O_3$ content. Refractory bricks are completely composed by mullite or mullite and silica, which significantly improves fire resistance. The use of pure alumina can obtain the highest refractory temperature. Sintered $Al_2O_3$ is used to manufacture laboratory glassware and the casted $Al_2O_3$ is used as refractory materials for glass tank furnace.

## 2. Structures and properties of Fe–C alloys

### a. Fe–Fe₃C phase diagram

Carbon steels and cast irons are the most widely used metal materials, and the iron-carbon phase diagram is an important tool to study the structure, properties, hot working, and heat treatment of iron and steel materials.

There are four kinds of forms of C element in irons and steels: C atom dissolves in $\alpha$-Fe and forms a solid solution called *ferrite* (BCC); or dissolves in $\gamma$-Fe and forms a solid solution called *austenite* (FCC); or combines with Fe and forms a complex structure compound Fe₃C (orthorhombic lattice) called *cementite*; C may also exist in the form of a stable free *graphite* (hexagonal structure). Generally, Fe–C alloys transform according to the Fe–Fe₃C system, where Fe₃C is only a metastable phase and could decompose into iron and graphite under certain conditions, namely Fe₃C→3Fe + C(graphite). Therefore, the iron–carbon phase diagram can have two forms: Fe–Fe₃C and Fe–C phase diagram. For the sake of convenience, they are drawn together, which is called the duplex iron–carbon phase diagram as shown in Fig. 7.50.

There are three three-phase isothermal phase transitions in the Fe–Fe₃C phase diagram, namely peritectic transformation at 1,495 °C: $L_B + \delta_H \rightarrow \gamma_J$, where the product is austenite; eutectic transformation at 1,148 °C: $L_C \rightarrow \gamma_E + Fe_3C$, where the product is a mechanical mixture of austenite and cementite, called *ledeburite*; and

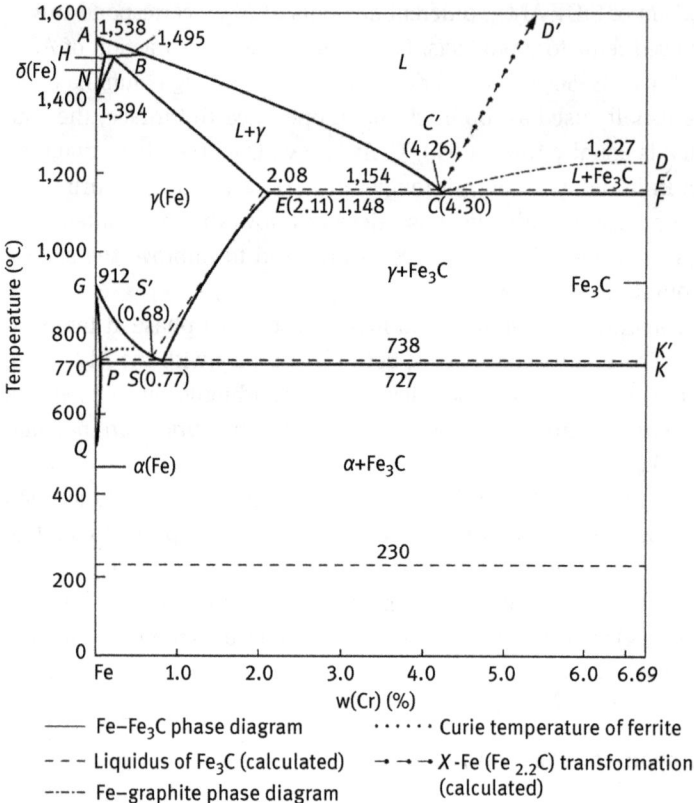

Fig. 7.50: The duplex Fe–C phase diagram.

eutectoid transformation at 727 °C: $\gamma_s \rightarrow \alpha_p + Fe_3C$, where the product is a mechanical mixture of ferrite and cementite, called *pearlite*. The eutectoid temperature is commonly marked as $A_1$ temperature.

In addition, there are another three important solid transformation curves:

(1) GS curve: The transformation line that ferrite begins to precipitate from austenite (upon cooling) or ferrite completely dissolves into austenite (upon heating), usually called $A_3$ temperature.

(2) ES curve: The solubility curve of carbon in austenite, often called $A_{cm}$ temperature. Cementite will precipitate from austenite below this temperature, which is also called secondary cementite indicated by $Fe_3C_{II}$ to differentiate it from primary cementite $Fe_3C_I$ that crystallizes from liquid across CD curve.

(3) PQ curve: The solubility curve of carbon in ferrite. Solubility limit of carbon in ferrite is $w(C) = 0.0218\%$ at 727 °C, thus, trace cementite will precipitate from ferrite upon cooling from 727 °C, termed tertiary cementite $(Fe_3C_{III})$ to differentiate it from the two other cases.

The horizontal line at 770 °C indicates the magnetic transition temperature of ferrite, which is often called $A_2$ temperature. Another horizontal line at 230 °C represents the magnetic transition of cementite.

## b. Equilibrium structures of typical iron–carbon alloys

Generally, iron–carbon alloys can be classified into three categories according to their carbon content and equilibrium structures at room temperature: industrial pure iron, carbon steel, and cast iron. Carbon steels and cast irons are distinguished by whether there exists eutectic transformation or not. Alloys without eutectic transformation, that is, without ledeburite, are called carbon steels, which can further be divided into hypoeutectoid steels, eutectoid steels, and hypereutectoid steels, while those with eutectic transformation are named cast irons.

Iron–carbon alloys with different carbon contents fall into seven categories according to their different structures obtained on the Fe–Fe$_3$C phase diagram, as shown in Fig. 7.51, which are:

① industrial pure iron, $w(C) < 0.0218\%$;
② eutectoid steel, $w(C) = 0.77\%$;
③ hypoeutectoid steel, $0.0218\% < w(C) < 0.77\%$;
④ hypereutectoid steel, $0.77\% < w(C) < 2.11\%$;
⑤ eutectic white cast iron, $w(C) = 4.3\%$;
⑥ hypoeutectic white cast iron, $2.11\% < w(C) < 4.30\%$;
⑦ hypereutectic white cast iron, $4.30\% < w(C) < 6.69\%$.

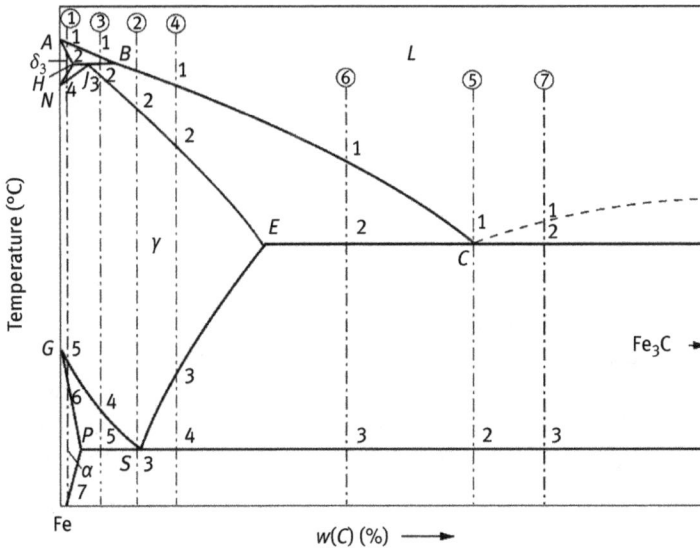

**Fig. 7.51:** Analysis of structure evolution of typical iron–carbon alloys upon cooling.

Next, we will choose one alloy from each category to analyze the transformation process and the structures at room temperature during equilibrium solidification.

### (1) Alloy with w(C) = 0.01% (industrial pure iron)

Position of this alloy is indicated by ① on the phase diagram as shown in Fig. 7.51. When the melt is cooled to points 1–2, the $\delta$ solid solution crystallizes via isomorphous transformation: $L{\rightarrow}\delta$. The single-phase solid solution $\delta$ forms between points 2 and 3. On cooling to points 3–4, a polymorphous transformation $\delta{\rightarrow}\gamma$ occurs, and austenite phase gradually nucleates and grows at grain boundaries of $\delta$ phase until point 4. A complete single-phase austenite forms and maintains above point 5. When cooling to points 5–6, another polymorphous transformation $\gamma \rightarrow \alpha$ takes place and produces ferrite $\alpha$ phase, which also preferentially nucleates and grows at grain boundaries of austenite and maintains above point 7. Once the temperature is cooled below point 7, tertiary cementite $Fe_3C_{III}$ will precipitate from ferrite. The structure of the industrial pure iron at room temperature is shown in Fig. 7.52.

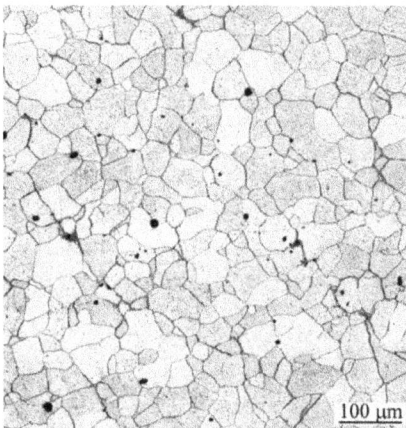

**Fig. 7.52:** Microstructure of industrial pure iron (Photograph by and courtesy of Mr. Bin Zong, Beijing University of Technology).

### (2) Alloy with w(C) = 0.77% (eutectoid steel)

Position of this alloy is indicated by ② on the phase diagram as shown in Fig. 7.51. Austenite crystallizes from the liquid alloy by isomorphous transformation between points 1 and 2. A single-phase austenite forms at the end of solidification at point 2 and maintains above point 3. As cooling to point 3 (727 °C), the eutectoid transformation $\gamma_{0.77} \rightarrow \alpha_{0.0218} + Fe_3C$ occurs, resulting in the formation of mixture which is alternating layers or lamellae of ferrite and cementite, called pearlite because it has the appearance of mother-of-pearl when viewed under the microscope at low magnifications. The cementite in pearlite is called eutectoid cementite. As the temperature continues to drop, a little $Fe_3C_{III}$ precipitates from ferrite, which usually grows together with the eutectoid cementite and cannot be recognized by optical microscope. Figure 7.53 shows the structure at room temperature.

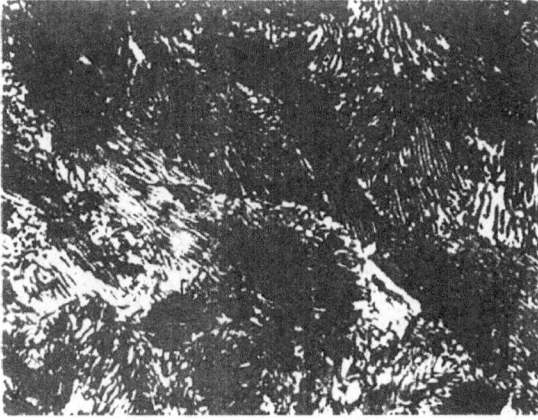

**Fig. 7.53:** Optical image of pearlite (600x).

At room temperature, the relative amounts of ferrite and cementite in pearlite can be computed by the lever rule as follows:

$$w(\alpha) = \frac{6.69 - 0.77}{6.69 - 0.0008} \times 100\% = 88\%$$

$$w(Fe_3C) = 100 - 88\% = 12\%$$

where $w(C) = 0.0008\%$ is the solubility limit of carbon in ferrite at room temperature. In Fig. 7.53, the thick bright lamellae are ferrite and the thin dark lamellae are cementite. This contrast of bright and dark is due to different corrosion rates of ferrite and cementite and their phase boundaries etched by metallographic corrosive agents. Cementite protrudes because it is less vulnerable to corrosion. Many cementite layers are so thin that adjacent phase boundaries are so close together that they are indistinguishable at the magnification of optical microscope; therefore, two phase boundaries of a cementite layer appear as a dark line. The morphology and width of cementite lamellar structure can be clearly shown by transmission electron microscopy (TEM), which has much higher resolution than the optical microscope. As shown in Fig. 7.54, there are no dark lines of cementite, and cementite exhibits bright contrast and ferrite does dark contrast in the TEM image of pearlite in a single-stage plastic replica, which is reverse to optical image.

The alternating ferrite and cementite layers in pearlite form similarly as the eutectic structure (Fig. 7.20 Pb–Sn phase diagram). The carbon content of the austenite phase (0.76 wt% C) is different from either of the ferrite (0.022 wt% C) and cementite (6.69 wt% C). During the eutectoid-phase transformation, a redistribution of the carbon

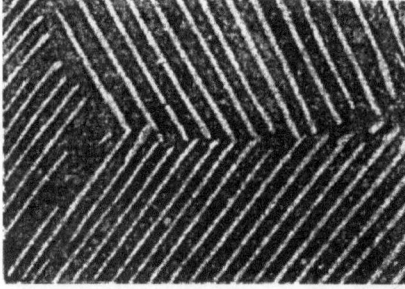

**Fig. 7.54:** The TEM image of pearlite in a single-stage plastic replica (8000x).

is required by lateral diffusion ahead of the phase interface. The pearlite first appears at the grain boundary and extends into the unreacted austenite grain, and carbon atoms diffuse away from the ferrite regions (0.022 wt% C) to the cementite layers (6.69 wt% C). The formation of layer structure of perlite is because carbon atoms can diffuse quickly along the lateral direction through minimal distances. The eutectoid lamellae in pearlite are much finer than the eutectic lamellae in lead-tin eutectic alloy because iron and carbon atoms must diffuse through solid austenite rather through a liquid.

In fact, the ferrite and cementite in pearlite do not form simultaneously during the eutectoid transformation. When the eutectoid transformation starts, an arbitrary constituted phase in pearlite—ferrite or cementite preferentially nucleates at grain boundaries of austenite and grows into lamellar morphology. Usually, cementite nucleates and grows up at the grain boundaries of austenite as a leading phase, resulting in a carbon deficient region nearby, which benefits the formation of ferrite nuclei at both sides of the cementite. As a result, a nuclei of pearlite constituted by ferrite and cementite forms. Because the solubility of carbon in ferrite is much less than that in austenite, the formation of ferrite promotes carbon in ferrite to partition (diffuse) into nearby untransformed austenite. Therefore, once the carbon content at these locations reaches a specific value (6.69%), the second layer of cementite will appear. Such a process continues to occur alternatively, resulting in the formation of pearlite colony (domain, or grain). Pearlite colonies have different orientations among colonies, and each colony nucleates and grows independently. Within each colony, the orientation lamellae in eutectoid microconstituent are identical. The orientation changes on crossing colony boundary. Therefore, there are specific crystallographic orientation relationships between ferrite and cementite:

$$
(A) \begin{bmatrix} (001)_{Fe_3C} \mathbin{/\!/} (2\bar{1}\bar{1})_\alpha \\ [100]_{Fe_3C} \mathbin{/\!/} [01\bar{1}]_\alpha \\ [010]_{Fe_3C} \mathbin{/\!/} [111]_\alpha \end{bmatrix} \quad (B) \begin{bmatrix} (001)_{Fe_3C} \mathbin{/\!/} (5\,\bar{2}\,\bar{1})_\alpha \\ [100]_{Fe_3C} \& [13\,\bar{1}]_\alpha, \text{ difference } 2.6° \\ [010]_{Fe_3C} \& [1\,1\,3]_\alpha, \text{ difference } 2.6° \end{bmatrix}
$$

In addition, lamellar spacing of pearlite decreases with increasing cooling rate, and the finer lamina of pearlite is, the higher strength and better toughness.

Lamellar pearlitic steels have poor machining characteristics. If they experience proper annealing treatment, such as about 30 °C below $A_1$ temperature holding for several hours, eutectoid cementite morphology changes into large, spherical particles in order to reduce boundary area (energy), as shown in Fig. 7.55. The microstructure, knowing as spheroidite, has a continuous matrix of soft, machinable ferrite. Such a treatment is called spheroidizing. Pearlitic steels with spherical particles of cementite have lower strength than lamellar pearlitic steels, but better plasticity and toughness.

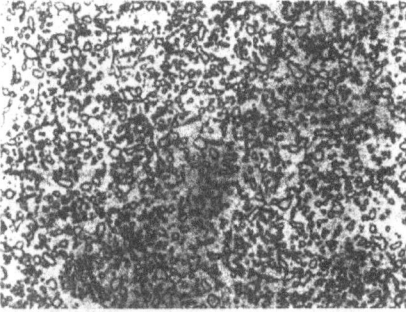

**Fig. 7.55:** Globular pearlite (400x).

### (3) Alloy with $w(C) = 0.40\%$ (hypoeutectoid steel)

Position of this alloy is indicated by ③ on the phase diagram as shown in Fig. 7.51. The δ solid solution crystallizes by isomorphous transformation between points 1 and 2. Upon cooling to point 2 (1,495 °C), a peritectic transformation $L_{0.53} + \delta_{0.09} \rightarrow \gamma_{0.17}$ takes place. Since the carbon content of the alloy is higher than that the peritectic point corresponds to, there exist some remaining liquid phase after the peritectic transformation. Between points 2 and 3, the remaining liquid continues to crystallize into austenite and forms a complete austenite with $w(C) = 0.4\%$ at point 3. Subsequent cooling brings no change of a single-phase austenite until cooling to point 4. With further cooling, ferrite starts to precipitate and increase, during which the carbon concentrations of ferrite and remaining austenite vary along the GP curve and GS curve, respectively. Once the temperature reaches point 5 (727 °C), the $w(C)$ of remaining austenite reaches 0.77% and eutectoid transformation occurs to form pearlite. Below point 5, tertiary cementite will precipitate from proeutectoid ferrite, but its amount is so little that it can be usually omitted. Structure of this alloy at room temperature is constituted by proeutectoid ferrite and pearlite as shown in Fig. 7.56.

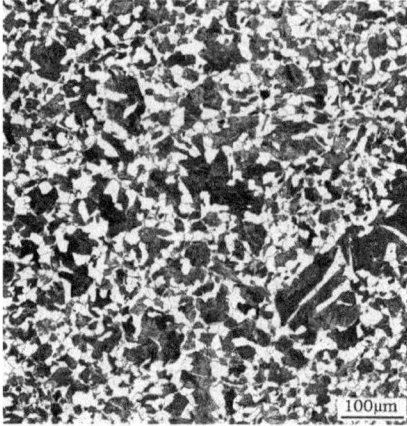

**Fig. 7.56:** Structure of hypoeutectoid steel at room temperature (Photograph by and courtesy of Mr. Bin Zong, Beijing University of Technology).

### (4) Alloy with $w(C)$ = 1.2% (hypereutectoid steel)

Position of this alloy is indicated by ④ on the phase diagram as shown in Fig. 7.51. Austenite crystallizes by isomorphous transformation between points 1 and 2. Secondary cementite starts to precipitate from austenite when cooling to point 3 and its precipitation ends until point 4. The composition of austenite varies along the ES curve. $Fe_3C_{II}$ precipitates along grain boundaries of austenite and hence appears as a network-like distribution. As cooling to point 4 (727 °C), the composition of austenite $w(C)$ decreases to 0.77% and then it experiences an isothermal eutectoid transformation, resulting in a final structure of pearlite with the distribution of network-like secondary cementite, as shown in Fig. 7.57.

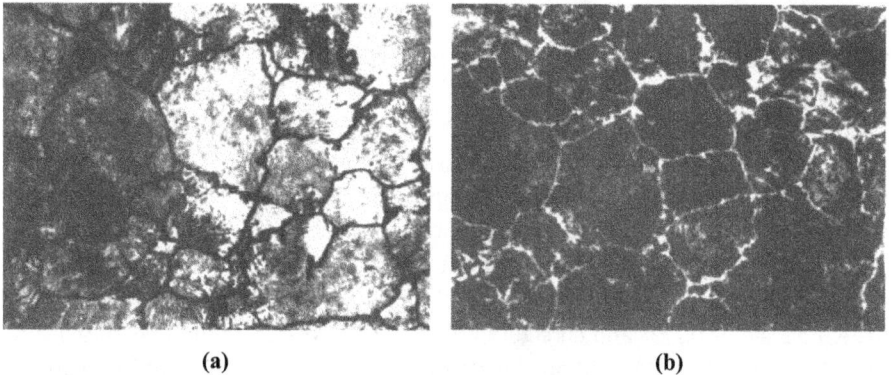

| (a) | (b) |

**Fig. 7.57:** Structure of hypereutectoid steel with $w(C)$ = 1.2% after slow cooling (500x). (a) etched by nitric acid alcohol solution, where network-like secondary cementite shows bright contrast and pearlite shows dark contrast; (b) etched by sodium picrate, where secondary cementite shows dark contrast and pearlite shows bright contrast with some gray.

### (5) Alloy with $w(C) = 4.3\%$ (eutectic white cast iron)

Position of this alloy is indicated by ⑤ on the phase diagram as shown in Fig. 7.51. When cooling to point 1 (1,148 °C), a eutectic transformation $L_{4.30} \rightarrow \gamma_{2.11} + Fe_3C$ occurs, and the produced eutectics is called ledeburite. Upon cooling to point 1~2, secondary cementite continuously precipitates from the austenite in the eutectics, which cannot be distinguished as it usually attaches on eutectic cementite. The relative amount of secondary cementite is up to 11.8% calculated by the lever rule. When the temperature drops to point 2 (727 °C), the carbon content of eutectic austenite decreases to the eutectoid point of 0.77% and an isothermal eutectoid transformation occurs to form pearlite. Neglecting the precipitated $Fe_3C_{III}$ as cooling below point 2, the final structure is room temperature ledeburite, which is also called variant ledeburite indicated by $L_d'$. The variant ledeburite keeps the morphology of original ledeburite, but the eutectic austenite has transformed into pearlite, as shown in Fig. 7.58.

**Fig. 7.58:** Structure of eutectic white cast iron at room temperature (the bright contrast is eutectic cementite, and the dark contrast is pearlite transformed from eutectic austenite).

### (6) Alloys with $w(C) = 3.0\%$ (hypoeutectic white cast iron)

Position of this alloy is indicated by ⑥ on the phase diagram as shown in Fig. 7.51. Austenite crystallizes from the liquid alloy between points 1 and 2, where the compositions of liquid and austenite vary along BC and JE curves during this period, respectively. When the temperature reaches point 2 (1,148 °C), the $w(C)$ of primary austenite and liquid are 2.11% and 4.3%, respectively, and eutectic transformation occurs immediately to produce ledeburite. Below point 2, secondary cementite precipitates from both primary austenite (or called *proeutectic austenite*) and eutectic austenite, and the composition of austenite varies along the ES curve. Once the temperature drops below point 3 (727 °C), all austenite will transform into pearlite through eutectoid transformation. Figure 7.59 shows the structure of the alloy at room temperature, where the dark dendritic constituents are pearlite transformed from both proeutectic austenite and the remaining is the variant ledeburite. Secondary

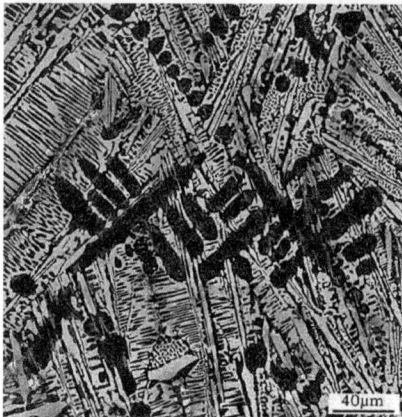

**Fig. 7.59:** Structure of hypoeutectic white iron at room temperature (the dark dendritic constituents are pearlite, and the remaining is the variant ledeburite) (Photograph by and courtesy of Mr. Bin Zong, Beijing University of Technology).

cementite precipitated from proeutectic austenite is attached on eutectoid cementite and difficult to identify.

### (7) Alloys with $w(C) = 5.0\%$ (hypereutectic white cast iron)

Position of this alloy is indicated by ⑦ on the phase diagram as shown in Fig. 7.51. Cementite crystallizes from the liquid alloy between points 1 and 2. This proeutectic phase is primary cementite, which grows in a strip morphology manner rather than a dendritic shape. Except for this, other transformations are the same as those of eutectic white iron. The room temperature structure of hypereutectic white cast iron is composed by primary cementite and the variant ledeburite as shown in Fig. 7.60.

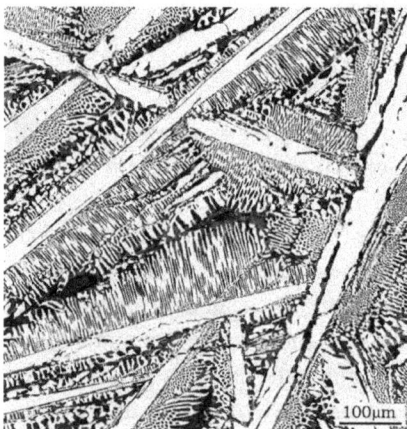

**Fig. 7.60:** Structure of hypereutectic white cast iron at room temperature (the bright strips are primary cementite, and the remaining is the variant ledeburite) (Photograph by and courtesy of Mr. Bin Zong, Beijing University of Technology).

Based on analysis of transformation processes of various iron–carbon alloys mentioned above, phase regions in the phase diagram can be marked according to the structures as shown in Fig. 7.61.

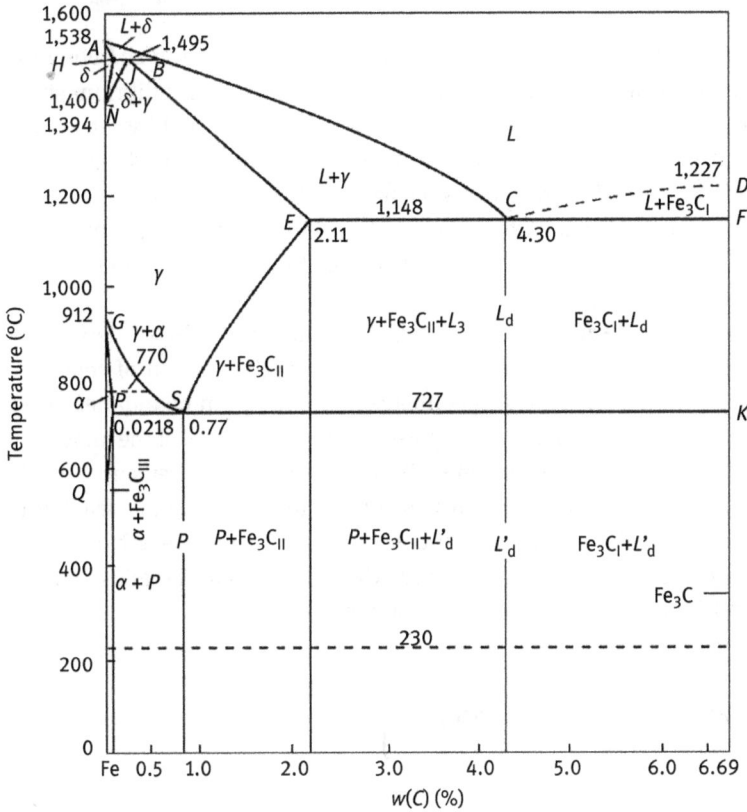

**Fig. 7.61:** Phase diagram of iron–carbon alloy showing different structures.

It is worthy to point out that white cast iron is a hard, brittle alloy containing massive amounts of $Fe_3C$. A fractured surface of this cast iron appears white, hence the name. White cast iron is composed of $Fe_3C$ and pearlite. The Fe–$Fe_3C$ system, as shown dash line in Fig. 7.50, is a metastable phase diagram. Under truly equilibrium conditions, the eutectic reaction is: $L \rightarrow \gamma + $ graphite. The Fe–C phase diagram is shown as solid line in Fig. 7.50. When the stable eutectic reaction mentioned above occurs at 1,148 °C, a ductile, or compacted-graphite cast iron forms, and its fractured surface appears gray, hence such a cast iron is called gray cast iron.

**c. Effect of carbon content of iron–carbon alloys on the structure and properties**

With increasing carbon content in Fe–Fe$_3$C system, structures of iron–carbon alloys change as follows:

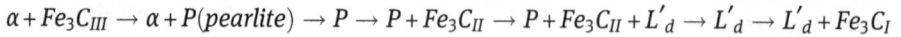

$$\alpha + Fe_3C_{III} \rightarrow \alpha + P(pearlite) \rightarrow P \rightarrow P + Fe_3C_{II} \rightarrow P + Fe_3C_{II} + L'_d \rightarrow L'_d \rightarrow L'_d + Fe_3C_I$$

The carbon content affects mechanical properties of steels mainly by changing the microstructures and relative amounts of the constituted phases. Equilibrium structures of all iron–carbon alloys at room temperature are constituted by ferrite and cementite. Ferrite is a soft and tough phase while cementite is a hard and brittle phase. Pearlite is composed by ferrite and cementite, hence its strength is higher than ferrite but lower than cementite while the plasticity and toughness are lower than ferrite but higher than cementite. Moreover, the strength of pearlite increases with decreasing its lamellar spacing.

Cementite is a strengthening phase in steels. If the matrix of alloys is ferrite, the strength of alloys increases with the carbon content since the amount of cementite increases as well. However, if brittle cementite distributes at grain boundaries, especially forms a network-like distribution, the plasticity and toughness of the alloys will significantly decrease. For example, when $w(C) > 1\%$, the secondary cementite exhibits a continuous network-like distribution with the increase of amount, leading to high brittleness, low ductility, and deleterious ultimate tensile strength of steels. When pearlite becomes the matrix (such as in white cast irons), the alloys will be harder and more brittle with increasing the amount of cementite.

## 7.4 Solidification theory of binary alloys

Except for following the general principles of pure metal crystallization, solute elements will be redistributed between the solid/liquid phases in binary alloys owing to the addition of the second element. As a result, the solidification manner and the crystallization morphology of alloys would be affected significantly by the redistribution of solute elements, and the macrosegregation and microsegregation may also be induced in the alloy ingots. In the following section, the solidification theory of a binary alloy with isomorphous transformation and eutectic transformation will be discussed. Finally, the structure and defects of alloy ingots (castings) will be briefly described.

### 7.4.1 Solidification theory of solid solution alloys

**1. Normal freezing**

The level of solute redistribution during solidification of alloys can be characterized by the equilibrium distribution coefficient $k_0$, which is defined as the ratio of the mass concentration in the solid phase to the mass concentration in the liquid phase:

$$k_0 = \frac{w_S}{w_L} \tag{7.8}$$

The two cases of isomorphous transformation process are illustrated in Fig. 7.62. When $k_0 < 1$, it indicates that both the start and finish solidification temperatures decrease with the increase of solute content. Conversely, for the case of $k_0 < 1$, it indicates that both the start and finish solidification temperatures increase with increasing solute content. When $k_0$ is close to 1, the solute content in the solid is close to the original content in the alloy, which means that the extent of solute redistribution (segregation) is less. Assumed that the liquidus and solidus line is straight, it can be proved geometrically that $k_0$ is constant for different alloy compositions.

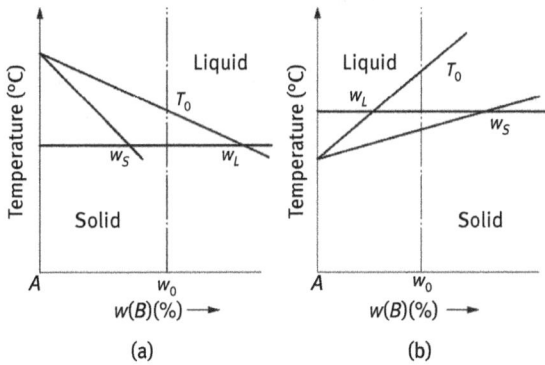

Fig. 7.62: Two cases of $k_0$: (a) $k_0 < 1$ and (b) $k_0 > 1$.

As shown in Fig. 7.63(a), a single-phase solid solution alloy is melted in a cylindrical crucible, and then directionally solidified from left to right. Under the condition of equilibrium solidification, the solute concentration in the solid is always uniform and corresponding to the solidus concentration at any specific temperature. At the end of solidification, the solid concentration is exactly the composition of the alloy, as shown in Fig. 7.63(b).

Under the condition of nonequilibrium solidification, however, the solute concentration in the solid is not uniform. It varies with the solidification distance, $x$. To deduce the analytic expression on the variation of the mass concentration $\rho_S$ with the solidified distance, five conditions are assumed:

① The liquid concentration is uniform at any time.
② The solid–liquid interface is planar.
③ The local equilibrium is maintained at the solid–liquid interface, that is, $k_0$ is a constant at the interface.

④ The solute diffusion is ignored in the solid.
⑤ The solid and the liquid phase have the same density.

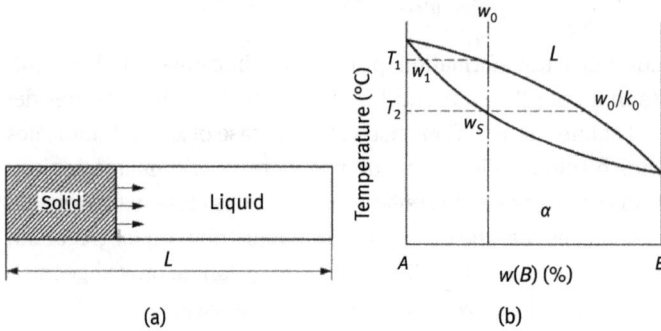

(a) (b)

**Fig. 7.63:** A billet ingot of L in length (a) and equilibrium solidification (b).

Assume the cross-sectional area and length of the alloy rod is $A$ and $L$, respectively. When a volume element $Adx$ with mass $dM$ solidifies, indicated by the shadow zone in Fig. 7.64, the mass change after solidification can be expressed as

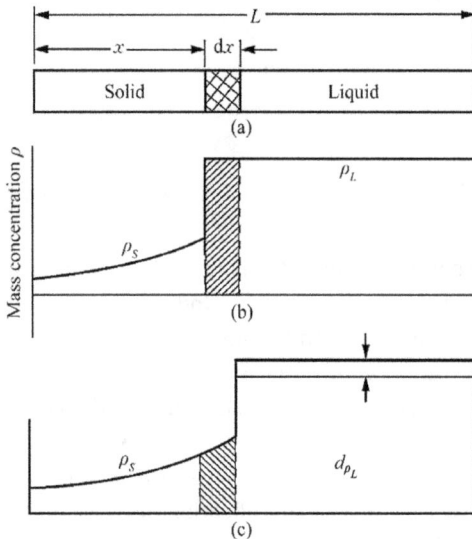

**Fig. 7.64:** Solidification of a volume element (a), the solute distribution before solidification (b) and after solidification (c).

$$dM(\text{before } solidification) = \rho_L A dx$$

$$dM(\text{after } solidification) = \rho_S A dx + d\rho_L A(L - x - dx)$$

where $\rho_L$ and $\rho_S$ are the mass concentration in the liquid and solid, respectively. According to the conservation of mass:

$$\rho_L A dx = \rho_S A dx + d\rho_L A(L - x - dx)$$

Ignoring the high-order small quantity $(d\rho_L dx)$, it can be obtained:

$$d\rho_L = \frac{(\rho_L - \rho_S) dx}{L - x}$$

Both sides are divided by the density of the liquid or solid phase. Considering the assumption that the solid and the liquid phases has the same density, so $\rho_S/\rho_L = w_S/w_L = k_0$. Now, integrating from $\rho = \rho_0$ at $x = 0$ to $\rho = \rho_L$ at $x = x$, it yields the mass concentration of liquid phase at the solid-liquid interface $\rho_L$ as a function of solidified length $x$:

$$\int_{\rho_0}^{\rho_L} \frac{d\rho_L}{\rho_L} = \int_0^x \frac{1 - k_0}{L - x} dx$$

$$\rho_L = \rho_0 \left(1 - \frac{x}{L}\right)^{k_0 - 1} \tag{7.9}$$

where $\rho_0$ is the initial mass concentration in the alloy. Considering the local equilibrium at the solid–liquid interface $\rho_L = \rho_S/k_0$, the mass concentration in the solid $\rho_S$ can be obtained:

$$\rho_S = \rho_0 k_0 \left(1 - \frac{x}{L}\right)^{k_0 - 1} \tag{7.10}$$

Equation (7.10) is called the *normal freezing equation*, or the *Scheil equation*, which expresses the variation of solid mass concentration with the freezing length.

The distribution of mass concentration of entire ingot (rod) after normal freezing was illustrated in Fig. 7.65 (for $k_0 < 1$), which is consistent with the general case of casting ingots, thus termed normal freezing. Such a nonuniform distribution of mass concentration that solute concentration increases gradually from the surface to center in the ingot is called *positive segregation*. The positive segregation is a kind of macrosegregation, which is difficult to be eliminated even by diffusion annealing.

## 2. Zone melting

In the above normal freezing process, the solid solution alloy with mass concentration $\rho_0$ is melted entirely before directionally solidified. Normal freezing produces positive segregation, in other words, normal freezing produces a purification in the first

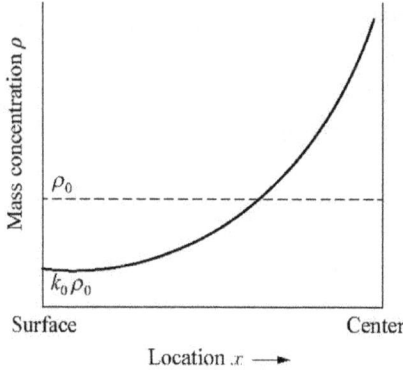

Fig. 7.65: The distribution of solute concentration in the ingot after normal freezing.

portion of the rod do solidify, as shown in Fig. 7.65. If this alloy is locally melted from the left to the right, what will the mass concentration distribution changes along with the solidification length? In 1952, W. G. Pfann published a famous paper showing that a much more effective way of purification could be obtained by the technique of zone melting.

In the following part, the mathematical formula about the mass concentration distribution of solute after one pass ($n = 1$) of the zone down the rod will be derived. The assumption is similar as the case of normal freezing. The mass concentration in raw material is assumed to be $\rho_0$ and uniformly distributed in the rod. Let the cross-sectional area $A = 1$, thus the volume of volume element is equal to $dx$ and the mass concentration in the solidified volume is $\rho_S = k_0\rho_L$, where $\rho_L$ is the mass concentration in the liquid. The mass of solute in the solidified volume is $\rho_S dx$ or $k_0\rho_L dx$, then:

$$\rho_L = \frac{\text{mass of solute in the liquid}}{\text{volume of the liquid}} = \frac{m}{V} = \frac{m}{l}$$

where $l$ is the length of melting zone.

After the melting zone moves forward a distance of $dx$, the increment of solute mass in the liquid (melting zone) can be obtained (see Fig. 7.66):

$$dm = m(x + dx) - m(x) = (\rho_x l - \rho_S dx + \rho_0 dx) - \rho_x l = \rho_0 dx - \rho_S dx = \left(\rho_0 - \frac{k_0 m}{l}\right) dx$$

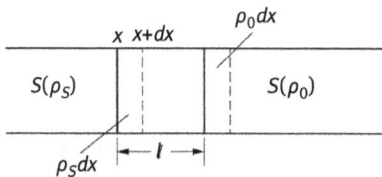

Fig. 7.66: The change in the solute after the melting zone moves forward a distance of $dx$.

Transposition and integrate, the above equation is transformed to

$$\int \frac{dm}{\rho_0 - \frac{k_0 m}{l}} = \int^d x$$

$$\left(-\frac{l}{k_0}\right) \ln\left(\rho_0 - \frac{k_0 m}{l}\right) = x + A$$

where $A$ is a constant to be determined. At $x = 0$ (the start point of zone melting), the total mass of solute in the liquid zone $m = \rho_0 l$, thus:

$$A = -\frac{l}{k_0} \ln \rho_0 (1 - k_0)$$

Substitute $A$ into the original equation, the mass concentration distribution in the solid can be expressed as

$$\rho_S = \rho_0 \left[1 - (1 - k_0)e^{\frac{-k_0 x}{l}}\right] \tag{7.11}$$

Equation (7.11) is the zone melting equation, which describes the variation of mass concentration in the solid solution with the solidification length after the zone has passed down the rod the first time ($n = 1$). Note that eq. (7.11) cannot be used to describe the result of more than one pass of the zone, because the mass concentration in the rod is not uniform after one pass of the zone. And this equation is also not satisfied with the last melting zone, because the length of melt is less than $l$ after a forward distance $dx$ and the expression of $dm$ cannot be obtained at the end of the rod.

The purification after one pass is less than that obtained in normal freezing. However, zone melting may repeatedly pass the zone down the rod and an increased purification is obtained after each pass, until after many passes an ultimate distribution is reached. The quantitative equation of the distribution of mass concentration after multipass zone melting ($n > 1$) had been derived by different authors. Figure 7.67 gives the schematic diagram of this distribution when $k_0 < 1$. With the increase of zone pass, the mass concentration of the impurity in the front end of the rod is reduced gradually, while that in the rear end of the rod increased gradually, resulting in the increasing purity of solid solution in the front end. Therefore, the zone melting process is also termed the zone purification. Figure 7.68 shows the results derived by Lord, where the distribution coefficient of impurity $k_0 = 0.1$. After eight zone passes of purification, the mass concentration of impurity in the solid is decreased by $10^4$–$10^6$ times at the length of 8 $l$. Zone melting is used extensively for various high-purity materials. For example, high purity germanium containing less than one impurity atom among ten million Ge atoms can be obtained from zone melting, which can be used as a semiconductor rectifier components. It can be concluded that the zone purification technology is an outstanding achievement in the

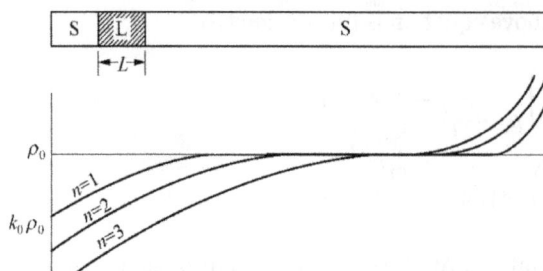

**Fig. 7.67:** Variation of composition along a rod produced by multipass zone melting ($n > 1$, $n$ equals the number of zone passes).

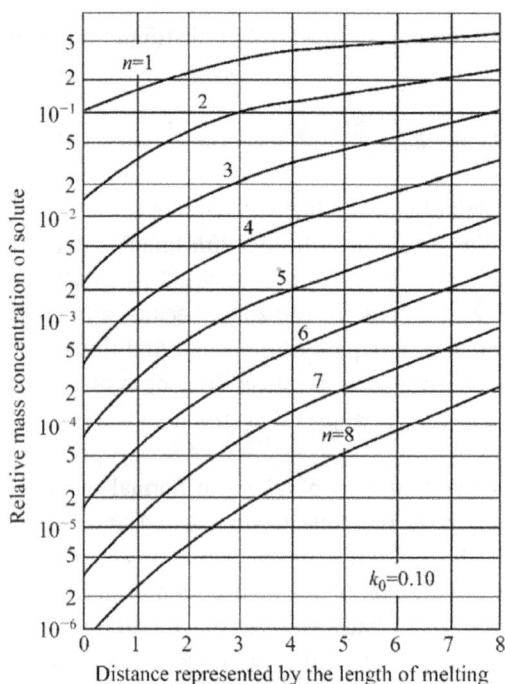

**Fig. 7.68:** Variation of composition along a rod produced by multipass zone melting based on Lord's calculation.

application of solidification theory of solid solution alloys. A typical zone melting setup is illustrated in Fig. 7.69, where purification process is realized by moving the rod to enter inside of interval induction heaters for melting. Therefore, multipass zone melting technique can be easily realized by mounting multiple induction heaters in parallel at some distance (interval) each other, and the rod is driven to slowly enter into induction heaters in the horizontal direction.

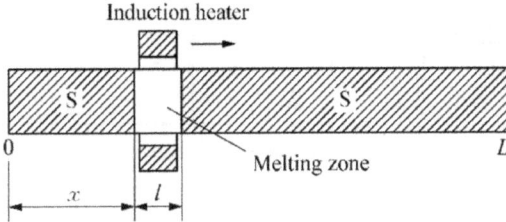

**Fig. 7.69:** Schematic view of a zone purification setup.

In principle, the normal freezing can also play a role of purification. However, the alloy is entirely melted during normal freezing, which can destroy the purification effect of previous freezing. Therefore, the purification efficiency of the normal freezing is far less than the zone melting.

### 3. Effective distribution coefficient $k_e$ and mixing in the liquid

In the derivation of both the normal freeze equation and the zone melting equation, the assumption of uniform liquid composition was employed. This assumption is usually reasonable, since there exist both diffusion and particularly convection in the liquid to homogenize the composition. However, this assumption in fact turns out to be a very severe restriction.

There exists natural convection in the liquid during solidification due to its low viscosity and high density. This natural convection tends to produce a uniform liquid composition. However, there is a fundamental characteristic of fluid flow that prevents this. When a liquid flows through a pipe at a low velocity, as we know, the fluid flows parallel to the pipe wall at all points, which is called laminar flow. The flow velocity is a maximum at the center of the tube and drops parabolically to zero as the wall is approached. As a result, there exists a very thin boundary layer of laminar flow in the vicinity of the wall. This kind of boundary layer also exists at the $S/L$ interface during solidification, which inhibits a uniform liquid composition.

The solute is continually being rejected from the solid into the liquid at the $S/L$ interface when $k_0 < 1$. In order to obtain a uniform liquid composition, this solute must be transported very quickly throughout the liquid. In the boundary layer at the $S/L$ interface, there can be no convective transport normal to the interface because of the laminar flow parallel to the interface. Solute can only be transported through the boundary layer into the convecting liquid by the slow mechanism of diffusion. Therefore, a buildup of solute is obtained in the boundary layer region, as shown by the dashed curve in Fig. 7.70(a). Beyond the boundary layer the bulk liquid composition is uniform at the value $(\rho_L)_B$ due to mixing. Since local equilibrium is approached at the $S/L$ interface we have $(\rho_S)_i = k_0 (\rho_L)_i$ (where mass concentrations are used and constant density of the solid and the liquid is assumed). The solute buildup causes $(\rho_L)_i$ to rise rapidly and, hence, $(\rho_S)_i$ must also rise rapidly.

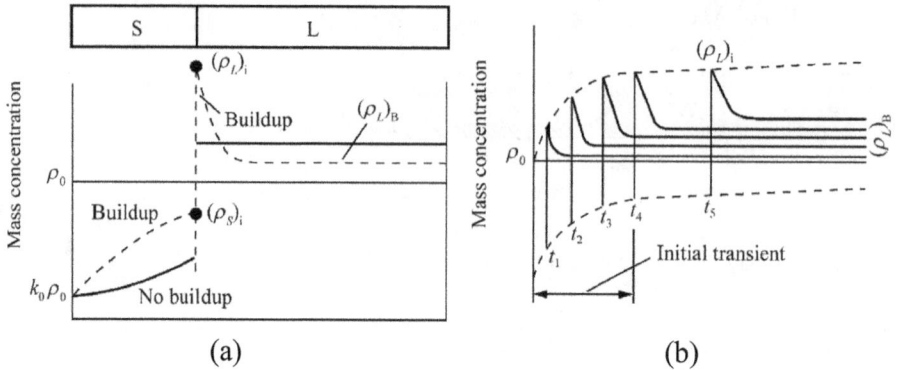

Fig. 7.70: Effect of solute buildup on the composition in the rod (a) and the establishment of solute buildup during initial transient.

Therefore, the solid composition rises more rapidly than for the case of no buildup, as shown in Fig. 7.70(a). As solute continues to build up, its concentration gradient across the boundary layer becomes steeper, and the rate of transport through the boundary layer by diffusion increases until eventually a balance is obtained between the input and output to the boundary layer region. At this point, the buildup ceases to rise so that $(\rho_L)_i/(\rho_L)_B$ becomes constant. The region over which the buildup occurs is called the initial transient, or the initial transition zone, as shown in Fig. 7.70(b).

In order to characterize the degree of mixing in the liquid, it is useful to define a term called the effective distribution coefficient, $k_e$:

$$k_e = \frac{(\rho_S)_i}{(\rho_L)_B} \tag{7.12}$$

After the initial transient is established, the effective distribution coefficient is constant. Since the value of $k_e$ tells a great deal about the effect of liquid mixing upon the solute profile when liquid is frozen into the solid, we now derive an equation for its dependence upon measureable parameters.

We take the $S/L$ interface as the reference point. The liquid flows toward an observer on the interface so that the flux of solute caused by liquid flow at any point in the liquid is $-R\rho_L$, where $\rho_L$ is the mass concentration of local liquid and $R$ is the velocity of the liquid flowing toward the observer ($R$ is the interface rate if taking any point in the liquid as the reference point). The minus sign of $-R\rho_L$ indicates that the direction of flow is opposite to the direction of diffusion ($z$), as shown in Fig. 7.71. The total flux of solute resulting from diffusion and fluid flow can be determined from:

$$J = -R\rho_L - D\frac{d\rho_L}{dz}$$

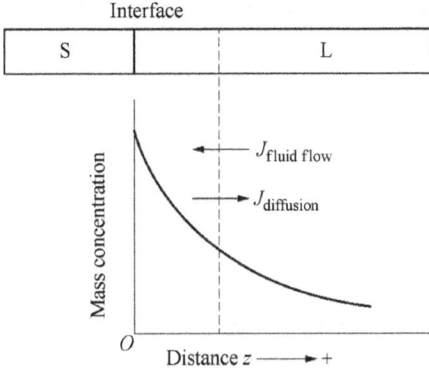

Fig. 7.71: The fluxes of solute in the liquid relative to the S/L interface.

Taking the partial derivative of above equation with respect to $z$, and considering the equation $\frac{\partial \rho_L}{\partial t} = -\frac{\partial J}{\partial z}$, which was an equation before deriving Fick's Second Law, we obtain:

$$D\frac{\partial^2 \rho_L}{\partial z^2} + R\frac{\partial \rho_L}{\partial z} = \frac{\partial \rho_L}{\partial t}$$

After the initial transient has been established, the amount of solute in the boundary layer remains relatively constant, and so we assume $\partial \rho_L/\partial t = 0$. Experiments have shown this assumption does not give rise to large error. Therefore, the above second-order partial differential equation becomes a second-order ordinary differential equation, and $\rho_L(z,t)$ becomes $\rho_L(z)$:

$$\frac{d^2 \rho_L}{dz^2} + \frac{R}{D}\frac{d\rho_L}{dz} = 0 \tag{7.13}$$

The general solution is

$$\rho_L = P_1 + P_2 e^{-Rz/D} \tag{7.14}$$

where the undetermined coefficients $P_1$ and $P_2$ can be obtained from the boundary conditions.

In a time interval $dt$, the S/L interface moves a distance $dz$ (equal to $Rdt$). The total amount of solute in the solid at one side of interface is $(\rho_S)_i ARdt$, where $A$ is the cross-sectional area of the rod, while that in the liquid adjacent to the interface is $(\rho_L)_i ARdt$. The difference between them should be equal to the solute rejected by the solid to the liquid outside the boundary layer by diffusion, the amount of which is $-AD\frac{d\rho_L}{dz}\,dt$, and thus:

$$(\rho_L)_i ARdt - (\rho_S)_i ARdt = -AD\frac{d\rho_L}{dz}\,dt \tag{7.15}$$

Then, we obtain:

$$\frac{d\rho_L}{dz} = \frac{R}{D}\left[(\rho_S)_i - (\rho_L)_i\right] = \frac{R}{D}(k_0 - 1)(\rho_L)_i \tag{7.16}$$

The solute distribution near the $S/L$ interface is illustrated in Fig. 7.72. Substituting the boundary condition $\rho_L = (\rho_L)_i$ at $z = 0$ into eq. (7.14), we have

$$(\rho_L)_i = P_1 + P_2 \tag{7.17}$$

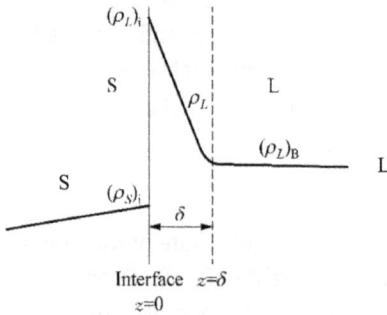

Fig. 7.72: The distribution of solute in the solid and liquid near the interface after the initial transient.

Taking the derivative of eq. (7.14) and $z = 0$, we obtain

$$\frac{d\rho_L}{dz} = -P_2\frac{R}{D} \tag{7.18}$$

Substituting eqs. (7.17) and (7.18) into eq. (7.16), we obtain

$$P_1 = \frac{k_0}{1-k_0}P_2 \tag{7.19}$$

Substituting eq. (7.19) and boundary condition $\rho_L = (\rho_L)_B$ at $z = \delta$ (the thickness of boundary layer, see Fig. 7.72) into eq. (7.14), we obtain

$$(\rho_L)_B = P_2\left(\frac{k_0}{1-k_0} + e^{-R\delta/D}\right) \tag{7.20}$$

So the undetermined coefficient $P_2$ can be obtained

$$P_2 = (\rho_L)_B \bigg/ \left(\frac{k_0}{1-k_0} + e^{-R\delta/D}\right) \tag{7.21}$$

Substituting eqs. (7.21) into (7.19), we have

$$P_1 = (\rho_L)_B \left(\frac{k_0}{1-k_0}\right) / \left(\frac{k_0}{1-k_0} + e^{-R\delta/D}\right) \tag{7.22}$$

Substituting $P_1$ and $P_2$ into eq. (7.14), we obtain

$$\rho_L = (\rho_L)_B \left(\frac{k_0}{1-k_0} + e^{-Rz/D}\right) / \left(\frac{k_0}{1-k_0} + e^{-R\delta/D}\right) \tag{7.23}$$

Considering $\rho_L = (\rho_L)_i$ at $z = 0$, eq. (7.23) transforms to

$$(\rho_L)_i = (\rho_L)_B \left(\frac{k_0}{1-k_0} + 1\right) / \left(\frac{k_0}{1-k_0} + e^{-R\delta/D}\right)$$

It can be simplified to

$$(\rho_L)_i = (\rho_L)_B / [k_0 + (1-k_0)]e^{-R\delta/D} \tag{7.24}$$

On account of the local equilibrium established at the $S/L$ interface, we have $k_0 = (\rho_S)_i/(\rho_L)_i$. Along with the definition of the effective distribution coefficient, $k_e$, given by eq. (7.12), we substitute them into eq. (7.24) and obtain the mathematic expression of $k_e$

$$k_e = \frac{k_0}{k_0 + (1-k_0)e^{-R\delta/D}} \tag{7.25}$$

It is a well-known equation first derived in 1953 by Burton, Prim, and Slichter, which indicates that the effective distribution coefficient $k_e$ is a function of the equilibrium distribution coefficient $k_0$ and the dimensionless quantity $R\delta/D$. A plot of Burton–Prim–Slichter equation (eq. (7.25)) is shown in Fig. 7.73 for a specific value of $k_0$. With the increase of $R\delta/D$, $k_e$ increases from a minimum value of $k_e$ to a maximum of 1. The following three cases of liquid mixing are discussed.

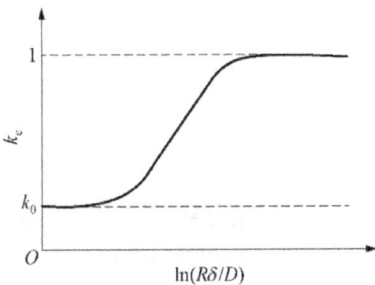

Fig. 7.73: A plot of the efficient distribution coefficient with $\ln(R\delta/D)$.

(1) When the solidification rate is extremely fast, $k_e$, that is, $k_e \to 0$. Hence, $k_e$ and $k_e$, as shown in Fig. 7.74(a). It represents a state of no mixing in the liquid, because the liquid convection is completely suppressed outside the boundary layer and the mixing of solute (a uniform distribution) cannot be achieved by diffusion only. In this case, the thickness of boundary layer has a maximum value of about 0.01–0.02 m generally.

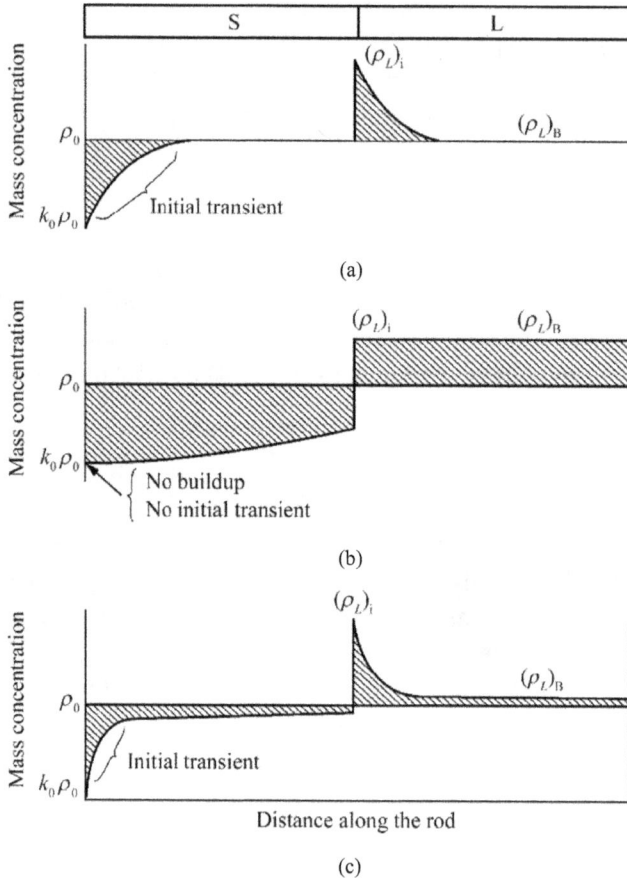

Fig. 7.74: The solute profile for different cases of $k_e$: (a) $k_e = 1$, (b) $k_e = k_0$, and (c) $k_0 < k_e < 1$.

(2) When the solidification rate is extremely slow, $k_e$, that is, $k_e \to 1$. Hence, $k_e$ and $k_e$, as shown in Fig. 7.74(b). The solute in the liquid is completely mixed, because sufficient convection eliminates the boundary layer and results in a complete mixing of the solute.

(3) When the solidification rate is between the two cases mentioned earlier, that is, $k_e$, $k_e$ is a constant after the initial transient, which corresponding to a state of

partially mixing as shown in Fig. 7.74(c). It means that it has time for the liquid beyond the boundary layer to experience insufficient convection to achieve some degree of mixing of the solute. In this case, the boundary layer is thinner with respect to the no-mixing state and its value is about 0.001 m generally.

Considering the degree of mixing in the liquid, $k_e$ in normal freeze equation and zone melting equation above-mentioned should be replaced by $k_e$. But it needs to be emphasized that these equations are still restricted to the assumption of a planar interface. Figure 7.75 shows the solute profile obtained after normal freeze at $k_e$ and $k_e$, respectively. Thus, $k_e$ should be as close to $k_e$ as possible if the greatest degree of purification is desired, that is, $k_e$ should be as small as possible. Hence, a smaller interface velocity $R$ and a higher degree of mixing are required to reduce the thickness of the boundary layer. On the contrary, $k_e$ is required if a rod with a uniform composition is desired, that is, a higher velocity $R$ and no mixing are required to achieve a maximum $\delta$. In this way, once the initial transient is over, a uniform distribution of solute can be obtained (except for the initial transient and the final melting zone at both ends of the rod).

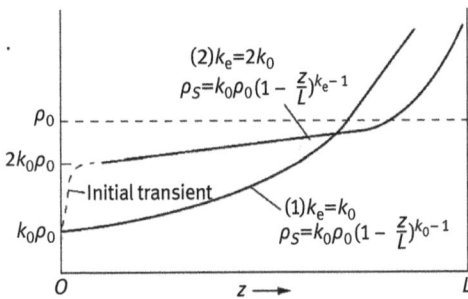

**Fig. 7.75:** Solute profiles for $k_e = k_0$ (complete mixing) and $k_e = 2k_0$ (partial mixing).

## 4. Constitutional supercooling during solidification of alloys

### a. The concept of constitutional supercooling
The theoretical freezing temperature $k_e$ is fixed for pure metals. Undercooling is generated when the actual temperature is lower than $T_m$, which is called thermal undercooling. During the solidification of alloys, the freezing temperature is changed by the solute redistribution in the liquid, which can be determined from the liquidus line on the phase diagram. When actual temperature in the liquid at the front of the $S/L$ interface is lower than the real freezing temperature dependent on solute redistribution, an additional undercooling is produced, which is called constitutional supercooling. Whether the undercooling can happen and its extent

depends on the solute and real temperature distribution in the liquid at the front of the $S/L$ interface.

Figure 7.76 illustrates the case of constitutional supercooling when $k_e$. Figure 7.76 (a) gives a corner of the binary phase diagram and a specific alloy composition $w_0$. Figure 7.76(b) gives the real temperature distribution in the liquid at the front of the $S/L$ interface ($z = 0$). Figure 7.76(c) gives the mass concentration profile in the liquid at the front of the interface under no-mixing condition ($k_e = 1$). By substituting the boundary conditions $z = 0$, $\rho_L = \rho_0/k_0$; $z = \infty$, $\rho_L = \rho_0$ into eq. (7.14), we obtain

$$\rho_L = \rho_0 \left( 1 + \frac{1 - k_0}{k_0} e^{-Rz/D} \right)$$

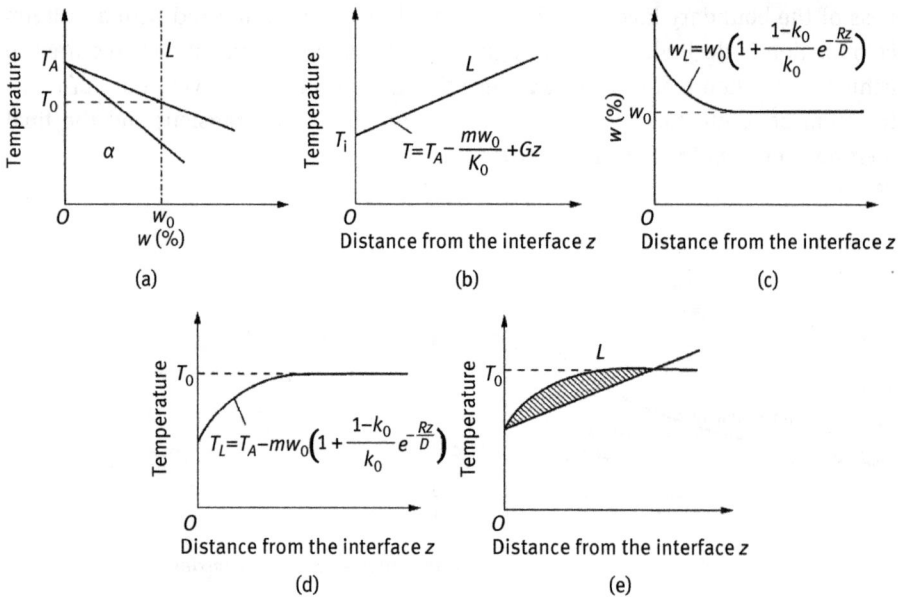

Fig. 7.76: Schematic of constitutional supercooling for an alloy of $k_0 < 1$.

Both sides of the equation are divided by alloy density $\rho$, we have

$$w_L = w_0 \left( 1 + \frac{1 - k_0}{k_0} e^{-Rz/D} \right) \tag{7.26}$$

The freezing temperature corresponding to the mass concentration of solute $w_L$ at each point of the curve in Fig. 7.76(c) can be found on the phase diagram and is shown in Fig. 7.76(d). Then, a plot of the real temperature is superimposed, and the constitutional supercooling zone is obtained as shown by the shaded area in Fig. 7.76(e).

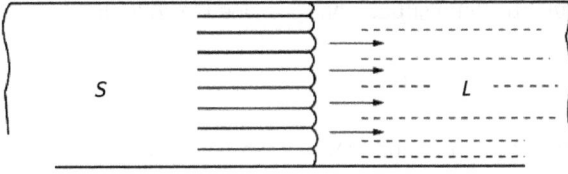

**Fig. 7.77:** Schematic view of cellular growth.

### b. The critical condition for constitutional supercooling

Assume that $k_0$ is a constant, thus the liquidus is a straight line with a constant slope $m$. From Fig. 7.76(a), the liquidus temperature of the alloy can be expressed by

$$T_L = T_A - m\omega_L \tag{7.27}$$

where $T_A$ is the melting point of pure A component. Substituting eqs. (7.26) into (7.27), we have

$$T_L = T_A - m\omega_0 \left(1 + \frac{1-k_0}{k_0} e^{-Rz/D}\right) \tag{7.28}$$

which is the mathematical expression of the curve given in Fig. 7.76(d).

As shown in Fig. 7.76(b), the real liquid temperature at a distance of $z$ from the S/L interface can be expressed by

$$T = T_i + Gz \tag{7.29}$$

where $T_i$ is the interface temperature, $G$ is the temperature gradient in the liquid relative to the S/L interface. During steady-state solidification after the initial transient, the interface temperature $T_i$ is equal to the liquidus temperature $T_L$ at $z = 0$. Under no-mixing condition in the liquid, the mass concentration in the solid and liquid at $z = 0$ are $\omega_0$ and $\omega_0/k_0$, respectively. Thus, the interface temperature $T_i$ is the liquidus temperature corresponding to the mass concentration of $\omega_0/k_0$:

$$T_i = (T_L)_{z=0} = T_A - \frac{m\omega_0}{k_0} \tag{7.30}$$

Substituting eqs. (7.30) into (7.29), we obtain

$$T = T_A - \frac{m\omega_0}{k_0} + Gz \tag{7.31}$$

Obviously, the constitutional supercooling is produced only when $T < T_L$, that is, the real temperature in the liquid is lower than the liquidus temperature.

The critical condition for constitutional supercooling is (see Fig. 7.76e):

$$\frac{dT_L}{dz}\Big|_{z=0} = G \tag{7.32}$$

Taking the derivative of eq. (7.28), the expression for $z = 0$ can be obtained:

$$\frac{dT_L}{dz}\Big|_{z=0} = mw_0 \frac{1-k_0}{k_0} \frac{R}{D} \tag{7.33}$$

Based on eqs. (7.32) and (7.33), the critical condition for constitutional supercooling is obtained

$$G = \frac{Rmw_0}{D} \frac{1-k_0}{k_0} \tag{7.34}$$

This is a well-known equation first derived by Tiller, Jackson, Rutter, and Chalmers in 1953. A large number of experiments have confirmed that it predicts the stability of solidifying planar interfaces remarkably well. Obviously, the condition for producing the constitutional supercooling is $G < (dT_L/dz)|_{z=0}$. Hence, we have

$$\frac{G}{R} < \frac{mw_0}{D} \frac{1-k_0}{k_0} \tag{7.35}$$

Otherwise, constitutional supercooling does not occur.

Based on the geometrical relationship in Fig. 7.76(a), the critical condition for constitutional supercooling can be expressed as

$$\frac{G}{R} = \frac{\Delta T}{D} \tag{7.36}$$

where $\Delta T = mw_0(1-k_0)/k_0$ is the liquidus-solidus temperature interval (freezing temperature range) at $w_0$.

The right hand of eq. (7.35) consists of the parameters associated with the properties of the alloy. Increasing $w_0$ and $m$, or decreasing $k_0$ all lead to the increase of $\Delta T$, which favors the constitutional supercooling. In addition, the smaller the diffusion coefficient $D$ is, the easier the solute enrichment in the boundary layer becomes, which also favors the constitutional supercooling. The left hand of eq. (7.35) consists of the parameters controlled by external conditions. The smaller the real temperature gradient $G$ is, the larger the shaded area in Fig. 7.76(e) becomes for a certain alloy composition and freezing rate, which means the tendency for constitutional supercooling is greater. Besides, increasing the freezing rate can reduce the degree of liquid mixing and enhance solute buildup in the boundary layer, which also promotes the constitutional supercooling.

The above derivation assumes that the liquid is completely no mixing, that is, $k_e = 1$. For the case of $k_0 < k_e < 1$ with partial mixing in the liquid, the basic conclusion mentioned above still is valid but needs modification. When $k_e = k_0$ with complete mixing in the liquid, there is no solute buildup at the front of the $S/L$ interface, and hence, the constitutional supercooling does not appear.

### c. The effect of constitutional supercooling on crystal growth morphology

Both the normal freezing and zone melting mentioned earlier require a planar solid/liquid interface. For this reason, very slow solidification rates and low solute content are required. Generally, the mass concentration of solute needs to be less than 1%. But the solidification rate $R$ in real alloy ingots or castings is usually larger than $2.5 \times 10^{-5}$ m/s, while the real temperature gradients are usually less than 300 $\sim$500 °C/m. Assume an alloy with $D = 10^{-9}$ m²/s at the liquidus temperature, the liquidus slope $|m| > 1°C/w$, and $R \approx 2.5 \times 10^{-5} m/s$. Based on eq. (7.34), if $k_0 = 0.1$, the critical temperature gradients for alloys with $w = 0.1\%$ and $w = 0.01\%$ are 225,000 and 2,250 °C/m, respectively. And if $k_0 = 0.4$, the critical temperature gradients for above alloys are 37,500 and 375 °C/m, respectively. These data are much larger than the real temperature gradients in ingots or castings, indicating that constitutional supercooling is generally inevitable during alloy solidification. Planar interface will be unstable when there is a small constitutional supercooling zone ahead of the growing $S/L$ interface. If a bump protrudes ahead of the interface, it enters the supercooled zone and will be expected to propagate in the liquid due to the increasing constitutional supercooling. But since the constitutional supercooling zone is small, it is impossible to have a large extension of the bump, which promotes the formation of a cellular interface, as shown in Fig. 7.77. If the constitutional supercooling increases, the bump can further grow and propagate into the undercooled liquid phase, and side branches will also develop by forming secondary and ternary axis, resulting in the formation of dendrites. The typical cellular and dendrite structures are shown in Figs. 7.78 and 7.79, respectively. There are also transitional forms between the two kinds of structures, namely the planar cells between the planar and the cellular morphology, and the cellular dendrites between the cellular and the dendritic morphology. For the alloy of $w_0$, the temperature gradient in the liquid and the solidification rate are the main factors that affect the constitutional supercooling. Figure 7.80 summarizes their effects on the growth morphology of the solid solution based on experiments.

From the above analysis, it is noted that for alloys dendritic structure can be obtained under a positive temperature gradient due to the constitutional supercooling, but for pure metals dendrites form only in the presence of negative temperature gradients. Therefore, constitutional supercooling is the main feature for alloy solidification that is different from pure metals.

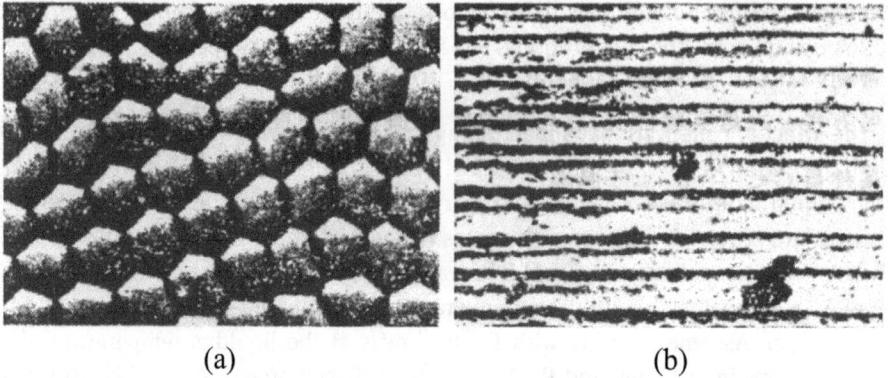

(a)                                                    (b)

**Fig. 7.78:** Regular cellular structure (unpolished, unetched): (a) transverse view and (b) longitudinal view.

00018590                  ———— 5 μm

**Fig. 7.79:** Typical dendritic structure of alloys.

### 7.4.2 Solidification theory of eutectics

For binary alloys, the eutectic structure forms by simultaneous growth of two solid phases from the liquid. There are many different kinds of morphology of eutectic structures. In this section, the classification of eutectics is briefly discussed, and the discussions are focused on the mechanism and growth kinetics of lamellar eutectic structure.

**1. The classification and formation mechanisms of eutectics**

For more than 100 years, a variety of eutectic structures have been discovered. They are roughly classified into lamellar, rod-like (fibrous), spherical, needle-like, and

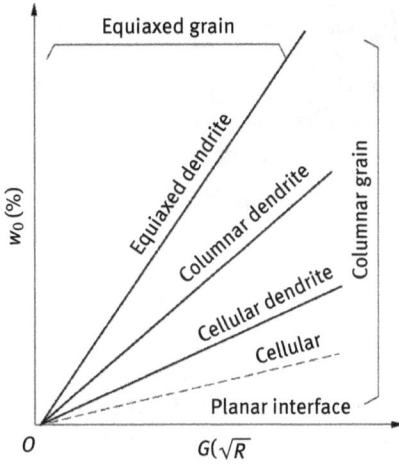

**Fig. 7.80:** Effect of $G/\sqrt{R}$ and $w_0$ on the growth morphology of solid solution alloys.

helical shapes, as shown in Fig. 7.81. The three-dimensional models of some eutectics are shown in Fig. 7.82, which are helpful to understand the morphology of eutectic microstructure in different two-dimensional sections.

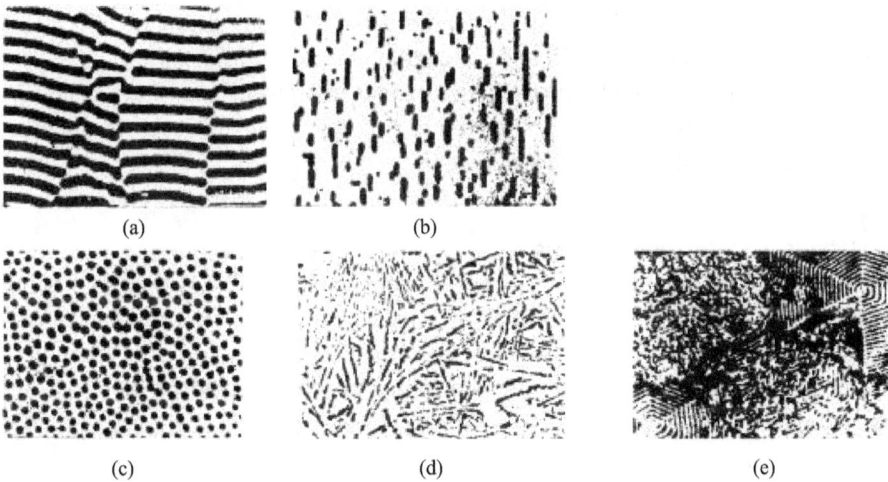

**Fig. 7.81:** Typical morphologies of eutectic structure: (a) lamellar, (b) rod-like (stripe or fibrous), (c) spherical, (d) needle-like, and (e) helical.

In the past, the classification of eutectics was mainly based on the morphology of the two phases or the morphology of the eutectic domain with the same orientation (the region growing from the same eutectic nucleus). However, this classification

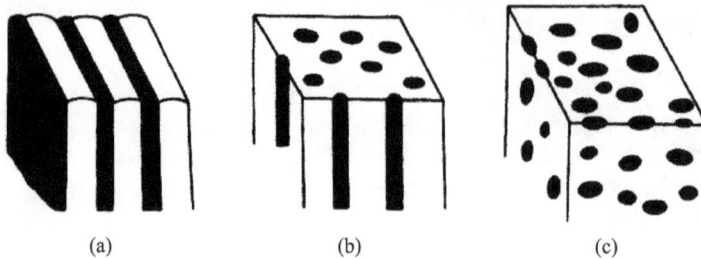

**Fig. 7.82:** Three-dimensional models for some eutectics: (a) lamellar, (b) rod-like, and (c) spherical.

cannot reflect the nature of formation of various types of eutectic alloys. Later, it was suggested that the eutectic microstructures can be classified into three categories according to the characteristics of the $S/L$ interface during eutectic growth, that is, the factor $\alpha$ presented by Jackson (see Chapter 6).

(1) Metallic–metallic type (rough–rough interface). It includes metallic–metallic eutectics, like Pb–Cd, Cd–Zn, Zn–Sn, and Pb–Sn, and many metallic–intermetallic eutectics, such as Al–Ag$_2$Al and Cd–SnCd.

(2) Metallic–nonmetallic type (rough–planar interface). The two constituent phases are metallic–nonmetallic or metallic–semimetallic. The $S/L$ interface of the non-metallic or semi-metallic phase in the eutectics is planar during solidification, such as Al–Ge, Pb–Sb, Al–Si, and Fe–C (graphite).

(3) Nonmetallic–nonmetallic type (planar–planar interface). There have been few researches on this category, and it does not belong to the scope of alloy research. Hence, it is not discussed hereafter.

### a. Metallic–metallic-type eutectic

Most of these types of eutectic structures are lamellar or rod-like. Although to some extent, it is affected by other factors, such as growth rates and temperature gradients ahead of the advancing interface, the formation of lamellar or rod-like eutectics is mainly controlled by the interface energy and determined by the following two factors:

### (1) The relative amount (volume fraction) of the two constituent phases in a eutectic

If the spacing between the lamellae or rods $\lambda$ is same, and the volume of one phase (e.g., $\alpha$ phase) is less than 27.6%, it is advantageous to form a rod-like eutectic. On the contrary, it is beneficial to form a lamellar eutectic. Detailed mathematical derivations are as follows.

The derivation model is shown in Fig. 7.83, where $r$ is the radius of the $\alpha$ rod or the thickness of the $\alpha$ lamellae, and $l$ is the length of the rod or lamellae. If the $\alpha$ rods

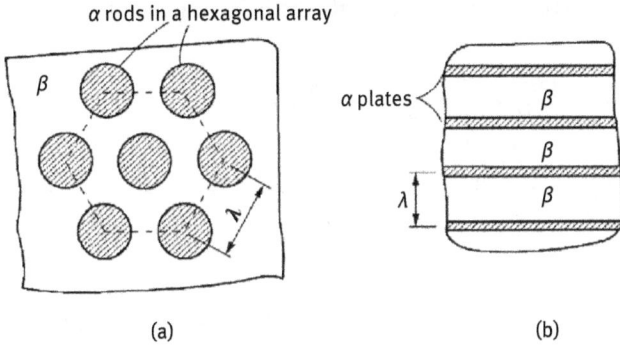

**Fig. 7.83:** Model for derivations of formation conditions of (a) the rod-like and (b) lamellar eutectic.

are distributed in a hexagonal array as shown in Fig. 7.83(a), the volume of the hexagonal eutectic is obtained

$$V_{\alpha+\beta} = 6 \times \frac{1}{2}\lambda \times \frac{\sqrt{3}}{2}\lambda \times l = \frac{3\sqrt{3}}{2}\lambda^2 l$$

The volume of the $\alpha$ rods in the hexagon is $V_\alpha = 3\pi r^2 l$

And the $\beta-\alpha$ interface area in the hexagon is

$$AA_R = 3 \times 2\pi r \times l = 6\pi r l$$

Assume the volume of the lamellar eutectic is equal to that of rod-like eutectic. Noting that the diagonal width of the rod-like hexagon is $2\lambda$, the spacing of the lamellar eutectic should be taken as $2\lambda$ to calculate the corresponding eutectic volume. Therefore, the width of the lamellar eutectic $x$ can be obtained from

$$\frac{3\sqrt{3}}{2}\lambda^2 l = 2\lambda x l$$

$$x = \frac{3\sqrt{3}\lambda}{4}$$

As shown in Fig. 7.83(b), there are four $\beta-\alpha$ interfaces existing in the spacing of $2\lambda$, and hence, its total area is obtained

$$4xl = 4 \times \left(\frac{3}{4}\sqrt{3}\lambda\right)l = 3\sqrt{3}\lambda l$$

If the $\beta-\alpha$ interface area in the rod-like eutectic is smaller than that in the lamellar eutectic, namely, $6\pi l r < 3\sqrt{3}\lambda l$, then:

$$r < \frac{\sqrt{3}}{2\pi}\lambda$$

In this case, the volume fraction of $\alpha$ phase must satisfy the following inequality:

$$\varphi = \frac{V_\alpha}{V_{\alpha+\beta}} = \frac{3\pi r^2 l}{\frac{3\sqrt{3}\lambda^2 l}{2}} < \frac{3\pi \left(\frac{\sqrt{3}}{2\pi}\lambda\right)^2 l}{\frac{3\sqrt{3}\lambda^2 l}{2}} = \frac{\sqrt{3}}{2\pi} = 27.6\%$$

The above result indicates that when the volume fraction of $\alpha$ or $\beta$ phase in the eutectic is less than 27.6%, the $\beta$–$\alpha$ interface area in the rod-like eutectic is smaller than that in the lamellar eutectic, which is in favor of the formation of the rod-like eutectic. On the contrary, it can be proved that when $\varphi > 27.6\%$, the formation of lamellar eutectic will be advantageous. This theoretical calculation has been confirmed by many experiments.

### (2) Interfacial energy per unit area between two constituent phases in eutectic

When the two eutectic phases are in a certain orientation relationship with each other, such as $(111)_{Al} \parallel (211)_{CuAl_2}$, $[\bar{1}01]_{Al} \parallel [\bar{1}20]_{CuAl_2}$, the interfacial energy per unit area will be lowered. To maintain this favorable orientation, the two phases can only be distributed in lamellar form.

Thus, when the volume of one phase is less than 27.6%, the morphology of eutectic depends on the relative contribution to the decrease of system energy by the reduction of interface area and the reduction of interfacial energy per unit area. If the former is dominant, it tends to form a rod-like eutectic. Otherwise, a lamellar eutectic forms.

Now we take the lamellar eutectic as an example to illustrate the formation mechanism of the eutectic microstructures.

For metallic–metallic type eutectic, the $S/L$ interfaces of these two eutectic phases are both atomically rough and approximately isothermal. The liquid phases at the front of the $S/L$ interfaces are undercooled to a small temperature (such as 0.02 °C) below the eutectic temperature. Thus, the solidification interface can be regarded as a planar interface on the macro scale.

During eutectic crystallization, the two solid phases do not appear at the same time. Instead, a solid phase preferentially nucleates and grows in the liquid, termed the leading phase. As shown in Fig. 7.84, the leading phase $\alpha$ nucleates and grows at the undercooling $\Delta T_E$ with a mass concentration of $w_\alpha^S$ for $B$ constituent. As the $\alpha$ phase continues to grow, owing to $w_\alpha^S < w_e$, excessive $B$ atoms will be rejected to liquid phase. As a result, the $B$ atoms in the liquid ahead of the $S/L$ interface are enriched and have a mass concentration of $w_\alpha^L$, which is higher than the content needed to form the $\beta$ phase, $w_\beta^L$. Therefore, the nucleation and growth of $\beta$ phase occur on the $\alpha$ leading phase. According to local equilibrium at the $S/L$ interface, the mass concentration of crystallized $\beta$ phase and the liquid are $w_\beta^S$ and $w_\beta^L$, respectively. At this time, the liquid ahead of $\beta$ phase is rich in $A$ atoms and its content of $A$ atoms is larger than the content needed to form $\alpha$ phase, which promotes the nucleation and growth of $\alpha$ phase on the $\beta$ phase again. In this iterative process of

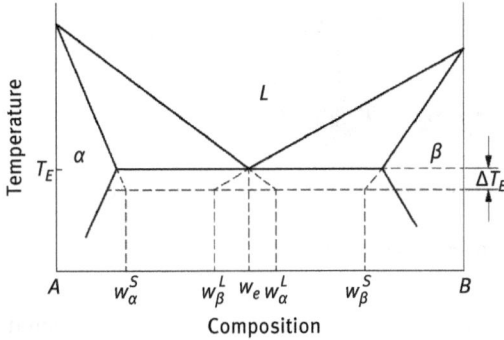

**Fig. 7.84:** Extrapolation of the phase boundaries to the supercooled interface temperature (often called the Hultgren extrapolation after A. Hultgren, Hardenability of Alloy Steels. American Society for Metals, Cleveland, 1938, p.55.).

alternative nucleation and growth, the microstructure of alternative $\alpha$ and $\beta$ phase arrangement forms in the end. When the $\alpha$ and $\beta$ phases grow forward together, the concentrations in the liquid ahead of the $\alpha$ and $\beta$ phases are $w_\alpha^L$ and $w_\beta^L$, respectively, and hence the transverse mass concentration difference between the two liquids is $w_\alpha^L - w_\beta^L$. The liquid far away from the $\alpha$ phase has a composition of $w_e$, thus the longitudinal mass concentration difference in the liquid ahead of $\alpha$ phase is $w_\alpha^L - w_e$, which is roughly half of that along the transverse direction. In addition, the diffusion length along the transverse direction is shorter than that along the longitudinal direction. Thus, as shown in Fig. 7.85, the alternative growth of $\alpha$ and $\beta$ is achieved mostly through the lateral diffusion process. The above analysis method is usually called the Hultgren extrapolation method, which is commonly used.

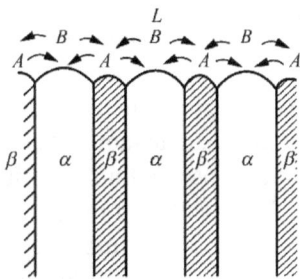

**Fig. 7.85:** Schematic view of lateral diffusion during solidification of a lamellar eutectic.

The alternative growth of the two phases in lamellar eutectic does not need repetitive nucleation, instead, it may depend on the bridging mechanism as shown in Fig. 7.86. The X-ray diffraction and the selected area electron diffraction have confirmed that the plates of each phase are grown from a same crystal in a eutectic colony and the two eutectic phases often have a certain crystallographic orientation relationship.

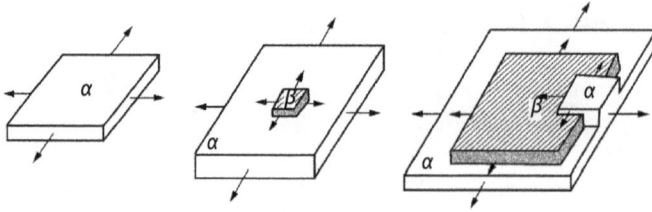

**Fig. 7.86:** The bridging mechanism of nucleation for a lamellar eutectic.

Generally, the thickness of the eutectic lamellae is represented by the lamellar spacing $\lambda$. The relationship between $R$ and $\lambda$ is

$$\lambda = \frac{k}{\sqrt{R}} \tag{7.37}$$

where $k$ is a constant that varies with different alloys. As the solidification rate, $R$, increases with the increase of undercooling $\Delta T_E$, a lager undercooling or a lager solidification rate leads to the decrease of the lamellar spacing and hence refine the eutectic structure.

The lamellar spacing of eutectics significantly affects the mechanical properties of alloys. The strength of eutectic alloys increases with the refinement of eutectic structures, which can be described by the Hall–Petch formula:

$$\sigma = \sigma^* + m\lambda^{-1/2} \tag{7.38}$$

where $\sigma$ is the yield strength, $\sigma^*$ is the constant relative to materials, and $m$ is a constant. The above discussion can also be applied to rod-like eutectics.

**b. Metallic–nonmetallic type eutectic**
Such eutectics are usually complex shaped, showing needle-like or bone-like. By SEM observation, however, the needles or plates within each eutectic domain are not isolated, instead they are interconnected into a whole structure. There are different viewpoints on the explanation of the reasons why the metallic–nonmetallic eutectic and the metallic–metallic eutectic are different. Some people think that it may be caused by the different kinetic undercoolings of the smooth and rough interfaces. The kinetic undercooling in the liquid at the front of a rough interface is about 0.02 °C, while that for a smooth interface is about 1–2 °C. A small kinetic undercooling in the liquid is enough to stimulate the nucleation and arbitrary growth of the metallic phase, which in turn causes the nonmetallic phases to branch out or forces them to stop growing, leading to the irregular microstructures. However, the kinetic undercooling theory cannot explain the formation of some metallic-nonmetallic eutectics. For example, in the case of Al–Si eutectic alloys, the leading phase at the front of the interface is not the metallic $\alpha$-A1 phase but the

nonmetallic $\beta$-Si phase, which is just opposite that predicted by the kinetic under-cooling theory. Experimental results have shown that the undercooling of Al–Si eutectic is mainly from the constitutional supercooling instead of the kinetic under-cooling, and the growth mode of Al–Si eutectic is determined by the difference of the mass concentration between the two phases and the constitutional supercool-ing. According to the phase diagram, the eutectic composition $w(Si)$ of the Al–Si alloy is about 12.6%, and the solid solubility of the $\alpha$-A1 and $\beta$-Si are approximately 1%. Thus, the ratio of mass fraction of $\alpha$ to $\beta$ in eutectics is about 9:1, which leads to a wide $S/L$ interface for $\alpha$-A1 phase and a narrow $S/L$ interface for $\beta$-Si phase. When the $\alpha$ phase grows, the Si atoms rejected from the interface diffuse into the front of narrow $S/L$ interface of the $\beta$-Si phase, which rapidly increases the Si content at the interface and the tendency of constitutional supercooling. Thus, the $\beta$-Si phase grows rapidly. Owing to the strong crystallographic anisotropy, different oriented needles or dendrites of $\beta$-Si phase form at the front of $S/L$ interface. As the $\beta$ phase grows, Al atoms rejected from the interface diffuse into the front of nearby $\alpha$ phase. Owing to the wide interface of $\alpha$ phase, the growth rate of $\alpha$ adjacent to $\beta$ phase is larger than that away from the $\beta$ phase, which results in a concave $S/L$ interface at the front of $\alpha$. Figure 7.87(a) and (b) show the schematic view of Al–Si eutectic growth and its secondary electron image of SEM, respectively.

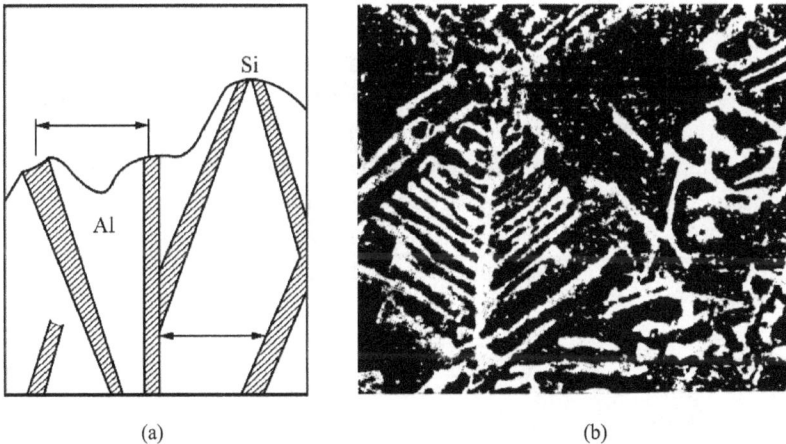

(a)                                    (b)

**Fig. 7.87:** Schematic view of Al–Si eutectic growth (a) and the secondary electron image of directionally solidified Al–Si eutectics after deep-etched by 2% NaCl aqueous solution (500x) (b).

In the metallic–nonmetallic eutectic alloys, the addition of a third component may change the solidified structure significantly. For example, adding a small amount of sodium salt to Al–Si alloy can refine the $\beta$-Si phase and increase its branch; add-ing a small amount of magnesium and rare earth elements can spheroidize the graphite flake, which is called modification treatment. Modification is an economical and practical process that can effectively improve the structure and properties of eutectic alloys.

## 2. Growth kinetics of lamellar eutectic

From the formation mechanism of a lamellar eutectic described earlier, one phase nucleates first from the liquid which is termed the leading phase. We assume the A-based $\alpha$ solid solution as the leading phase. The formation of $\alpha$ rejects excessive B atoms into the liquid ahead of the S/L interface, and the enrichment of B atoms at the front promotes the formation of B-based $\beta$ solid solution. Once the $\beta$ phase grows, excessive A atoms are rejected, which in turn promotes the nucleation and growth of $\alpha$ phase on the $\beta$ phase. Such an iterative process results in eutectic microstructures of alternative $\alpha$ and $\beta$ phases. During eutectic growth, due to the small kinetic undercooling and the strong lateral solute diffusion, the effective constitutional supercooling cannot be established at the front of S/L interface, and hence the interface is planar. For further understanding the moving rate of the S/L interface, the growth kinetics of lamellar eutectic need to be studied.

During solidification of a eutectic alloy, the free energy per volume released can be expressed as

$$\Delta G_B = \Delta S_f \Delta T_E \tag{7.39}$$

where $\Delta S_f$ is the solidification entropy per volume of the eutectic liquid, $T_E$ is the undercooling of the liquid at the front of S/L interface during eutectic solidification.

Considering the interfacial region shown in Fig. 7.88, which has a lamellar spacing $s_0$ and a depth into the page of 1 unit. The interface of Fig. 7.88 is advanced a distance $dz$ and an energy balance is carried out on the volume ($s_0.1.dz$): the free energy released $\Delta G_B \cdot s_0 \cdot 1 \cdot dz$ is used to supply energy for the production of two $\alpha$–$\beta$ interfaces $2\gamma_{\alpha\beta} \cdot 1 \cdot dz$ and for driving solute diffusion $\Delta G_d \cdot s_0 \cdot 1 \cdot dz$:

$$\Delta G_B = \frac{2\gamma_{\alpha\beta}}{s_0} + \Delta G_d \tag{7.40}$$

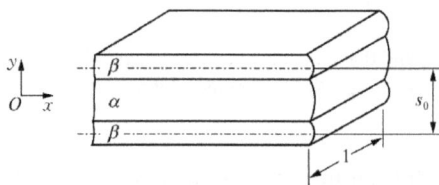

Fig. 7.88: Model for lamellar eutectic growth.

Suppose that all of the free energy released were used to generate $\alpha$–$\beta$ interface. $\Delta G_d$ would be zero in the above equation, and the spacing would have to be a minimum $s_0 = s_{min}$, because we have the maximum interface area. We obtain

$$s_{min} = \frac{2\gamma_{\alpha\beta}}{\Delta G_B} \qquad (7.41)$$

This equation gives the possible minimum spacing of the eutectic plates.

Substituting eqs (7.39) and (7.41) into eq. (7.40), we obtain the free energy to drive diffusion

$$\Delta G_d = \Delta S_f \Delta T_E \left(1 - \frac{s_{min}}{s_0}\right) \qquad (7.42)$$

The total undercooling $\Delta T_E$ determines the amount of free energy available for the eutectic reaction, which can be divided into two parts: $\Delta T_s$ that provides the free energy to generate the $\alpha$–$\beta$ interface and $\Delta T_d$ that provides the free energy to drive diffusion. Thus,

$$\Delta G_B = \Delta G_d + \Delta G_s$$

$$\Delta S_f \Delta T_E = \Delta S_f \Delta T_d + \Delta S_f \Delta T_s \qquad (7.43)$$

This result may be displayed geometrically on a phase diagram as shown in Fig. 7.89. The concentration difference available for diffusion $\Delta w$ is shown on Fig. 7.89 and it can be calculated from the slopes of the two liquidus $m_\alpha$ and $m_\beta$, as

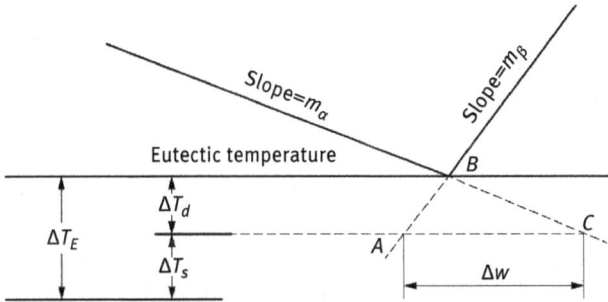

**Fig. 7.89:** Relation of total undercooling $\Delta T_E$ to the phase diagram.

$$\Delta w = \Delta T_d \left(\frac{1}{|m_\alpha|} + \frac{1}{m_\beta}\right) \qquad (7.44)$$

The relationship between the concentration difference for diffusion and the rate of the $S/L$ interface motion is established below.

The flux of B atoms that is rejected from the $\alpha$ plate may be written as

$$J(\text{reject}) = R(\rho_\alpha^L - \rho_\alpha^S) \approx R(\rho_E - \rho_\alpha) \tag{7.45}$$

where $\rho_\alpha^L$ and $\rho_\alpha^S$ are the corresponding concentrations of the extrapolation of the $\alpha$ liquidus and solidus at undercooling $\Delta T_E$, respectively (see Fig. 7.84); $\rho_E$ and $\rho_\alpha$ are the corresponding concentrations of the liquid (eutectic point) and the $\alpha$ solid at $\Delta T_E = 0$, respectively; and $R$ is the rate of interface motion. In order to calculate the lateral diffusion of this rejected solute, it is necessary to determine the three-dimensional distribution of the solute in the liquid. To simplify the calculation, we assume the lateral diffusion between $\alpha$ and $\beta$ plates will occur mainly in the $y$ direction on Fig. 7.88. Thus, the first-order approximation results in

$$J(\text{diffusion}) = \frac{D\Delta\rho}{s_0/2} \tag{7.46}$$

where $\Delta\rho$ is the average concentration difference in the liquid between the $\alpha$ and $\beta$ fronts, and $s_0/2$ is the diffusion length. At steady state, the flux of B atoms rejected from an $\alpha$ plate should be equal to the lateral diffusion flux that transports these atoms into the $\beta$ plates, that is, $J(\text{reject}) = J(\text{diffusion})$, which results in $R(\rho_E - \rho_\alpha) = \frac{D\Delta\rho}{s_0/2}$

Dividing both sides of the above equation by the density of the alloy $\rho$, we have

$$R = \frac{2D\Delta w}{s_0(w_E - w_\alpha)} \tag{7.47}$$

Considering $\Delta T_d = \Delta G_d / \Delta S_f$, eq. (7.42) can be expressed as

$$\Delta T_d = \Delta T_E \left(1 - \frac{s_{\min}}{s_0}\right) \tag{7.48}$$

Substituting eqs. (7.48) into (7.44), it gives

$$R = \left(\frac{1}{|m_\alpha|} + \frac{1}{m_\beta}\right) \frac{2D\Delta T_E}{(w_E - w_\alpha)s_0} \left(1 - \frac{s_{\min}}{s_0}\right) \tag{7.49}$$

In above equation, parameters are constant except for $s_0$. The spacing would adjust so as to minimize the undercooling at the interface. Based on this optimization principle, the optimal spacing corresponding to the minimum $\Delta T_E$ is found by taking the first derivative of eq. (7.49) equating to zero ($dR/ds_0 = 0$) with respect to variable $s_0$. One obtains

$$s_{\text{opt}} = 2s_{\min} \tag{7.50}$$

Substituting $s_{min} = s_{opt}/2$, $s_0 = s_{opt}$, and $\Delta T_E = \Delta G_B/\Delta S_f = \frac{2\gamma_{\alpha\beta}}{\Delta S_f s_{min}} = \frac{4\gamma_{\alpha\beta}}{\Delta S_f s_{opt}}$ into eq. (7.49), we can obtain a rate equation of the $S/L$ interface:

$$R = \left(\frac{1}{|m_\alpha|} + \frac{1}{m_\beta}\right)\frac{4\gamma_{\alpha\beta}D}{\Delta S_f(w_E - w_\alpha)} \cdot \frac{1}{s_{opt}^2} \tag{7.51}$$

The above equation can be simplified in the form

$$s_{opt} = \frac{k}{\sqrt{R}} \tag{7.52}$$

where $k$ is a constant. This result indicates that the observed spacing should vary inversely with $R$.

Equation (7.52) agrees well with a large number of experiments in the investigation of eutectic solidification. The results for experiments on Pb–Sn are shown in Fig. 7.90. Experiments show that there appears to be a limited range of the spacing $s$, in which it is possible to obtain regular eutectic structures. It is generally found that $s$ is over 10 μm, the lamella plates become very bending and begin to break up. When $s$ is below ~0.5 μm, it is very difficult to keep the regular structure from breaking up. The usual spacing is around 1–3 μm for regular eutectic.

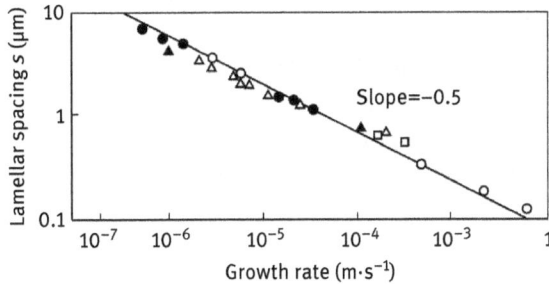

**Fig. 7.90:** Relation between lamellar spacing and rate of solidification in Pb–Sn alloys.

### 3. Stability of eutectic interfaces

The constitutional supercooling theory gives a good judgment on the stability of the planar $S/L$ interface during growth of a single-phase solid solution. The stability of the planar interface will be broken with the occurrence of the constitutional supercooling. With the increase of the constitutional supercooling, the interface morphology changes from cellular to dendritic, which has been experimentally proved. For analysis of eutectic alloys, the constitutional supercooling theory does not work very well, but it can still explain the interface stability of eutectic in some cases.

## a. Pure binary eutectic

For the solidification of a pure binary eutectic, when the leading phase $\alpha$ crystalli-zes, the excessive B atoms will be rejected into liquid at the front of the $\alpha$ phase, and the equilibrium composition of the liquid with $\alpha$ phase is eutectic composition according to the phase diagram. After the $\beta$ phase forms, A atoms are rejected into liquid at the front of the $\beta$ phase, and the equilibrium liquid still remains the eutec-tic composition. Therefore, there is no solute buildup in the liquid at the front of the $S/L$ interface and no constitutional supercooling from such a buildup. If the undercooling $\Delta T_E$ exists during eutectic solidification, and such solute profiles can produce the constitutional supercooling. However, the undercooling $\Delta T_E$ is very small (<0.02 °C) for metallic-metallic eutectic, so the caused constitutional super-cooling is negligible. Thus, the planar $S/L$ interface can be stable under a positive temperature gradient, and generally dendrites do not appear. For metallic-nonme-tallic eutectic, however, possibly owing to the relatively large kinetic undercooling (1–2 °C) for nonmetallic phase, significant solute buildup can be generated, and in turn obvious constitutional supercooling can be produced under a small positive temperature gradient accompanying with the formation of dendrites.

## b. Impure binary eutectic

If the binary eutectic contains impurities, the impurity element will have a certain average distribution coefficient $\bar{k}$ between the two solid phases and the liquid. If $\bar{k} < 1$, the impurity element will build up in front of the eutectic liquid interface and may give rise to constitutional supercooling. If the amount of impurity is small, the constitutional supercooling will be small, which can cause the planar interface to become cellular in a manner similar to that for single-phase growth. The lamellae tend to grow perpendicular to the $S/L$ interface, and consequently, each cell is eas-ily distinguished in a transverse section. If the amount of impurities is sufficient, it is possible to form dendrites, and the dendrites are usually composed of pure $\alpha$ phase, pure $\beta$ phase, or an impurity phase.

## c. Binary pseudoeutectic

The eutectic microstructure can be obtained in off-eutectic composition, which is called a pseudoeutectic, as shown in Fig. 7.27. The composition of a pseudoeutectic must be an average composition over the two eutectic phases, that is, $\bar{w} = f_\alpha w_\alpha + (1 - f_\alpha) w_\beta$, where $f_\alpha$ is the volume fraction of the $\alpha$ phase in the pseudoeutectic, the $w_\alpha$ and $w_\beta$ are the mass fraction of the $\alpha$ and $\beta$ phases as determined from the phase diagram, respec-tively. For the eutectic with eutectic composition $w_E$, the value of $f_\alpha$ can be determined from the phase diagram as $\frac{w_\beta - w_E}{w_\beta - w_\alpha}$. However, if one can make the $\alpha$ plate thicken, then the relative amount of $\alpha$ phase will increase and the average composition $\bar{w}$ of the pseu-doeutectic must be lower than $w_E$. Therefore, a pseudoeutectic composition can, in principle, have any average composition between $w_\alpha$ and $w_\beta$ depending only on the relative amount of the two phases.

The next question then is how to obtain such a pseudoeutectic structure at off-eutectic compositions. Assume the pseudoeutectic alloy composition is $w_0$ that is less than eutectic composition $w_E$, and the equilibrium distribution coefficient $k_0 < 1$. In order to maintain a regular eutectic microstructure after solidification, two conditions must be satisfied: the liquid in front of the $S/L$ interface is close to the eutectic composition $w_E$; and the planar interface must be stable (no dendrites). The first condition is met by the presence of a solute buildup at the interface. When the cooling rate is very fast, and if the liquid is completely no-mixing, the solute builds up rapidly at the interface to reach $w_E$ at steady-state solidification, and the solid phase also increases to the original alloy composition $w_0$. Consequently, the liquid with eutectic composition can form pseudoeutectic structure by eutectic reaction. For the growth of pseudoeutectic, according to eq. (7.34), the criterion for the stability of planar interface can be obtained: $G_{CR} = mR(w_E - w_0)/D$

The planar interface is stable when the constitutional supercooling does not appear. Once the constitutional supercooling forms, experiments show that the above equation agrees well when $w_0$ is close to $w_\alpha$; however, the planar interface can still keep stable even at very low temperature gradients below $G_{CR}$ when $w_0$ is close to $w_E$. So far, except for experiments, there is no practical method used to predict the conditions for the pseudoeutectic structure. Under the conditions to determine the pseudoeutectic region, the temperature gradient is very low (probably less than 0.01–0.05 °C/cm). Therefore, the pseudoeutectic region on the diagram shows the compositions where pseudoeutectic will grow at very low temperature gradients and rather high cooling rates.

### 7.4.3 The structure and defects of alloy castings (ingots)

Generally, there are two ways to make industrial parts. One is called castings which are directly solidified from alloy melts in a mold with a certain geometry and size. The other method needs a series of processes: at first, the alloys were cast into ingots, then they were machined slabs or billets; following hot forging or hot rolling; finally, shaped by machining and heat treatment, or even welding. Obviously, the former can save more energy, time, and manpower, thus reduce production costs, but its application is limited within a range. For castings, as-cast structure and defects directly affect their mechanical properties. For ingots, as-cast structure and defects directly affect their processing performance, and may also affect the mechanical properties of final products. Therefore, the quality of the alloy castings (or ingots) is important not only in foundry practice but also in almost all metal products.

## 1. Macrostructure in ingots or castings

The solidified grains of metals and alloys are relatively coarse and generally macroscopically visible. As shown in Fig. 7.91, the macrostructure of the ingot consists of three distinct zones, that is, the surface fine-grain zone, the columnar zone, and the center equiaxed zone. The formation mechanisms for these zones are discussed as follows.

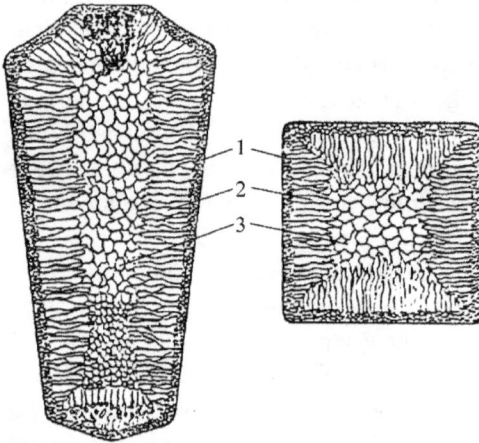

**Fig. 7.91:** The three-zone structure of a steel ingot: 1, surface fine-grain zone; 2, columnar zone; 3, center equiaxed zone.

### a. Surface fine-grain zone

During casting process, owing to the low temperature of the mold, a thin layer of melt in contact with the inner wall is rapidly chilled. Furthermore, the mold wall can serve as a substrate for heterogeneous nucleation. As a result, a large number of nuclei form immediately, and rapidly grow and contact with each other to form a fine-grain zone filled with fine, disoriented equiaxed grains. Therefore, surface fine-grain zone also is called chill zone.

### b. Columnar zone

As a shell of fine-grain zone forms, the inner wall of the mold is heated by the liquid and its temperature is increased, which causes the decrease in cooling rate of the remaining liquid. In addition, the latent heat of crystallization is released into the liquid. The above two factors result in a decrease in the undercooling in the liquid in front of the fine-grain zone, thus further nucleation is getting difficult while existing crystals can continue to grow into the liquid. In this case, the grains with the primary dendrites (i.e., preferred orientation with fast grow rate) perpendicular to the mold wall (the direction with fast heat dissipation) tend to crowd out the other grains, so that the columnar zone consists of relatively large long grains crystallographically oriented with their dendrite directions parallel to the heat flow direction. Since the growth direction of each columnar crystal is the same, for example, the preferred growth direction of the cubic systems is <100>. The as-cast structure with the same crystallographic orientation is called casting texture or crystallization texture.

For pure metals, since there exists a positive temperature gradient in the liquid in front of the $S/L$ interface and there is no constitutional supercooling, the columnar crystals grow with a planar $S/L$ interface. For alloys, the columnar grains tend to grow into columnar dendrites as there exist significant constitutional supercoolings in the liquid; but the primary axis of the columnar dendrites is still perpendicular to the inner wall, that is, is along the heat flow direction.

### c. Center equiaxed zone

When columnar dendrites grow to a certain extent, owing to the difficulty of heat dissipation in the liquid away from the mold wall, the liquid (melt) is cooled slowly and the temperature difference decreases gradually, which inhibits the growth of columnar dendrites. As the temperature of liquid drops below the liquidus temperature, many nuclei appear in the liquid and grow in random orientation, leading to the formation of center equiaxed zone. There are several different opinions about the formation of center equiaxed grains.

(1) Constitutional supercooling: With the growth of columnar dendrites, the constitutional supercooling occurs and the undercooled zone is extended from the $S/L$ interface to the center of the liquid, leading to formation of a large number of nuclei and random growth into equiaxed grains. Therefore, the columnar zone is inhibited and the center-equiaxed zone develops.

(2) Liquid convection: During casting process, the liquid adjacent to the mold wall is cooled rapidly, leading to large temperature differences in the liquid while forming the surface fine-grain zone. The outer cold metal is denser than the hot metal at the center of the ingot and hence it tends to sink while the hot metal rises, producing intensive natural convection, as shown in Fig. 7.92. The existing dendrites can be torn off by the convection currents, and then the resulting dendrite fragments are swept into the center of the ingot to act as seeds for nucleation of equiaxed crystals.

(3) Seed crystals formed by local remelting of dendrites: In the growth of columnar dendrites, narrowing of the side branches generally occurs at their bases. The fine "necks" of the side arms can be easily melted off due to temperature fluctuations or solute enrichment. These fragments drift to the center of the liquid, and sever as seed crystals to form equiaxed grains in the center.

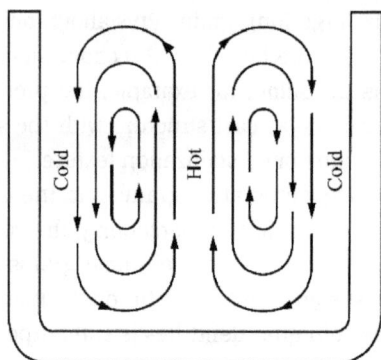

**Fig. 7.92:** Convection currents in the liquid metal in a mold during casting.

It should be emphasized that the macrostructure of the ingot or casting is closely related to the casting conditions. By changing these conditions, the relative thickness and grain size of the three crystal regions can be changed, or even only two crystal zones appear. Usually a fast cooling rate, high pouring temperature and directional heat dissipation is beneficial to the formation of columnar crystals. On the contrary, a slow cooling rate, low pouring temperature, addition of effective nucleating agent or stirring, and so on promote the formation of equiaxed grains in the center.

The advantage of columnar grains is that the structure is dense and this casting texture can be utilized in some cases. For example, the <001> direction of the cubic metal is parallel to the long axis of columnar crystals, and this characteristic can be used to produce iron alloys for magnets. The magnetic induction is anisotropic and it is usually higher along the <001> direction. Casting texture can be used to improve the mechanical properties and subsequent processing performance of alloys. The columnar grains have drawbacks that the parallel contact areas and particularly perpendicular interfaces among the columnar grains are relatively weaker, and low melting point impurities and nonmetallic inclusion often segregate there. Thus, cracks are easily propagated along these boundaries in hot working process and in service. By comparison, there are no preferred growth direction and weak boundaries in equiaxed grains, and the grains with random orientations have good combination with each other, which can prevent the crack propagation. Therefore, the microstructure with fine equiaxed crystals can improve the mechanical properties of castings. Relative to the columnar grains, however, the density of the equiaxed structure is lower. The surface fine-grain zone has little effect on the properties of the castings, and since it is very thin, it is usually removed during machining process.

## 2. Defects in ingots or castings

### a. Shrinkage

When the melt (liquid) is cast in a mold, the melt contact with the mold wall first solidifies, and then the central part of the liquid starts to solidify. Since most of metals, except for a few metals, such as Sb, Ga, and Bi, contract during freezing, a shrinkage cavity forms in the ingot or casting, also known as shrinkage or porosity.

Shrinkage can be divided into concentrated shrinkage and dispersed shrinkage, and the latter is also called porosity. As shown in Fig. 7.93, there are a variety of different forms of concentrated shrinkage, such as pipe shrinkage, cavity shrinkage, and surface sink, and the porosity can be divided into common porosity and central porosity.

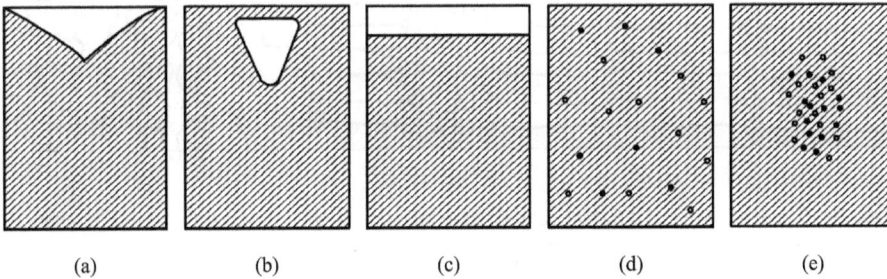

Fig. 7.93: Different types of shrinkage (a) pipe shrinkage, (b) cavity shrinkage, (c) surface sink, (d) common porosity, and (e) central porosity.

In general, the concentrated shrinkage can be controlled in the riser of the steel ingot or casting and then be removed subsequently. Improper feeding or riser design may lead to a deeper shrinkage that is difficult to remove, and the remained shrinkage will have great harm on subsequent hot working process and service performance. Porosities are an inevitable result during the solidification of alloys with dendritic structures. At the end of solidification, the arrangement of the growing dendrites can lead to a lot of isolated liquid pools. Once there is no sufficient liquid to compensate the volume contraction, tiny dispersed porosities will form. Therefore, even with the correct riser design, the porosity will also exist.

The type of shrinkage or porosity is closely related to the solidification manner of alloys.

Similar to pure metals, alloys with eutectic compositions freeze at a constant temperature. If the cooling rate and temperature gradient can be properly controlled, the constitutional supercooling effect is negligible in front of solidification interface, thus the planar interface keeps stable. Therefore, solidification starts from the mold wall, then the columnar crystals grow into the center liquid, which is called skin-forming solidification, as shown in Fig. 7.94(a). In this manner of solidi-fication, the fluidity of the liquid is very good and the feeding is easy, thus the shrinkage is concentrated in the riser. As a result, the total volume of dispersed shrinkage porosities is markedly decreased, and the casting is very dense.

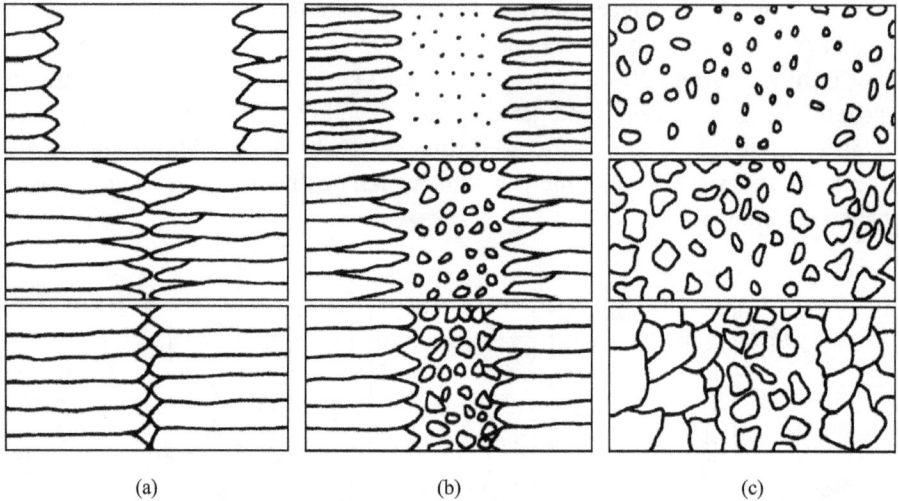

(a)                              (b)                              (c)

**Fig. 7.94:** Schematic view of different solidification manners (a) skin-forming, (b) an intermediate case, and (c) mushy forming.

For solid solution alloys with a wide temperature range of solidification and a small solute distribution coefficient $k_0$, the constitutional supercooling is easily formed in front of the $S/L$ interface, which promotes the growth of equiaxed dendrites. There exists a wide mushy zone, where solid and liquid phases coexist, between the com-plete solid phase zone and complete liquid phase zone. This manner of solidifica-tion is called mushy forming solidification, as shown in Fig. 7.94(c). Apparently, the liquid fluidity is poor in this manner of solidification. In addition, multiple den-drites in the mushy zone usually intersect with each other. As a result, the feeding of the shrinkage at the last solidified section is not sufficient, which causes the for-mation of dispersed porosities and decreases the compactness of the casting.

In order to improve the feeding property of mushy forming alloy, grain refining method is usually employed, which can restrict the formation of fully developed dendrites, weaken the crossing dendrite network and improve the fluidity of the liq-uid. In addition, since the porosity is mainly dispersed in the interdendritic region, grain refining can directly reduce the volume of these voids, and thus increase the compactness of the casting. This principle is often used in aluminum and magne-sium based alloys. As shown in Fig. 7.94(b), the manner of solidification for real alloys is generally between the skin solidification and mushy solidification.

The gas dissolved in the supersaturated liquid may precipitate to form bubbles during solidification and also leads to porosity in the casting and decreases the compactness of the casting. Therefore, it is also important to pay attention to the gas content in the liquid in order to reduce the porosity within the casting.

## b. Segregation

The term of segregation refers to the nonuniform distribution of chemical composition in metals or alloys. Segregation always exists in alloy casting to some extent, which is determined by the characteristics of solidification process. In the normal freezing, when a rod is directionally solidified with a planar $S/L$ interface from one end, significant segregation occurs along the length of the rod. When the equilibrium distribution coefficient $k_0 < 1$, the first part of the rod contains less solute, while the last part contains more solute. However, there is usually a constitutional supercooling in the liquid in front of the $S/L$ interface, thus the interface is mostly dendritic, and this completely changes the type of segregation. When the dendritic interface extends into the liquid, the solute is rejected in the longitudinal and lateral directions from the freezing solid. The longitudinal solute transport gives rise to a macrosegregation along directions parallel to the dendrite axes, and the lateral solute transport gives rise to a microsegregation along directions perpendicular to the dendrite axes. Macrosegregation can be observed by naked eyes or in low magnification after etched, while the microsegregation can only be seen by a microscope.

## (1) Macrosegregation

Macroscopic segregation is also called regional segregation. According to the different phenomena presented in castings or ingots, the macrosegregation can be divided into three categories, including normal segregation, inverse segregation and gravity segregation.

① *Normal segregation* (positive segregation). When $k_0 < 1$, the content of solute in the first solidified outer layer is lower than that in the last solidified inner layer, which gives rise to a high solute concentration in the center of the ingot. This is a normal phenomenon of solidification and thus termed the normal segregation.

The extent of normal segregation is related to the casting size, the cooling rate, and the degree of mixing in the liquid during solidification. Generally, the normal segregation is severe in the center part of large-sized castings or ingots, which is the last solidified part and has a relatively high solute concentration. Sometimes, the nonequilibrium second phases may appear, such as carbides. In the center of some high alloy tool steel ingots, the nonequilibrium ledeburite may even occur due to segregation.

The normal segregation is usually inevitable and it deteriorates casting performance that can hardly be eliminated by subsequent hot working and diffusion annealing treatment. Thus, the appropriate methods should be taken during the casting process to alleviate the normal segregation.

② *Inverse segregation.* Contrary to the normal segregation, inverse segregation is that the solute concentration at the surface is higher than that in the center of the casting for $k_0 < 1$.

It has been demonstrated that inverse segregation can only be formed when alloys undergo a volume contraction during solidification and porosities exist in the center of castings. The presence of columnar dendrites, a large temperature range of solidification and gas dissolved in the liquid are in favor of the formation of inverse segregation. According to experiments, it is generally believed that the formation of inverse segregation is related to the interdendritic liquid flow. Owing to the volume contraction, cavities form at the base of the dendrite array (negative pressure there). As the temperature decreases, gas precipitates from the liquid in the central part and generates pressure. Consequently, the liquid with high solute concentrations (for $k_0 < 1$) flows through the interdendritic channels to fill cavities at the skin layer, which gives rise to the inverse segregation. As the melting point decreases with increasing solute concentrations, the "sweating" phenomenon usually occurs at the surface of castings, such as Cu–Sn alloy, which is an obvious sign of inverse segregation.

The inverse segregation can be inhibited by expanding the central equiaxed zone or reduce the gas content in the liquid, both of which can inhibit the interdendritic liquid flow from the center toward the surface.

③ *Gravity segregation.* Gravity segregation usually takes place in the early stage of solidification. The primary phases may float up or settle down due to their large density difference relative to the bulk liquid, causing the solute concentration difference along the gravity direction. For example, in the solidification of Pb-15 wt.% Sb alloy, the lighter primary-phase Sb solid solution floats on top of the liquid and the heavier (Pb + Sb) eutectics sink at the bottom, leading to gravity segregation. The floating of primary graphite is also a kind of gravity segregation.

There are several methods to alleviate the gravity segregation. Increasing the cooling rate of castings to inhibit the settling or floating of the primary phases; or adding the third element to form dendritic compounds with high melting points and similar density to the liquid phase, which develops a dendritic skeleton to block the settling or floating of other phases in the early solidification stage. For example, the addition of Ni or S can be used to form Cu–Ni solid solution phase or $Cu_2S$ intermetallic phase in Cu–Pb alloys, and the addition of Cu can be used to form $Cu_6Sn_5$ or $Cu_3Sn$ intermetallics in Sb–Sn alloy to prevent the gravity segregation.

## (2) Microsegregation

Microsegregation also can be divided into three categories: cellular segregation, dendritic segregation, and grain boundary segregation.

① *Cellular segregation.* As previously mentioned, the solid solution crystals grow in a cellular manner when the constitutional supercooling is small. If the equilibrium distribution coefficient $k_0 < 1$, the solute enrichment will occur at the cell wall; otherwise, the solute depletion will occur at the cell wall for $k_0 > 1$; and this is termed cellular segregation. Since the size of the cells is relatively small, that is, the range of composition variations is relatively narrow, the cellular segregation can be easily eliminated by homogenization annealing.

② *Dendritic segregation.* As discussed earlier, dendritic segregation is resulted by nonequilibrium solidification, which causes the nonuniform solute distribution on the dendritic scale. Generally, alloys freeze in dendritic manner and one dendrite forms one grain in the final microstructure. Therefore, the dendritic segregation is within a grain, and it is also known as intragranular segregation. The degree of dendritic segregation depends on several factors. The higher the solidification rate is, the smaller the diffusion ability of the segregating element in the solid solution is, or the wider the solidification temperature range is, the more serious the intragranular segregation will become.

③ *Grain boundary segregation.* For alloys with $k_0 < 1$, grain boundary segregation is caused by the solute atoms rejected at grain boundaries at the final stage of solidification. During solidification, the solute concentration in the liquid is gradually increased. Once the adjacent grains grow to impinge on each other, the solute will be enriched in the remaining liquid which is finally solidified into grain boundary.

The factors affecting the grain boundary segregation include the solute concentration in the alloy, the morphology of solidification interface, and solidification rate. The higher the solute concentration is, the greater the degree of grain boundary segregation becomes. Nondendritic growth increases the degree of grain boundary segregation, in other words, dendritic segregation can decrease the grain boundary segregation. The decrease of solidification rate provides the solute enough time to diffuse into the interdendritic liquid, and thus increases the degree of grain boundary segregation.

Grain boundary segregation tends to cause intergranular fracture, and thus it should be eliminated as far as possible. In addition to controlling the content of the solute, a suitable third element can be added to reduce grain boundary segregation. For example, the addition of carbon in iron can reduce the grain boundary segregation of oxygen and sulfur, and adding molybdenum can weaken the grain boundary segregation of phosphorus; the addition of iron in copper can reduce the segregation of antimony at grain boundaries.

### 7.4.4 Casting and secondary processing of alloys

Figure 7.95 summarizes four types of casting processes employed in engineering. In some processes, the same mold can be repeatedly used, such as die casting; in others the mold is used only once to be expendable, such as sand casting and investment casting. Sand casting, also known as sand molded casting, uses typically silica sand ($SiO_2$) as the mold material to produce metal castings. The investment casting process is also known as the lost wax process, in which the pattern that has the shape of the casting is made from wax with a low melting temperature. The pattern is dipped into a slurry, coated with ceramics particles, and then dried to form a ceramic shell around the patterns. The shell is then heated, such that the wax pattern is melted out, leaving a hollow ceramic shell having the desired shape. This technique is beneficial for metal castings with high melting temperatures, high dimensional accuracy, complex geometries, and excellent surface finish, such as blades and cases for jet engines. In another casting process known as the lost foam process, the expendable pattern is made from molded polystyrene (PS) foam and embedded in sand. As the molten metal is poured into the mold, it replaces the pattern which vaporizes and precisely duplicates all of the features of the pattern. Lost foam casting process is also advantageous for very complex castings with high-dimensional accuracy and

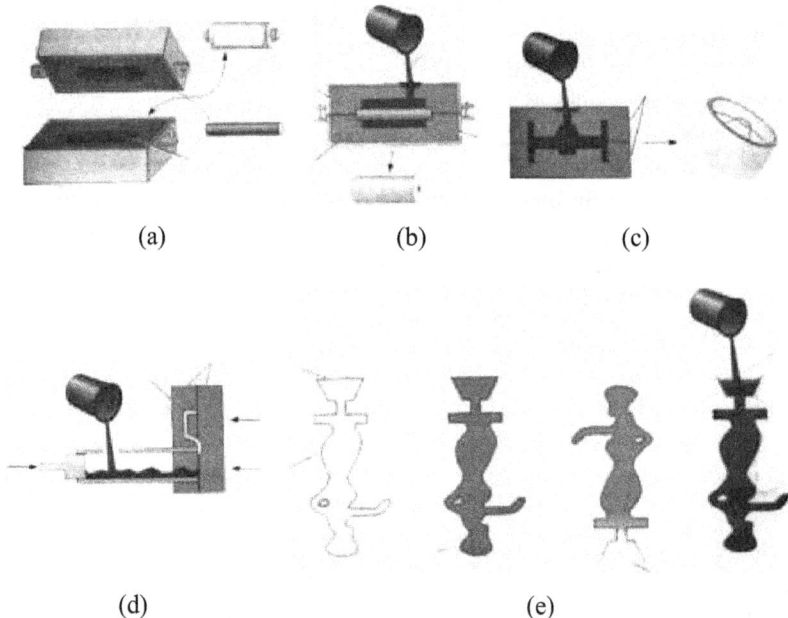

(a)          (b)          (c)

(d)                              (e)

**Fig. 7.95:** Four typical casting processes (a) and (b) sand casting, (c) permanent mold casting, (d) pressure die casting, and (e) investment casting.

excellent surface finish. Lost foam is generally more economical than investment casting because it removes the need to melt the wax out of the mold.

The permanent mold and pressure die casting processes employ reusable molds, usually made from metallic material (steel or cast iron). The liquid metal is poured into the mold under gravity or injected into the mold under tons of pressure. Once the metal is solidified, the casting is removed from the mold and the mold is reused. The processes using metallic molds tend to give superior mechanical properties, such as higher tensile strength, because the solidification rate is higher than ceramic or sand molds. The permanent mold casting can produce castings with good surface finish and dimensional accuracy, but the tooling costs of making the permanent die are high. Die casting is ideal for high volume production of castings with thinner walls, smoother surface, great dimensional tolerance and little post-machining. Many nonferrous metals, such as aluminum, magnesium and zinc based alloys, are processed using permanent mold and pressure die casting to produce truck and car pistons, mobile phone cases, clothes zippers, and so on.

Except for the manufacturing of shape castings, casting is also used for producing ingots that can be further processed into different shapes (e.g., rods, bars, and wires) by secondary processes.

In ingot casting, metals or alloys are melted in a furnace and refined in a tundish, and then cast into ingots of different shapes (e.g., rods, billets, and slabs) in a continuous or semicontinuous way. The resultant ingots are then processed into desired shapes via thermomechanical processing, such as hot mill, cold mill, forging, extrusion, drawing, bending, and stretch forming. The secondary processing steps in the processing of steels and other alloys are shown in Fig. 7.96. Currently, in the steel

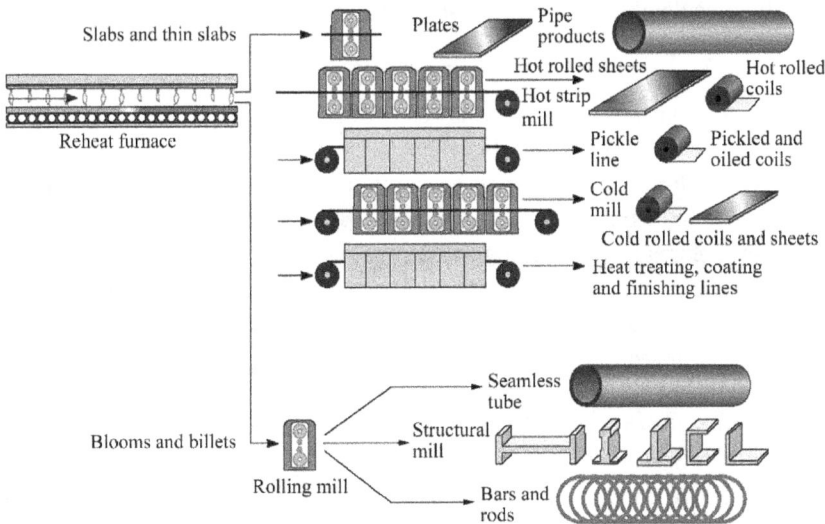

**Fig. 7.96:** Secondary processing steps in processing of steel and other alloys.

industry, the casting and rolling process have been combined to a continuous operation, namely the continuous casting and continuous rolling process. In the aluminum
industry, another process termed the continuous roll casting process, which combines the casting and rolling together, is widely used to produce aluminum strips.

## 7.5 Brief introduction to polymer alloys

Polymer alloy, which is known as multicomponent polymer, refers to the complex
system containing two or more polymer chains, including block copolymers, graft
copolymers, and a variety of blends and so on. The aim of mixing different polymer chains is to obtain a more diverse range of polymer materials and a higher
overall performance by combining the physical and chemical properties of the
macromolecules. Therefore, the polymer composite system is so called "polymer
alloy". When there are two components, it is called a binary polymer. This section
will summarize the compatibility criterion of polymer alloy and method for determination of alloy phase diagram, main preparation method of polymer alloy, morphology of polymer alloy, and general relationship between the properties and
the components in binary polymer, and finally make a brief introduction of main
types of polymer alloys.

### 7.5.1 Compatibility of polymer alloy

The compatibility criterion of two polymers is the same as the compatibility criterion of small molecule: the free energy of mixing is less than zero:

$$\Delta G = \Delta H - T\Delta S < 0 \tag{7.53}$$

where $\Delta H$ and $\Delta S$ are mixed enthalpy and mixed entropy, respectively. For polymer systems, if there is no special interaction between heterogeneous molecules, then the $\Delta H$
value is always greater than zero, which means it is endothermic during dissolution.
Therefore, the heat of mixing is always detrimental to the mixing. Thus, the degree of
entropy increase before and after mixing will determine whether the two kinds of polymers can be mixed. In fact, the entropy increase for the two polymers is much smaller
than that for the two low molecular blends, which can be explained visually in
Fig. 7.97. The small molecules $A$ and $B$, respectively, occupy a lattice, polymer can be
regarded as composed of several chain links, each chain link occupies a lattice.
Figure 7.97(a) shows the case where small molecules $A$ and $B$ are mixed. From the thermodynamic formula $S = k \ln w$, the entropy ($S$) is a function of the configuration number
($w$). The larger configuration number is, the larger the entropy. There are $N_A$ molecules
of $A$ and $N_B$ molecules of $B$ in the mixture, since any one of the $A$ and $B$ interchange
positions in the lattice model is a new arrangement, the number of permutations they

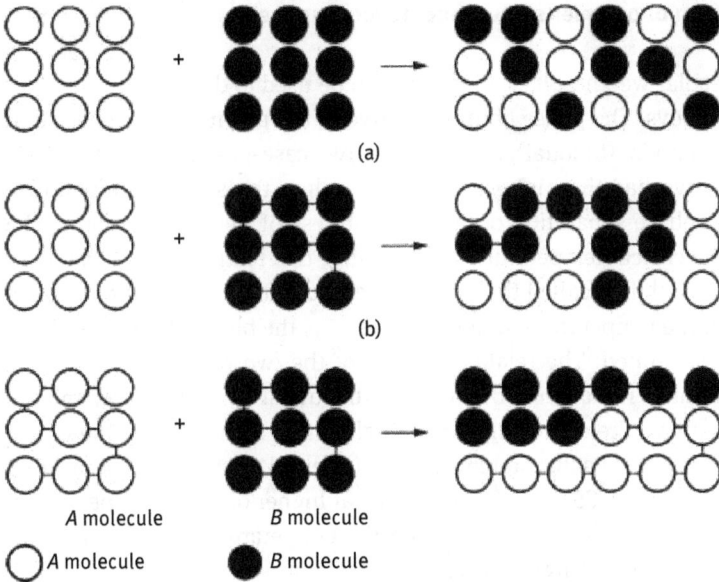

(a)

(b)

A molecule          B molecule

A molecule          B molecule

**Fig. 7.97:** Sketch of mixed entropy. (a) Two kinds of small molecules, (b) small molecules and macromolecules, and (c) two kinds of macromolecules.

may take is $w = (N_A + N_B)! / (N_A! N_B!)$. The $N_A$ and $N_B$ in the macroscopic system are a large number, and there are a lot of molecular arrangements in small molecule solutions, so the dissolution leads to a great increase in entropy. Figure 7.97(b) shows the case where the small molecule $A$ and the macromolecule $B$ are mixed. In the case because the same $B$-chain links must be linked together, $B$-chain link location cannot be arbitrarily exchanged with molecule $A$. In this way, the entropy still increases significantly compared to the cases of pure $A$ or pure $B$ ($w = 1$ for pure components, $S = 0$), but it is much smaller in the same volume compared to the case where small molecules $A$ and $B$ are blended. Figure 7.97(c) is the case of macromolecules blending. In this case, since each chain link on the same $A$ chain or on the same $B$ chain must be linked, respectively, then interchanged, and thus new arrangement is less so that the increase in the entropy of the polymer mixture is very limited. In most cases, the entropy increase is not enough to overcome the contribution of $\Delta H$, that is, $T\Delta S < \Delta H$, the mixing of polymers is unlikely to reach the level of small molecules, which always forms a heterogeneous system. This is why multicomponent polymers are often incompatible thermodynamically.

The mixture of polymers generally cannot be fully compatible to the homogeneous phase, so it will appear phase separation. There are two kinds of mechanism: spinodal decomposition and nucleation and growth mechanism. Spinodal decomposition is the final-phase separation through the uphill diffusion of the composition, while the nucleation and growth mechanism is the formation of two phases through the downhill diffusion of the component.

### 7.5.2 Phase diagram of polymer system and its measurement

Like the low molecular weight, phase diagrams can be used to describe the compatibility of polymer alloys. The phase boundary curve in the polymer alloy phase diagram is called the binodal (binodal), and there are two cases, as shown in Fig. 7.98. One is that the curve has the highest point ($T_C$), when the system temperature $T > T_C$, there is no phase separation occurred regardless of the composition of the blends, so $T_C$ is the critical temperature. And because this temperature is the highest point of the binodal, it is called the upper critical solution temperature (referred to as UCST). When the temperature of the system $T < T_C$, the blend of the $A/B$ within the curve will be separated. The relative amount of the two phases can be determined by the lever rule. The other is that there is the lowest point $T_C$ curve, and the two-phase area is above the curve, while the single-phase area is below the curve, so there is the lower critical solution temperature (referred to as LCST). Figure 7.98 shows a system with both UCST and LCST, that is, at higher or lower temperatures, the system is phase separated, and only when the temperature is in the range of UCST $< T <$ LCST, the blend is single phase.

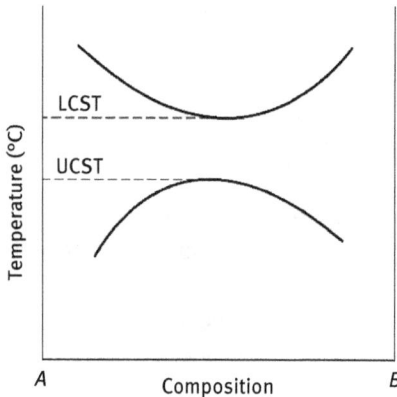

**Fig. 7.98:** Phase diagram of polymer alloys.

In principle, all methods which are sensitive to the phase structure of the system can be used to determine the boundary line, such as thermal analysis methods and dynamic mechanics; or methods which can directly observe phase structure, morphology, such as scanning electron microscopy and TEM. An example of determining the boundary line with the scattered light intensity I is given below. For a typical heterogeneous blend system, the size of the dispersed phase is usually in the range of several hundred nanometers to several tens of microns. The lower limit of this size has been comparable to the wavelength of visible light. Therefore, when light passes through such a material, strong light scattering occurs and turbidity occurs. For a material of single phase, there will not be the intensity mutation of scattered light. Using this property of light scattering, the boundary line can be determined. Figure 7.99(a) gives a typical result. The temperature at which a curve of the scattered light intensity changes with temperature is often referred to as a "cloud point" or "turbidity point". By plotting the cloud points of the blends of the different components against the elements, a phase boundary as shown in Fig. 7.99(b) can be obtained. The limitations of this method are three points: First, the refractive index of polymer components should be evident difference, moreover, the refractive index temperature coefficient should be taken in attention. Otherwise, if the temperature change makes the refractive index of the two components almost be equal, it can also cause the sharp drop of scattered light intensity, which may be mistaken for a phase transition; second, when the size of the dispersed phase is much smaller than the visible wavelength, such as the microphase separation of the block copolymer is only a few to several tens of nanometers, and the method is invalid; third, the mutation of scattering intensity is also affected by dynamic factors, such as cooling or heating rate, the change speed of the phase area size, so the measured phase boundary cannot be called solubility limits in the in equilibrium phase diagram. Therefore, the determination of the boundary line is usually used a variety of methods.

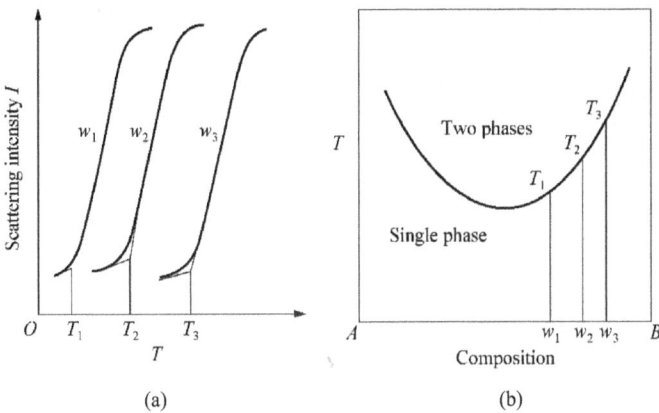

**Fig. 7.99:** Phase boundaries measured by light scattering method. (a) Variations of scattering intensity $I$ with temperature and (b) relationship between the cloud points and components.

The phase diagram of polymer alloys has been very little, there may be two reasons: First, for many incompatible systems, the LCST or UCST are not at temperature range which is easy to experiment, or not at temperature range polymer can withstand. Another reason is the experimental difficulty, for example, the light scattering method to determine the phase boundary is widely used, but it has many fatal weaknesses that limit its scope of application. Figure 7.100 shows the phase diagram of (chlorinated PE (CPE)) and organic glass (PMMA) blends with different

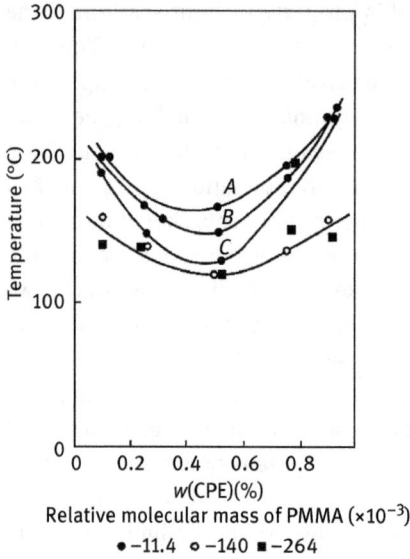

Fig. 7.100: Experimental phase diagram for CPE/PMMA mixture.

relative molecular weight were determined by the above method. It is shown that the lowest miscibility temperature of the PMMA/CPE alloy with different relative molecular mass indicates that these alloys are in the two-phase region above the $T_C$ temperature (the lowest point) and the single-phase region is below the $T_C$ temperature. It is clear that from Fig. 7.100 that with the increase of relative molecular mass of PMMA, the $T_C$ temperature will decrease.

### 7.5.3 Preparation of polymer alloys

The preparation methods of polymer alloys can be divided into two types: physical and chemical methods. Before discussing the preparation method of polymer alloy, at first let us review the polymer polymerization methods: addition polymerization reaction and polycondensation reaction. The polymer is a compound based on the long-chain structure formed by the covalent linkage of a large number of repeated structural units. For example, the molecular chain structure of polyethylene (PE) can be abbreviated as $-[-CH_2-CH_2-]-_n$, which is composed of many structural elements formed by repeated connections. This structural unit is called the PE chain link. $N$ is the repeat number of links, also known as degree of polymerization. In fact, the length of each polymer chain is not the same, so it represents the average number of chains. The structural units are connected with strong covalent bonds, while the long chains are connected with weak Van der Waals. The long-chain structure of the polymer can be obtained by two basic polymerization reactions.

## 1. Addition polymerization reaction

The formation process of PE by addition polymerization reaction is described below, taking ethylene as an example. A single ethylene molecule is called a monomer, in which two carbon atoms is covalently bonded together in an unsaturated bond, in addition, each carbon atom is bonded with two hydrogen atoms, and has eight stable electron shell. If an initiator is added, for example, adding a hydrogen peroxide $H_2O_2$, it can be divided into two OH* active initiating groups ($H_2O_2 \rightarrow 2HO$), and the carbon double bonds are broken, so that the ethylene monomer changes into the chain link. One of the OH groups is attached to the chain link, which is the initiation stage of the chain. Once it is triggered, the spontaneous, rapid growth stage of the chain will occur. The growth of 1,000 molecular links takes about $10^{-2}$–$10^{-3}$s. Because the carbon double bond is broken up into a single bond, therefore, both two ends of the carbon atom form a free radical, respectively. Since the valence electrons are unsaturation, the polymerization is easily carried out. Moreover, the energy released by this polymerization reaction is greater than the energy to destroy the double bond so that the polymerization process is carried out continuously until the chain growth is terminated. There are two ways to terminate the chain growth: one is that the chain activity end meets OH* and ends growth, as addition polymerization process shown in Fig. 7.101(n); the other is the two growth chains meet, combined into a longer chain, and the reaction is terminated, as shown in Fig. 7.101($n'$). It can be seen from Fig. 7.101 that the addition polymerization reaction undergoes three stages, namely chain initiation, chain growth, and chain termination. The initiation of the chain can be initiated by an initiator, thermal, or photoinitiation. The characteristic of the reaction product is that the composition of the polymer is the same as that of the monomer.

In the above addition, polymerization reaction monomer is only one kind. Such a reaction is called homopolymerization, and the product is called homopolymer. If in the polymerization reaction, there are two or more monomers, such a reaction is called copolymerization, and the product is called a copolymer. For example, butadiene monomer and styrene copolymerize into styrene–butadiene rubber:

$$n(NH_2(CH_2)_5COOH) \rightarrow H[NH(CH_2)_5CO]_nOH + (n\text{-}1)H_2O$$

The constituents of styrene–butadiene rubber obtained by addition polymerization are a combination of the two monomers, and no new components appear.

## 2. Condensation polymerization

Condensation polymerization is a chemical reaction by which one or more monomers join together to form a polymer, accompanying precipitation of some molecules with low molecular weight (such as water, ammonia, alcohol, and halide). The resulting polymer is called a polycondensate and has a different composition

(1)  $OH^* + $ (ethylene: H H on top, C=C, H H on bottom)  ⎫
(1')  $HO-C-C^{\bullet}$ (with H H top and bottom)  ⎬ Initiation

(2)  $HO-C-C-C-C^{\bullet}$ (H H H H)  ⎫
(3)  $HO-C-C-C-C-C-C^{\bullet}$ (H H H H H H)  ⎬ Growth
       ⋮

(n)  $HO-C-C-C-C-C-C- \cdots -C-C-C-C^{\bullet} + OH^{\bullet}$  ⎫
(n')  $HO-C-C-C-C-C-C- \cdots -C-C-C-C-OH$  ⎬ Termination

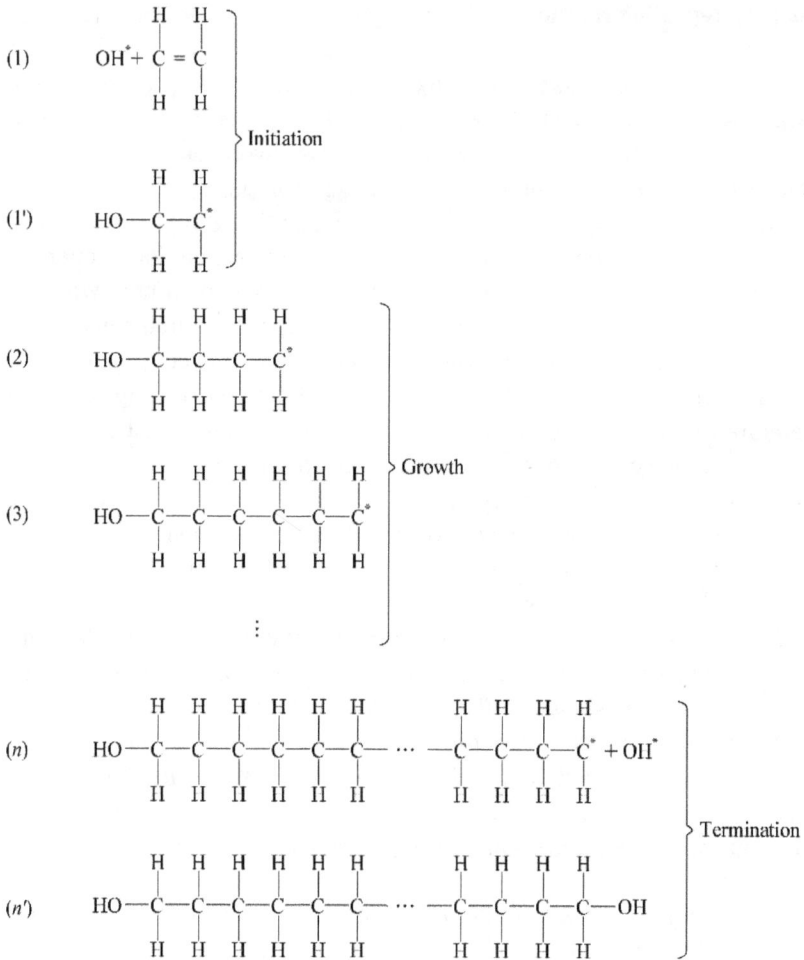

**Fig. 7.101:** Three stages of addition polymerization reaction.

from the monomer. The condensation polymerization is much more complex than the addition polymerization. The condensation polymerization is carried out in a step-growth way, and its polymerization rate is much slower than that of the addition polymerization. Figure 7.102 shows the process of a phthalic acid monomer and a ethylene glycol monomer forming a polyester fiber and a methanol with low molecular weight as by-product. With the repetition of this process, the relative molecular mass gradually increase, and eventually the polyester fibers with relative high molecular mass will be obtained.

The condensation polymerization is a reaction between the functional groups, and the above reaction between the two or more monomers is referred to as copolycondensation. While the condensation polymerization between the same monomers

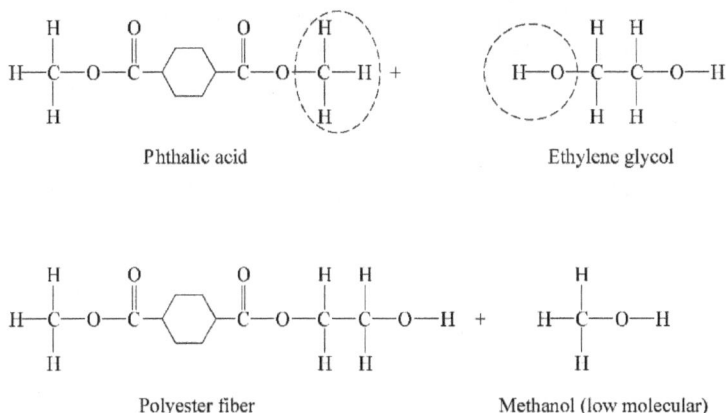

Fig. 7.102: The condensation polymerization of polyester fiber.

is called homopolymerization, the product is called homopolymers. For example, n aminocaproic acid ($NH_2(CH_2)_5COOH$) can produce a polyamide 6 (nylon 6) by a homopolymerization:

$$n(NH_2(CH_2)_5COOH) \rightarrow H[NH(CH_2)_5CO]_nOH + (n-1)H_2O$$

As mentioned earlier, the polymer chains have various molecular structures: linear, branched, cross-linked, or three-dimensional networks. The molecular structure depends on the number of bonds that a monomer can form in the polymerization, that is, the functionality of the monomer. Linear molecular chain is obtained by the reaction of the bifunctional monomer (monomer molecules that have two active points that can form molecular chain) such as vinyl chloride and ethylene glycol are bifunctional monomer. If the monomer with the functionality greater than 2 is involved in the reaction, a branched molecular chain is obtained, such as the branched structure of the phenol (trifunctional) and formaldehyde(trifunctional). With the degree of polymerization, branching can be developed into a three-dimensional network structure, that is, the formation of cured phenolic resin (the polymer) obtained by condensation polymerization of two monomers, which is often named after the monomer name with "resin" word. In addition, the "resin" is also commonly used in the word refers to the unprocessed polymer compounds synthesized by chemical plant or laboratory.

**3. Physical blending method**
Physical blending method, also known as mechanical blending method, is a method to blend different types of polymers in the mixing (or milling) equipment. Blending processes generally include mixing and dispersing. In the blending operation, due to the role of energy (mechanical energy and thermal energy) supplied by the various

mixing machines, mainly convection and shear, diffusion is less important, so that the size of mixed material particles are constantly decreased and dispersed each other, and ultimately achieve uniform dispersion and become a mixture. In the mechanical blending operation, generally only physical changes occur, but under a strong mechanical shear, a little amount of polymer may be degraded to form some macromolecular free radicals, followed by the formation of graft or block copolymer, and in the process some chemical reaction will occur.

Physical blending methods include dry blending, melt blending, solution blending, and emulsion blending. The most common methods are melt blending. The melt blending method is a process in which each polymer component is dispersed and mixed at a temperature above the viscous flow temperature to obtain a homogeneously dispersed mixture, and the process is shown in Fig. 7.103. This method has the advantages of good blending effect and wide application, so it is the most commonly used blending method.

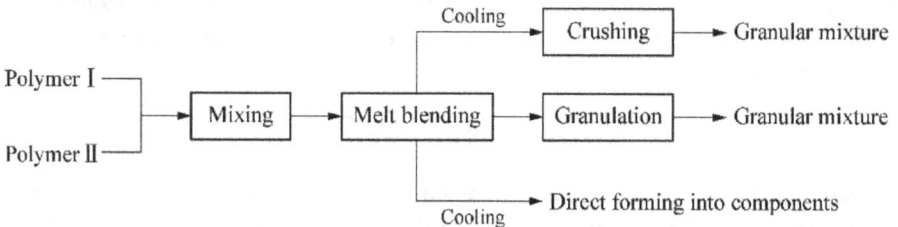

**Fig. 7.103:** Schematic view of melt blending process.

## 4. Chemical blending method
There are two main types of chemical blending: copolymerization-blending and interpenetrating polymer networks.

### a. Copolymerization blending
The copolymerization blending method includes graft copolymerization blending and block copolymerization blending. Graft copolymerization blending method is more important in the preparation of polymer. This method is characterized in that the monomer B is polymerized to "in situ" produce a blend in the presence of polymer A. Copolymerization blending method is different from method of polymer $A$ and polymer $B$ blending, that is, the former has the graft copolymer of polymers $A$ and $B$, and it plays an "emulsion" stabilizing effect on the two phases during the polymerization process; moreover, it plays a decisive role to enhance the cohesion of the two phases of the final product. For example, high-impact PS (HIPS) is a blend of styrene monomer "in situ" polymerizations in the presence of polybutadiene (PB), that is, PS

is a blend of styrene and graft copolymer of butadiene, because of the presence of graft copolymer, the impact toughness of PS significantly improves.

### b. Interpenetrating polymer networks

Interpenetrating polymer networks (IPN) are important new methods for preparing polymer alloys. The most common method for preparing IPN is to swelling the polymer $A$ with a monomer $B$ containing an initiator and a crosslinking agent (the solvent molecules penetrate into the polymer so that the volume of the polymer expands, called swelling), and polymerizes again, get $A$ and $B$ interpenetrating network, as shown in Fig. 7.104. The polymerization process mentioned above should be carried out independently. Usually one is addition polymerization; the other is condensation polymerization. IPN technology has been widely used in the preparation of damping materials, special coatings, dental materials, ion exchange resins, controlled release drugs, and so on.

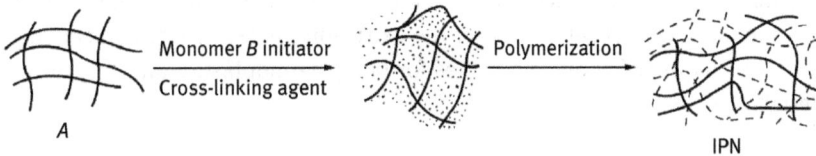

**Fig. 7.104:** Schematic view of IPN process.

### 5. Forming process of polymer alloy

Molding is the typical technique for forming plastic polymers, and a variety of molding methods are developed, including blow, compression, injection, transfer, and extrusion molding. In all cases, a finely pelletized plastic is forced to flow into, fill, and obtain the shape of a mold cavity at an elevated temperature and by pressure. There are many forming methods of polymer materials, and the choice of methods also depends on the nature of the material itself, that is, thermoplastic or thermosetting. For the thermoplastic material, the choice of forming methods is more, at first polymer material is heated to near or above its melting point, get the melt or elastomer, and then injected into a mold to get the desired shape. Thermoplastic elastomer also can be formed by this method. In these processes, the waste can be easily reused, minimizing waste. The forming method of thermosetting polymer material is relatively less, this is because once the polymer network structure is formed, the polymer material is difficult to re-shape. The thermoplastic elastomer needs to be shaped in a high shear apparatus, and it is heated to the state of viscoelastic deformation, and the material can form a permanent network structure. After the mixing step, the material is added with a curing agent, such as zinc oxide. Since the material from the mixing container is easy to bend, it can be small-scale extrusion, rolling or immersion coating and other treatment. The following is a variety of polymer-forming process, from which you can see that most of the process can only be used for thermoplastic materials.

### a. Extrusion forming

This is the most used thermoplastic polymer preparation process. Extrusion forming has two advantages: (1) it provides a means of continuously preparing a simple shaped material; (2) the extrusion method ensures that the additives (such as carbon black and fillers) evenly mixed; moreover, it can combine the subsequent process. Spiral processing equipment consists of one or a pair of screws (mating relationship), forcing the heated thermoplastic material (solid or liquid) and additives through the mold to produce a film, sheet, test tube, tubular material, and even produce plastic bags, as shown in Fig. 7.105(a). Extruders in the industry can have 60–70 ft (1 ft = 0.3048 m) in length, 2 ft in diameter, and is divided into hot and cold areas. Because of the shear shrinkage effect and the viscoelasticity of the thermoplastic material, it is critical to control the forming temperature and the consistency of the material viscoelasticity for the extrusion process. Extrusion can also be used for the coating of wires and cables for both thermoplastics and thermosetting materials.

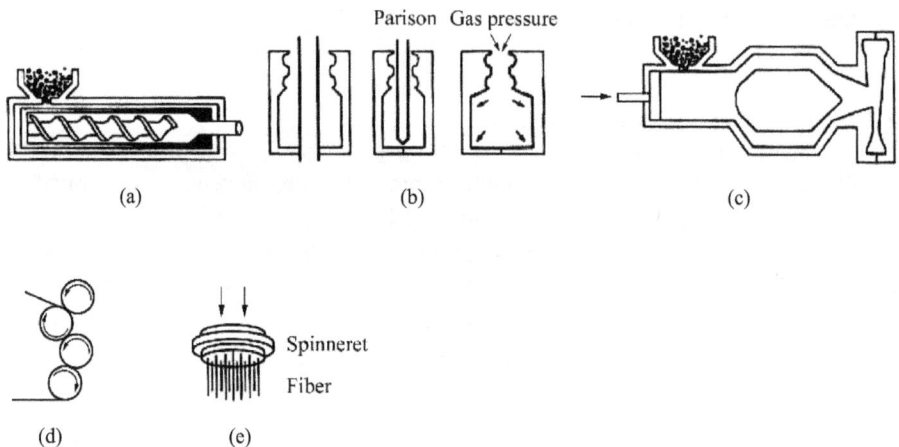

**Fig. 7.105:** Typical forming methods for thermoplastic polymer materials. (a) Extrusion forming, (b) blow molding, (c) injection molding, (d) film extrusion, and (e) spinning.

### b. Blow molding

Blow molding is a specific fabrication process by which hollow plastic parts are formed. It is designed to manufacture high volume, one-piece hollow objects and similar to that used for forming glass bottles or other hollow shapes. With blow molding, a plastic tube is heated and filled with air until it becomes a balloon of hot plastic called a parison. The parison is placed in a two-piece mold having the desired shape of the part while air continues to fill the parison into the shape of hollow piece under pressure, as shown in Fig. 7.105(b). During the blow molding process, the temperature and viscosity of the parison must be carefully controlled. This process is used to produce plastic bottles, containers, stadium seating and chairs, automotive fuel tanks, and other hollow-shaped devices.

### c. Injection molding
The polymer injection molding is the most widely used method for forming thermoplastic materials, as shown in Fig. 7.105(c). Crafted from die casting for metals, the thermoplastic material is melted to form a viscous liquid and injected into the cavity of molds by the motion of a plunger or ram at high temperatures under extreme pressure. Pressure is maintained until the molding has solidified and the molds are then cooled to release complete plastic pieces. Injection molding is ideal for high-volume and mass production requiring millions of the same plastic parts. A lot of products such as cups, combs, and can-opener are made in this way.

### d. Film extrusion
Many films are simply extruded through a thin slit die onto a polished chill roll, and pulled to a second set of rolls to cool the other side, as shown in Fig. 7.105(d). A thin cylindrical film can be produced by continuously reducing the wall thickness. Typical products created with film extrusion are polyvinyl chloride (PVC) tiles and rain-proof roller blinds.

### e. Spinning
Spinning is a process whereby bulk polymer material is shaped into a bundle of filaments, called a fiber. Usually, the polymer must be first brought into a liquid state, either by being melted (melt spinning) or dissolved in a suitable solvent (solvent spinning). In the case of melt spinning, the liquid thermoplastic material is pumped through a plate called a spinneret, which contains a lot of small holes, to produce a continuous filament fiber, as shown in Fig. 7.105(e). The spun fibers solidify rapidly by cooling in contact with air or water after extrusion form the spinneret. For some polymer materials, including nylon, these fibers may eventually be stretched in a direction parallel to the filament so as to increase their strength.

### f. Casting
Casting includes a number of processes that take a monomer, powder, or solvent solution, put them into a mold and allow them to solidify. The mold may be flat glass for the production of thinner sheets. Different from molding and extrusion, casting relies on atmospheric pressure to fill the mold rather than using external force to

push the plastic into the mold cavity. There is a vast variety of materials than can be cast, including thermoplastic and thermosetting plastics. For thermoplastics, solidification occurs upon cooling from the molten state while hardening occurs as a consequence of the actual polymerization or curing process for thermosets. Rotary casting is a special casting method in which the molten polymer material is poured into a mold which rotates about two axes. The centrifugal force makes the material better contact with the heated mold surface. A thin tent roof can be produced in this way.

### g. Compression molding

Compression molding is a method of molding in which the appropriate amounts of plastics materials and necessary additives are first placed in an open, heated mold cavity (female), or form (male) using a two-part mold system, as shown in Fig. 7.106(a). The plastics are generally preheated to reduce molding time and pressure, extend the die lifetime, and produce a more uniform finished part. The mold is closed with a top force and heat and pressure are applied, the plastics then become viscous and flow to conform to the mold shape. Once the material is set-hardened and has taken shape of the mold, an "ejector" releases the finished component. Both thermoplastic and thermosetting polymers are used in compression molding. But for thermoplastics, it is more time-consuming and expensive than the more commonly used extrusion or injection molding techniques mentioned above. Compression molding is one of the least expensive ways to mass-produce products due to its cost-effectiveness and efficiency. Small power boards, fenders, hoods, clothing fasteners, and car side panels can all be produced by this process.

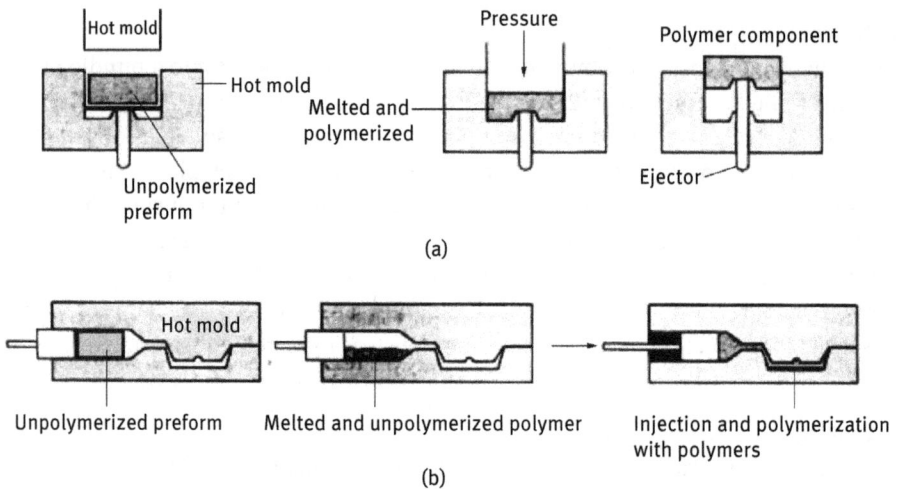

(a)

(b)

Fig. 7.106: Typical forming methods for thermosetting polymers. (a) Compression molding and (b) transfer molding.

## h. Transfer molding

Transfer molding is similar to compression molding; however, the material is first placed in a transfer chamber, and the mold is closed and clamped prior to the material entering the mold. As the molten material is injected into the mold cavity, the pressure is distributed more uniformly over all surfaces, as shown in Fig. 7.106(b). The mold remains closed until the material inside is cured (thermosets) or cooled (thermoplastics). This process provides certain advantages of the injection molding process typically for thermoset polymers and for pieces having complex geometries. One of the key advantages of transfer molding over compression molding is that different inserts, such as metal prongs, semiconductor chips, and ceramics, can be placed/positioned in the mold cavity before the polymer is injected into the cavity.

## i. Reaction injection molding

The liquid thermosetting polymeric resin is first injected into the mixing vessel and then directly forms a product, during which molding and curing takes place simultaneously. In the reinforced reaction injection molding process, reinforcing materials containing fine particles and short fibers are injected into the cavity to form a composite material. Automobile bumper, fender and furniture parts can be made by this method.

## j. Foaming molding

The foam materials can be made from PS, urethane, polymethylmethacrylate, and a large number of other materials. Polymeric raw materials are pellets, and a foaming agent, such as pentane needs to be added. In the pre-expansion process, the diameter of the pellets is expanded by about 50 times. The pre-expanded particles are then injected into the mold, and the particles are fused by steaming, resulting in an ultra-light material with a density of only about 0.02 $g \cdot cm^{-3}$. Foam cups, packaging materials and insulating materials are all application examples of this process. Besides, engine warmer is made of foamed PS material.

## 7.5.4 Morphological structure of polymer alloy

Polymer alloys may be single-phase (homogeneous) or may be multiphase (heterogeneous). The single-phase alloys correspond to a random copolymers, while the multiphase alloys correspond to polymer blends. The morphological structures of polymer alloys include the phase structures formed by different polymers, the size range of about 0.01–10 μm, and they have an important impact on the performance of the alloys. The same component, but the different morphological structure of the polymer alloys, may have significant performance differences.

According to the continuity of the phase, binary polymer alloys can be divided into single-phase continuous structure and two-phase continuous structure.

### 1. Single-phase continuous structure

Single-phase continuous structure in binary polymer alloys is that one of the two phases is a continuous distribution, which is called the matrix, the other phase dispersed in the matrix. According to the phase domain of the dispersed phase, it is divided into three kinds: the dispersed phase is irregular, the regular, and the cellular: (1) Dispersive phase consist of particles with irregular shape, different sizes of domains. The products obtained by the mechanical blending process generally have this morphological structure. (2) The domain morphology of the dispersed phase is regular, generally spherical or columnar. These particles do not contain or contain only a very little amount of continuous phase component. For example, styrene 80% mass fraction of styrene–butadiene block copolymer belongs to this category. The domains of some polymer alloys are not only regular in shape, uniform in size, but also arranged in space in the order of macroscopic order, for example, styrene-isoprene star-shaped block copolymer has a long-range-ordered structure. (3) The domains of the dispersed phase have a cellular structure, that is, the dispersed phase contains smaller particles composed of continuous-phase component, the dispersed-phase particles is considered as a cell, and the cell wall consists of continuous-phase component, and within cell there are more fine continuous-phase particles. Such a structure is named the cellular structure. Polymer alloys prepared by graft copolymerization blending method have mostly this structure. TEM image in Fig. 7.107 shows the typical cellular structure of high impact PS. In this polymer alloy, the continuous phase is PS, and the dispersed phase is rubber, which contains many dispersed PS-phase domain.

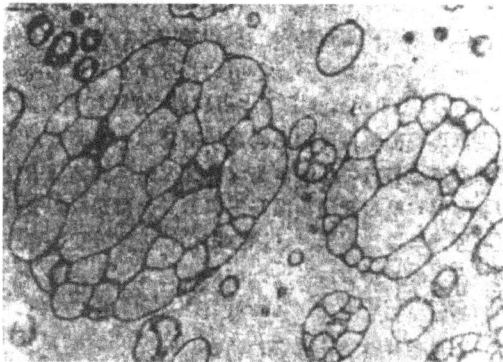

**Fig. 7.107:** Cellular structure of high-impact polystyrene.

## 2. Two-phase continuous structure

The interpenetrating polymer network has a typical two-phase continuous structure. The networks of the two components in the polymer alloy consist of continuous phases, and they interpenetrate each other, which make the whole specimen exhibit an interwoven network. If the compatibility of the two components in the alloy is not good, a certain degree of phase separation will occur. At this time, the two component networks are not molecular scale, but the domain scale, as shown in Fig. 7.108. The better the compatibility of the two components, the smaller the phase domain.

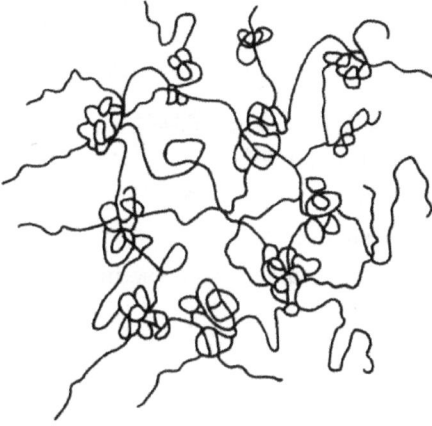

**Fig. 7.108:** Schematic view of interpenetrating network of two-phase continuous structure with a certain compatibility.

According to the morphological observation of the binary polymer alloy, some rules can be summarized as follows. The shape of the dispersed phase changes from spherical to rod to lamella with the increase of its amount, which is similar to the eutectic alloys in metals. The structural model is shown in Fig. 7.109. The electron microscopic photographs shown in Fig. 7.110 are the morphologies of styrene-butadiene-styrene triblock copolymer (SBS) with different ratio of styrene/butadiene. In the figure, the black portion is a PB rubber phase and the white portion is a PS plastic phase. It is shown in Fig. 7.110 that when the amount of butadiene is small, the

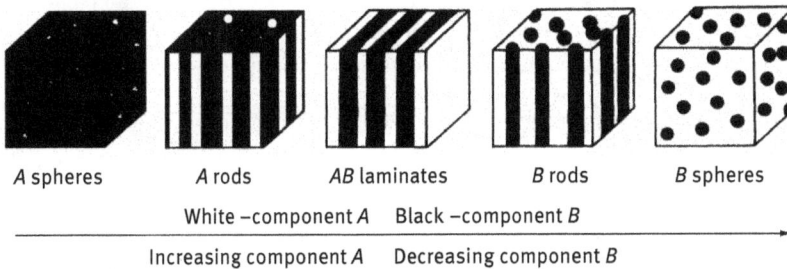

| A spheres | A rods | AB laminates | B rods | B spheres |

White –component A   Black –component B

Increasing component A   Decreasing component B

**Fig. 7.109:** Structural model showing the morphologies of polymer alloy.

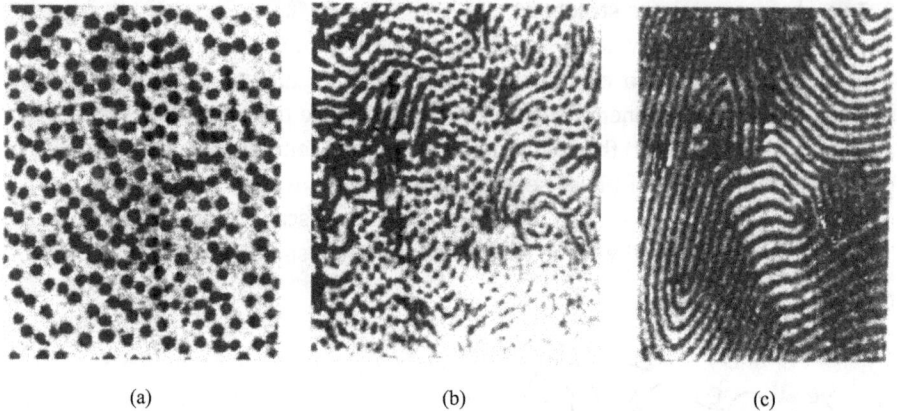

**Fig. 7.110:** SEM images showing morphologies of styrene–butadiene–styrene triblock copolymer (SBS) with different ratio of styrene/butadiene. (a) 80:20, (b) 60:40, and (c) 50:50.

rubber phase is dispersed phase, distributing in the plastic phase matrix. The shape of the rubber phase changes from spherical to rod to lamella with the increase of butadiene amount. When the amount of butadiene is over styrene, the rubber phase will be transformed into a continuous phase, as the matrix, while the plastic phase will be transformed into a dispersed phase. With the decrease of styrene amount, the plastic phase changes from lamellar to rod to globular. In polymer alloys, the transition from a continuous phase to a dispersed phase or from a dispersed phase to a continuous phase is called a phase inversion.

In a broad sense, crystalline polymers are also multiphase systems: one phase is a crystalline phase and the other is an amorphous phase. When the degree of crystallization is low, the crystal phase is the dispersed phase and the amorphous phase is the continuous phase. But when the degree of crystallization is high (more than 40%), the crystal phase is continuous phase and the amorphous phase is the dispersed phase. If one component of the binary polymer alloy can be crystallized, or the two components can both be crystallized, the basic structure of the alloy is shown in Figs. 7.111 and 7.112. Figure 7.111 illustrates the morphology of the crystalline-amorphous blend. When the amorphous phase is the matrix, the grains or spherulites will disperse in the amorphous region. When the crystallinity is large and the crystal phase is continuous phase (matrix), the amorphous phase will disperse in the spherulites, or the amorphous phase is aggregated into larger phase domains in the spherulites. Figure 7.112 illustrates the morphology of a crystalline-crystalline polymer alloy. When the crystallinity of the two components is low, amorphous is a continuous phase, and grains or spherulites will disperse in the amorphous region. When the two components have high crystallinity, two different spherulites or a mixture of spherulites may be formed, and in spherulites an amorphous phase exists.

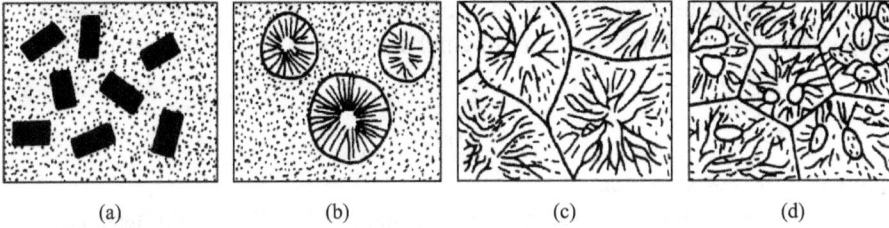

**Fig. 7.111:** Schematic view of the morphology of the crystalline-amorphous blend. (a) Grains dispersed in the amorphous region; (b) spherulites dispersed in the amorphous region; (c) amorphous phase dispersed in the spherulites; and (d) amorphous phase aggregated into larger phase domains in the spherulites.

**Fig. 7.112:** Schematic view of the morphology of the crystalline–crystalline polymer blend. (a) Two kinds of grains dispersed in the amorphous region; (b) spherulites and grains dispersed in the amorphous region; (c) two different kinds of spherulites; and (d) a mixture of spherulites.

### 7.5.5 The general relationship between properties and components in polymer alloys

The relationship between the component and property of binary polymer alloy can be approximately estimated by using "mixture rule," and the two most commonly equations are described as follows:

$$p = p_1 w_1 + p_2 w_2 \tag{7.54}$$

$$\frac{1}{p} = \frac{w_1}{p_1} + \frac{w_2}{p_2} \tag{7.55}$$

where $p$ is a certain property of the alloy, such as density, electrical properties, and modulus, $p_1$ and $p_2$ indicate properties of components 1 and 2, respectively, $w_1$ and $w_2$ are the mass fraction of components 1 and 2, respectively. In most cases, eq. (7.54) gives the upper limit of alloy property, while eq. (7.55) gives the lower limit. If the interaction of the two components has an effect on the property of the alloy, the interaction factor $I$ may be introduced, then the expression after modification of eq. (7.54) is

$$p = p_1 w_1 + p_2 w_2 + I w_1 w_2 \tag{7.56}$$

The interaction factor may be positive or negative, for example, the glass transition temperature of a random copolymer of vinyl acetate and vinyl chloride may be approximately expressed as

$$T_g = T_{g1}w_1 + T_{g2}w_2 - 28w_1w_2$$

where $w$ is the mass fraction.

For biphasic polymer alloys, the size and shape of the dispersed phase particles have an effect on certain properties. The relationship between the properties of the alloy and the components is

$$\frac{p}{p_1} = \frac{1 + AB\varphi_2}{1 - B\psi\varphi_2} \tag{7.57}$$

where $p$ and $p_2$ are the properties of the alloy and the continuous phase, respectively, $\varphi_2$ is the volume fraction of the dispersed phase, $\psi$ is the constant associated with the maximum stacking fraction of dispersed particles, $A$ is the shape factor of the dispersed phase. For uniform spherical particles, $A = 1.5$, and $B$ is a constant related to the property ratio of the two components.

For the polymer alloy that two phases are continuous, the relationship between the property and component can be expressed as follows:

$$p^n = p_1^n\varphi_1 + p_2^n\varphi_2 \tag{7.58}$$

where $\varphi_1$ and $\varphi_2$ are the volume fractions of component 1 and 2, respectively, and $n$ is a constant, such as in an IPN mixture, whose elastic modulus corresponds to $n = 1/3$.

As mentioned earlier, polymer alloys can exhibit "phase inversion" with variation of component amount, and in range of phase inversion, the elastic modulus of polymer alloy follows equation:

$$\lg p = \varphi_1 \lg p_1 + \varphi_2 \lg p_2 \tag{7.59}$$

It is worthy to note that the above equations give the basic guiding principle, the actual relationship between property and component is much more complex, and the specific relationship still needs to be summarized according to the experimental results.

### 7.5.6 The main types of polymers and their alloys

Polymer materials according to the use of materials can be divided into plastic, rubber, and fiber. Plastic has a wide range of performance change. Some plastics are hard and brittle, and some are very flexible, producing elastic–plastic deformation under external forces. Therefore, the plastic polymer chain can have crystallinity, but also can be linear, branched, and cross-linked. The elastomeric rubber must have sufficient cross-linking, which is usually achieved by vulcanization at elevated temperatures. The fiber macromolecule has the property of being drawn into filaments, and its length to diameter ratio is generally at least as high as 100/1. Since the fibers are subjected to various mechanical deformations in use, they must have high tensile strength, high elastic modulus, and abrasion resistance. This requires the fiber material to have a high relative molecular mass. Moreover, the structure and configuration of the molecular chains should have high crystallinity and can transform into symmetric linear and unbranched molecular chains. Above three types of plastics can be divided into two types of thermoplastic and thermosetting according to their mechanical properties at elevated temperature.

### (1) Thermoplastic polymer

The thermoplastic polymer will soften when it is heated and will harden when it is cooled, and this property can be repeated. These materials are usually prepared by the combination of heat and pressure. At the molecular level, when the temperature rises to below the temperature at which the covalent bond breaks, the secondary bond force between the molecular chains will weaken because molecular motion increases, which results in the relative motion between the molecular chains that becomes much easier under the action of external forces. Therefore, the thermoplastic material is relatively soft, good plasticity, and impact toughness. Most of the linear chain and some of the flexible structure of the branched polymer are thermoplastic polymers.

### (2) Thermosetting polymer

The thermosetting polymer exhibits a hard characteristic when cooled, but does not soften during the subsequent heating. During the initial heat treatment, covalent bond crosslinking structures between adjacent molecular chains have formed. These bonds anchor the chains so that the vibration and rotation of the molecular chain at high temperatures is prevented. And about 10% to 50% of the thermosetting polymer chains are crosslinked. When the temperature exceeds the temperature at which the crosslinking bond is broken and the polymer is degraded, the characteristics of the thermosetting polymer disappear. Thermosetting polymers are generally more rigid, stronger, and brittle than thermoplastic polymers, but have better dimensional stability. Most of the crosslinked and networked polymers are thermosetting polymers.

Polymer alloys also can be classified by polymer base with different properties as follows:

### 1. Polyethylene-based polymer alloys

PE is one of the most important plastics. Its advantages are good plasticity and easy processing. And its disadvantage is low softening temperature, low strength, easy to crack caused by stress, not easy to dye. Blending can effectively overcome these shortcomings.

Blends with PE as the main component have the following types: blends of PE with different densities; PE and ethylene–vinyl acetate copolymer (EVA); PE and acrylic vinegar blends; PE and CPE blends and so on. For example, blends of PE with different densities can widen the melting zone and retard crystallization during cooling, which is valuable for the preparation of PE foams. By changing the proportion of PE with different density, we can get a variety of foam with different properties.

### 2. Polypropylene-based polymer alloys

Polypropylene (PP) has a better heat resistance than PE and can be used below 120 °C for a long time. It also has the advantages of good rigidity, good folding, and excellent process ability. The main drawbacks are the large forming shrinkage, easy to brittle at low temperatures, poor wear resistance and light resistance, not easy to dye. Alloying is an effective way to overcome these drawbacks. These polymer alloys are PP/PE blends, PP/EPR (ethylene propylene copolymer), PP/BR (butadiene rubber) blends, and so on. For example, blends of PP with BR can significantly improve the toughness of PP. When blended with 15% BR, the impact strength of PP can be increased by over six times, and the brittle transition temperature decreases from 30 °C to 8 °C. The extrusion expansion of PP/BR is smaller than PP, PP/PE, PP/EVA, and so on, so the dimensional stability of product is better.

### 3. Polyvinyl chloride-based polymer alloys

PVC is a widely used polymer with good comprehensive properties. The main drawback is that the thermal stability is not good, and the decomposition of PVC starts at 100 °C, and thus processing properties is poor. PVC is hard and brittle, has a low impact strength, poor aging resistance, and cold resistance. Usually polymer alloys by alloying modification mainly include PVC/EVA, PVC/CPE, PVC/PB, PVC/ABS (A: acrylonitrile, B: butadiene, S: styrene, ternary copolymerization). For example, PVC/ABS alloy has high impact strength, good thermal stability, and excellent processing performance.

### 4. Polystyrene-based polymer alloys

PS is a high strength, but its shortcomings are brittle, poor toughness, easy to cracking and poor boiling water intolerance. Alloying modified polymer alloys mainly include PS (HIPS), ABS resin with high impact toughness and so on. For example, PS/styrene–butadiene rubber alloy overcomes the brittleness of PS and greatly enhances the impact strength.

## 7.6 Brief introduction of ceramic alloys

The engineering ceramics can be divided into two categories: structural ceramics and functional ceramics. The former is mainly concerned with the mechanical properties of ceramics; the latter mainly uses the optical, electrical, magnetic, thermal, and other physical properties of the ceramic. Most ceramics are composed of two components and multicomponents; therefore, these ceramic materials can also be called ceramic alloys. Ceramic materials can be divided into traditional ceramics and advanced ceramics according to quality. The main raw materials of traditional ceramics are quartz, feldspar, and clay minerals. The raw materials of advanced ceramics are a series of synthetic chemicals or refined chemical raw materials.

The atomic bonding force of the ceramic materials is mainly ionic bond, covalent bond, or ionic-covalent bond, which leads to the intrinsic brittleness of the ceramic materials, in applied stress ceramic materials only exhibit little deformation or no deformation before cracking. This intrinsic characteristic limits the preparation of ceramic materials that cannot be made using various processes commonly used in metallic materials, and must be accomplished by powder preparation, powder molding and sintering methods, similar to the powder metallurgy preparation of metallic cemented carbides. This section will briefly describe the ceramic powder synthesis, ceramic molding, and sintering methods and principles, the mechanical properties of ceramic materials and physical properties, the number of ceramic components is unlimited.

### 7.6.1 Synthesis of ceramic alloys

Oxygen, silicon, and aluminum account for 90% of the total elements in the earth's crust. They exist in the natural world in the status of silicate and aluminum silicate as the main raw materials of ceramic industry. Kaolinite-based minerals are the principal component of the high-duty clay, and aqueous talc is an important raw material for the manufacture of electrical, electronic components, and ceramic tiles. Anhydrous $SiO_2$ is the main component of glass, glaze, enamel, refractories, abrasives, and other white porcelain products. $SiO_2$ has a polymorphic form, and quartz is mainly used as a raw material.

In addition to clay and quartz, the feldspar is also the main raw material of traditional ceramics. Feldspar is an anhydrous aluminosilicate containing $K^+$, $Na^+$, or $Ca^{2+}$, which acts as a flux to promote the formation of a glassy phase. The main feldspar materials are potash feldspar $K(AlSi_3)O_8$, albite feldspar $Na(AlSi_3)O_8$, and anorthite $Ca(Al_2Si_2)O_8$, and so on.

The preparation of ceramic powder, also called the synthesis of powder, it can be divided into two categories, one is the mechanical crushing method and the other is the physical and chemical methods.

## 1. Mechanical crushing method

Mechanical crushing method can be used in a variety of ways, such as jaw crusher (coarse crushing equipment), wheel mill (grinding equipment which can also be used for mixing materials), ball mill (crushing equipment, which can also be used for mixing), jet mill (super fine crushing equipment, the minimum particle size of 0.1–0.5 μm), and airflow crushing (doing in nitrogen, carbon dioxide or an inert gas atmosphere).

## 2. Physical and chemical method

### (1) Solid-phase method

Solid-phase method use solid-phase reaction between solid substances to produce powder. The reactions commonly used in the preparation of ceramic powder materials include the combination reaction, the thermal decomposition reaction, and the oxide reduction reaction. These reactions often occur simultaneously in the actual process. The particle size of powder prepared by solid-phase method is sometimes too large. It needs to be further crushed in order to be used as raw materials. The following example illustrates the use of various methods.

The combination reaction is generally two or more than two kinds of solid materials mixing in the atmosphere and temperature conditions to generate one or more composite solid material, and sometimes may be accompanied with the escape of certain gases. Such as the synthesis of barium titanate ceramic powder, which mixes $BaCO_3$ and $TiO_2$ solid powder with the same molar amount, and it is heated between 1,100 and 1,150 °C for 2–4 h to form barium titanate raw material accompanying with release of carbon dioxide. The reaction formula is as follows:

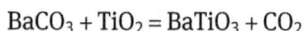

$$BaCO_3 + TiO_2 = BaTiO_3 + CO_2$$

The thermal decomposition reaction is to obtain a high-purity oxide powder with an excellent property by thermal decomposition during heating the metal sulfate, nitrate. For example: $\alpha$-$Al_2O_3$ powder prepared by ammonium aluminum sulfate $[Al_2(NH_4)_2 (SO_4)_2 \cdot 24H_2O]$ at different temperatures in the heating system in air, the specific thermal decomposition reaction is as follows:

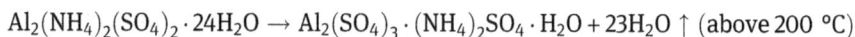

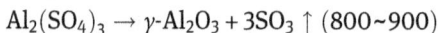

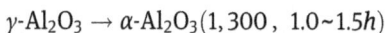

$$Al_2(NH_4)_2(SO_4)_2 \cdot 24H_2O \rightarrow Al_2(SO_4)_3 \cdot (NH_4)_2SO_4 \cdot H_2O + 23H_2O \uparrow \text{ (above 200 °C)}$$

$$Al_2(SO_4)_3 \cdot (NH_4)_2SO_4 \cdot H_2O \rightarrow Al_2(SO_4)_3 + 2NH_3 \uparrow + 2H_2O \uparrow \text{ (500~600)}$$

$$Al_2(SO_4)_3 \rightarrow \gamma\text{-}Al_2O_3 + 3SO_3 \uparrow \text{ (800~900)}$$

$$\gamma\text{-}Al_2O_3 \rightarrow \alpha\text{-}Al_2O_3 (1,300, \ 1.0~1.5h)$$

Silicon carbide and silicon nitride powders in advanced ceramic materials are typically prepared by an oxide reduction process. For example, SiC powder is quartz sand ($SiO_2$)

mixed with carbon powder, in the resistance furnace carbon is used to reduce $SiO_2$ to generate silicon carbide. The basic reactions taking place in the furnace are:

$$SiO_2 + 3C = SiC + 2CO \text{ (above 1,500 °C)}$$

The reaction process is:

$$SiO_2 + C = SiO + CO$$

Then, silicon monoxide reacts directly with carbon to produce silicon carbide:

$$SiO + 2C = SiC + CO$$

SiO can also be reduced to elemental silicon by carbon:

$$SiO + C = Si + CO$$

At this time, silicon vapor and carbon continue to react to produce silicon carbide:

$$Si + C = SiC$$

## (2) Liquid-phase method

Liquid-phase method is widely used in the preparation of ultrafine powder of advanced ceramic materials. It can better control the powder chemical composition and realize the ionic level of uniform mixing. The liquid-phase method mainly includes the precipitation method, the sol–gel method and so on.

Precipitation method is to add or generate precipitation agent in the metal salt solution and make the solution volatile, then the desired ceramic powder is obtained by heating the obtained acid salt and hydroxide. For example, $BaTiO_3$ powder is made by direct precipitation method. $Ba(OC_3H_7)_2$ and $Ti(OC_5H_{11})_4$ were dissolved in isopropanol or benzene and hydrolyzed with water to obtain stoichiometric $BaTiO_3$ crystalline powder with particle diameter of several nanometers.

The sol–gel method is to heat the metal oxide or hydroxide sol in 90 to 100 °C to form a gel material, and then it need filtering, dehydrating, drying, and sintering at suitable temperature to obtain the high purity superfine oxide powder. For example, the $ThO_2$ powder prepared by sol–gel method is sintered, which exhibits ultra-high density, reaching 99% of the theoretical density.

## (3) Gas-phase method

There are two methods for the preparation of ceramic powders by gas-phase method: physical vapor deposition (PVD) and chemical vapor deposition (CVD).

PVD method is used to heat the raw material at arc or plasma temperature to gasify, and then the gasifying material is condensed into powder particles by quenching (rapid cooling) in a large temperature gradient between the heating source and the environment.

The chemical vapor reaction method is a method of synthesizing the required powder by chemical reaction. For example, the chemical vapor reaction of $SiCl_4$ and $NH_3$ at 500–900 °C can generate amorphous powders with 0.1 μm, followed by heat treatment at higher temperature high purity ultrafine $Si_3N_4$ powder can be obtained.

### 7.6.2 Forming and sintering of ceramic powder

The basic steps for the preparation of the ceramic are shown in Fig. 7.113. In addition to the synthesis of the powders described above, the following steps include powder refinement and drying, casting forming, solidification sintering, secondary processing, and final product formation. For example, unidirectional compression forming is to obtain green body with desired shape by pressing powder into a mold under pressure, as shown in Fig. 7.114(a), where the powder particles displace and fill the voids under pressure, thereby reducing porosity and raising the density of green body. During sintering the grain boundary diffusion and bulk diffusion raise, the density of sintering ceramics, while surface diffusion and evaporation of the sintered ceramics make the grain growth. The microstructure of the sintered ceramics is generally equiaxed crystal, as shown in Fig. 7.114(b).

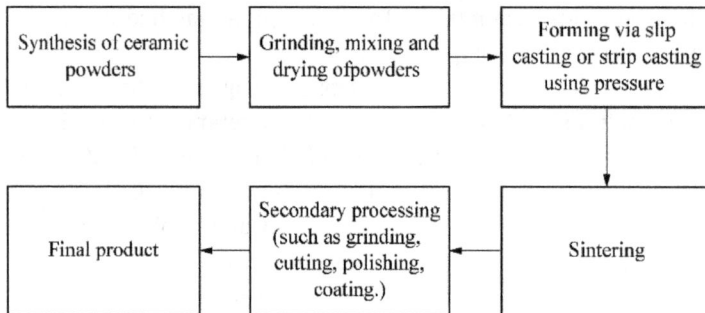

| Synthesis of ceramic powders | → | Grinding, mixing and drying of powders | → | Forming via slip casting or strip casting using pressure |
|---|---|---|---|---|

| Final product | ← | Secondary processing (such as grinding, cutting, polishing, coating.) | ← | Sintering |

**Fig. 7.113:** Basic procedure for preparing ceramics.

### 7.6.3 Preparation of glass

The manufacturing process of the glass is very special and different from the molding and sintering process of the ceramic. Sheet glass is produced in the molten state, which involves the molten glass through a water-cooled roll or floating over the tin surface, as shown in Fig. 7.115. Liquid tin treatment makes the glass surface extremely smooth. Float glass process is a breakthrough development in the field of glass processing.

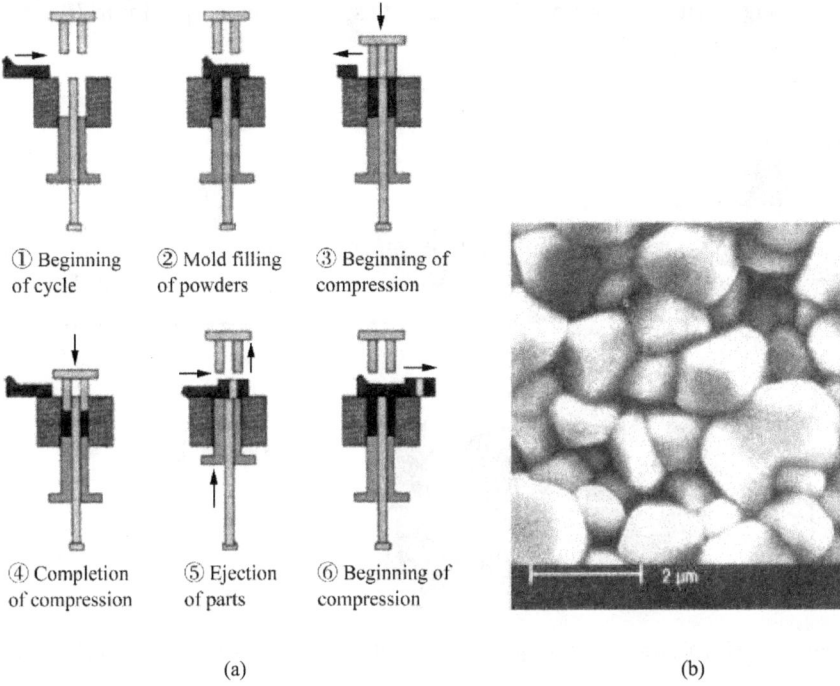

① Beginning of cycle   ② Mold filling of powders   ③ Beginning of compression

④ Completion of compression   ⑤ Ejection of parts   ⑥ Beginning of compression

(a)                     (b)

**Fig. 7.114:** Molding of ceramics (a) and the microstructure after sintering (b).

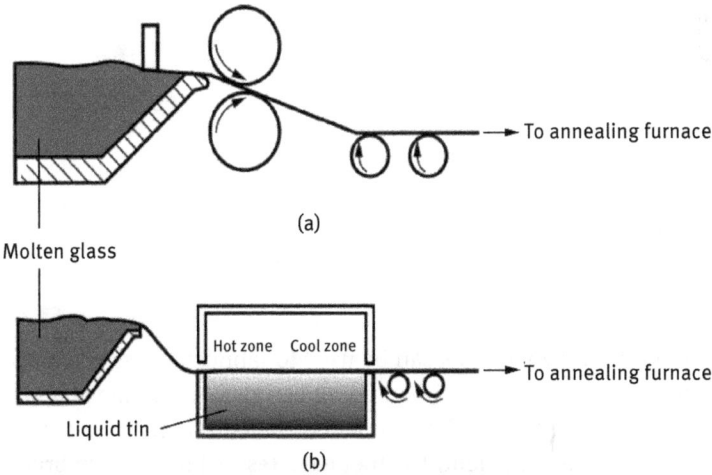

To annealing furnace

(a)

Molten glass

Hot zone   Cool zone

To annealing furnace

Liquid tin

(b)

**Fig. 7.115:** Production of glass plate. (a) Roll forming and (b) forming by floating on the surface of tin.

Some special shapes of glass, such as utensils and optical glass lenses, are produced by casting molten glass through a mold and then cooling it as slowly as possible to minimize residual (thermal) stress and to avoid cracking of glass pieces (shown in Fig. 7.116). Glass fiber is a liquid glass through the platinum (Pt) die drawn from the hole, as shown in Fig. 7.117.

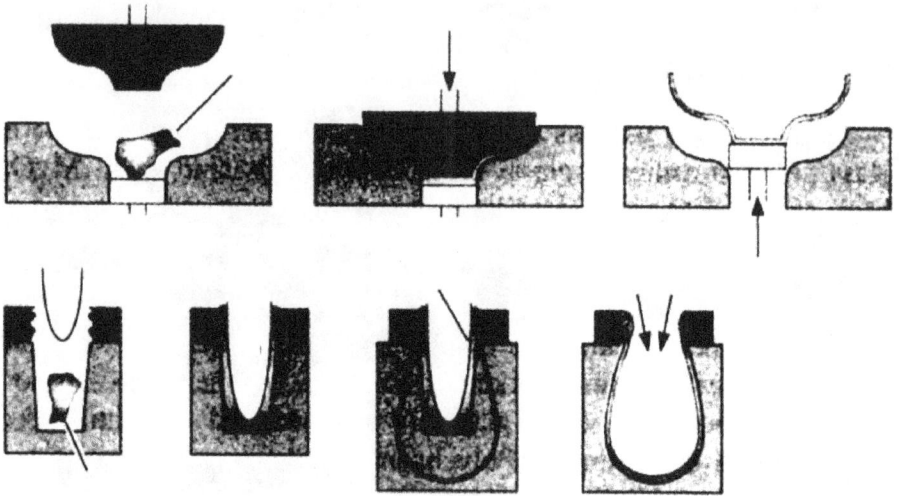

**Fig. 7.116:** Processing of glasses with particular shapes. (a) Extrusion and (b) extrusion with blowing.

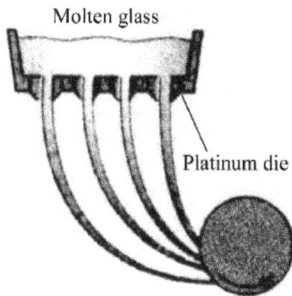

**Fig. 7.117:** Drawing technology of glass fibers.

In general, five different forming methods can be used to fabricate glass products: pressing, blowing, drawing, and sheet and fiber forming. When a ceramic material is cooled from an elevated temperature, internal stress, called thermal stress, may be introduced due to temperature gradient. The thermal stress is important in brittle ceramics, especially glasses, because they may weaken materials, even lead to fracture, which is termed thermal shock. Therefore, an annealing heat treatment as post-treatment of glass fabrication is necessary to eliminate or reduce such thermal stress.

### 7.6.4 Properties of ceramic materials

Ceramic properties include mechanical properties and optical, electrical, magnetic, thermal, and other physical properties. The following will give a brief introduction.

#### 1. Mechanical properties

Ceramic materials do not undergo plastic deformation at room temperature under tensile load, which means that brittle fracture occurs immediately after the end of elastic deformation, as shown in Fig. 7.118, it is different from metallic materials. The binding of atoms in ceramics is an ionic bond or a covalent bond, with obvious directionality and few slip systems. The same ion encounters a large repulsive force due to the electrostatic energy. Therefore, it is difficult to produce plastic deformation at room temperature for most of the ceramic materials, while obvious plastic deformation only can be seen only at high temperature.

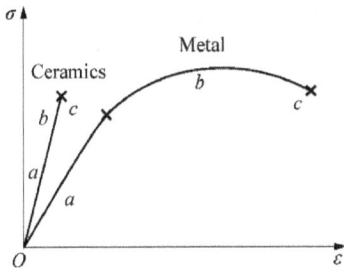

**Fig. 7.118:** Stress–strain curve of metals and ceramics.

Only a few ceramic materials with simple crystal structures, such as MgO, KCl with the NaCl structure has the plasticity at room temperature. Some ultrafine grain ceramics, such as $Al_2O_3$ and $ZrO_2$, at high temperatures will appear superplasticity, which is of engineering significance for ceramic materials difficult to machine.

For the brittle ceramic materials, the compressive strength replacing the tensile strength is used to represent of the mechanical properties of the material. Ceramic material compressive strength is much higher than the tensile strength. The compressive strength of the components is usually used to set prestress, which can enhance resistance under the tensile stress state of the material.

The current methods improving the brittleness and toughness of ceramic materials are the following: adding the dispersed whiskers, fibers or toughening second phase; For example, the addition of second phase $ZrO_2$ in brittle ceramics can enhance toughness by stress-induced martensitic transformation of $ZrO_2$.

The high hardness of ceramics makes them widely used at room temperature and high temperature. For example, diamond is the hardest material in the natural world. Industrial diamonds are used as abrasives for grinding and polishing. Diamond and diamond-like coatings prepared by CVD are used as abrasion-resistant coatings for cutting tools and, of course, as jewelry. SiC has excellent high temperature (higher than the melting point of steel) antioxidant capacity; therefore, it is often used for metal alloy cutting tool coating; SiC also is used in the metal matrix and ceramic matrix composites in the form of particles or fiber to reinforce matrix strength. $Si_3N_4$ has similar properties with SiC, but the high temperature strength is slightly lower. Borosilicate glass contains 15% $B_2O_3$, which exhibits very good physical and chemical stability, and it is used as laboratory glassware and high radioactive waste storage containers. Calcium aluminoborosilicate glass is usually used as a reinforcing fiber for composites, such as glass fibers. Aluminosilicate glass containing 20% $Al_2O_3$ and 12% MgO, and 3% $B_2O_3$ and high amount of silica can withstand high temperature and thermal shock. Concrete from cement and sand mixture has been widely used in construction.

Glass ceramics for the crystal material which is from the glass transition of amorphous glass, usually have a high degree of crystallinity ($\geq$70–99%). The special structure of glass-ceramic makes it have a high strength and toughness, and in general has a low coefficient of thermal expansion and good high temperature performance. Perhaps, the most important glass ceramics are based on the $Li_2O$–$Al_2O_3$–$SiO_2$ system, which is used in ceramic devices for cooking appliances and furnaces. Other glass ceramics are used in the fields of communication engineering, computer and optics.

## 2. Physical properties

The physical properties of the material refer to light, electric, magnetic, and thermal effects. The following briefly describes the main ceramic system involved in these properties.

### (1) Conductive and dielectric materials

The conductivity of conductive ceramics is far greater than the general ceramics. The majority of ceramics are insulators, but ceramics with the interstitial structure of the carbides show good electrical conductivity, and the electric conduction mechanism is electronic conductivity. While some ceramic materials in a certain temperature or pressure conditions show ionic conductivity characteristics, which is called fast ion conductor or fast ion ceramic. Because of the sensitivity of ion conduction to the activity, temperature, and pressure of the surrounding substances, a variety of solid-state ion-selective electrodes, gas-liquid, heat, humidity, and pressure-sensitive sensors can be fabricated by fast ion conductors. Color display can be produced by using oxidation-reduction coloring effect of some ions in the fast ion conductor, besides its charge and discharge effects can be used to produce batteries, electrochemical switches and memory devices. For example, non-stoichiometric $SnO_2$, $CuO_2$, SiC are electronic conductors. $ZrO_2$, $LaCrO_3$, and $LaMnO_3$ are the components of solid oxide fuel cell using ion conduction; $BaTiO_3$ and $SrTiO_3$ are heat-sensitive materials with a positive temperature coefficient; $SnO_2$ is a gas-sensitive material; spinel-type $MgCrO_4$–$TiO_2$ is a moisture vapor sensitive material; ZnO is a pressure sensitive material that the resistance value nonlinearly changes with the voltage; CdS and doped Cu, Ag, Au ceramic are optoelectronic materials in a visible light range; $SrTiO_3$ and $La_{2-x}Ba_xCuO_4$ are superconducting materials with a high superconducting transition temperature ($T_c$).

Dielectricity indicates an ability of the division and insulation of current. Dielectric ceramic is a functional material which has the ability of polarization under the action of electric field and can establish the electric field in the body, mainly including insulating ceramics, the capacitor ceramics and the microwave ceramics and so on. For example, $Pb(Mn_{1/3}Nb_{2/3})O_3$ and $Pb(Zn_{1/3}Nb_{2/3})O_3$ are high dielectric constant capacitor ceramics; $Pb(Zr_xTi_{1-x})O_3$(PZT) is a new type of piezoelectric ceramic; (Pb, La) (Zr, Ti)$O_3$ (PLZT) is a new type of electrostrictive ceramic.

## (2) Magnetic ceramics
Magnetic ceramics are usually called ferrite, it is mainly used for high-frequency technology, such as radio, television, computer, microwave, and ion accelerator. For example, $Fe_2O_3$- and $Fe_2O_3$-based composite oxide by adding MnO, MgO, CuO, and so on. are ferrite soft magnetic ceramics, while $BaFe_{12}O_{19}$ and $SrFe_{12}O_{19}$ are ferrite hard magnetic ceramics.

## (3) Heat storage and heat insulation ceramic
Heat capacity of most ceramic material increases with increasing temperature from low temperature. The low dielectric constant of $Al_2O_3$ reached 3 R (25 J $(mol K)^{-1}$) at about 1,000 °C. It can be used for electronic packaging materials of silicon chip. Alumina, zirconia, magnesium oxide, silicon carbide, boron nitride, and fused silica can be used as heat insulation materials under the temperature of 2,000 °C.

## (4) Optical glass
Glass is transparent to visible light and therefore has a wide range of uses. Pure $SiO_2$ needs to be heated to a very high temperature to obtain the viscous flow state for forming. Most commercial glasses are $SiO_2$-based ceramics; additives, such as soda ($Na_2O$), can break the network structure; so it can also reduce the melting point, meanwhile the addition of CaO can reduce the water solubility in the glass. Most common commercial glasses contain about 75% $SiO_2$, 15% $Na_2O$, and 10% CaO, called the green plate glass.

Special optical properties (e.g., photosensitivity) of the glass can also be obtained. Photochromic glass is opaque to ultraviolet (UV) light and can be used as a sunglass material. Multicolor decorative glass is not only sensitive to UV light, but also to all light, also including nanocrystalline semiconductor, such as the formation of nanocrystalline CdS by nucleated in $SiO_2$-based glasses. These glasses not only give us a rich color but also have many useful optical properties.

# Chapter 8
# Ternary phase diagrams

Most metallic or ceramic materials are composed of more than two components. The addition of the third component or the fourth component not only changes the solubility of other components, but also makes the diagram more complicated. For the purpose of better understanding of the relationships among the components, microstructure, and properties, it is necessary to obtain the knowledge of ternary phase diagrams or multicomponent phase diagrams. However, the phase diagrams that have more than three components are too complicated to analyze clearly in a three-dimensional space. They are often simplified as the pseudoternary phase diagrams. Therefore, it is obvious that the knowledge of ternary phase diagrams is essential for multicomponent materials.

Compared with the binary phase diagram, the ternary phase diagram has one more component. In a ternary diagram with components $A$–$B$–$C$, the sum of the mass (or mole) fraction is unity, that is, $X_A + X_B + X_C = 1$. As a result, there are two independent component variables in the diagram, which means that the component axis should be two and it needs to be represented as a plane. In addition, there is a temperature axis perpendicular to the component plane. Hence, a ternary phase diagram is exhibited as a three-dimensional graphics, with a series of spatial curved surfaces, unlike the curved lines in binary diagram separating the different phases.

It is a time-consuming project to prepare a complete ternary phase diagram experimentally. Meanwhile, it is also a difficult task to imagine the three-dimensional details correctly only according to two-dimensional graphics. Therefore, we often refer to the sections and projection drawings that have practical value in research and analysis of materials, such as the various isothermal sections, the variable temperature sections, and the projection drawings of each phase in the composition triangle. A complete three-dimensional ternary phase diagram is thus composed of many such sections and projections.

In this chapter, we mainly learn how to use the ternary phase diagrams, particularly focusing on the analysis of sections and projection drawings.

## 8.1 The basics of ternary phase diagrams

The difference between the ternary phase diagrams and the binary phase diagrams lies in the added component variable. Hence, a symmetrical composition representation of all three components can be obtained with an equilateral "composition triangle" as shown in Fig. 8.1. The compositions at the triangle corners correspond to the pure component. Compositions corresponding to the three binary subsystems $A$–$B$, $B$–$C$, and

https://doi.org/10.1515/9783110495379-008

$A$–$C$ are found along the edge of the triangle. The basic characteristics of ternary phase diagrams are as follows.

(1) A complete ternary phase diagram is a three-dimensional space pattern.

(2) A four-phase equilibrium reaction can occur in a ternary system at constant pressure. It can be determined by the Gibbs phase rule that the maximum number of equilibrium phases is three in binary systems, while the maximum number of equilibrium phases is four in ternary systems. The four-phase equilibrium zone in the ternary phase diagram is an isothermal horizontal plane, that is, composition and temperature of equilibrium phases are fixed and unchanged since degrees of freedom, $f$, is equal to zero.

(3) In addition to the single-phase regions and the two-phase equilibrium regions, the three-phase equilibrium regions in ternary phase diagrams occupy a certain space. According to the phase rule, three-phase equilibrium of ternary systems has one degree of freedom, so the three-phase equilibrium transformation is a temperature-changing process. Reflecting on the phase diagrams, the three-phase equilibrium regions will occupy a certain space rather than horizontal lines in the binary phase diagrams.

## 8.1.1 Representation of components in ternary phase diagrams

The composition of the binary systems can be expressed by a point in a straight line, while the composition points of the ternary systems are arranged in a triangle defined by two coordinate axes. The triangle is called composition triangle or concentration triangle. The common composition triangle is an equilateral triangle, in some special cases a right triangle or an isosceles triangle is also sometimes used to represent the composition.

### 1. The equilateral composition triangle

Figure 8.1 is the equilateral triangle representation. Three corners ($A$, $B$, $C$) of triangle represent three components and the three triangle sides ($AB$, $BC$, $CA$), respectively, indicate the component coordinates of three binary subsystems, so any point within the triangle represents a composition of a ternary system. For example, the composition of point $S$ in the composition triangle $ABC$ can be obtained by the following method.

Every set of the equilateral triangle side length is 100% and the sides $AB$, $BC$, $CA$, respectively, represent the content of the component $B$, $C$, $A$. From this point $S$, draw three lines that are parallel to $BC$, $CA$, $AB$, respectively, and intersect three sides at points $c$, $a$, $b$, and thus obtain $Sa$, $Sb$, and $Sc$, as shown in Fig. 8.1. According to the nature of equilateral triangle, we can know that

$$Sa + Sb + Sc = AB = BC = CA = 100\%$$

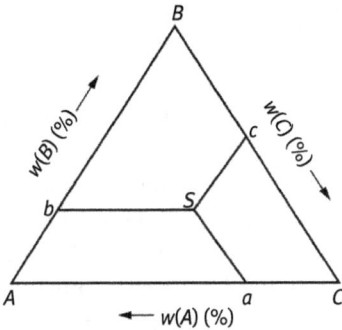

**Fig. 8.1:** The composition presentation of ternary alloy in equilateral composition triangle.

Among them, $Sc = Ca = \omega_A$ (%), $Sa = Ab = \omega_B$ (%), $Sb = Bc = \omega_C$ (%). Then, the segments $Ca$, $Ab$, $Bc$, respectively, represent the mass fraction of the components $A$, $B$, $C$. As a result, the point $S$ within the composition triangle is determined.

### 2. The special lines in equilateral composition triangle

The following lines have specific meanings in the equilateral composition triangle.

(1) All the ternary phases whose component points are in a line being parallel to one side of the equilateral triangle have same mass fraction of this component. As shown $ef$ line paralleling to $AC$ and facing toward the triangle vertex – component $B$ in Fig. 8.2, namely, the mass fraction of component $B$ of all ternary phases in the line $ef$ is $Ae = \omega_B$ (%).

(2) All the ternary phases whose components are in the line from a triangle vertex (corner) to the opposite side with an intersection point have same mass fraction ratio of the two components, which are represented by the ratio of spacings of two vertices to the intersection point. As shown in Fig. 8.2, the mass fraction ratio of component $A$ and $C$ of all ternary phases in line $Bg$ is same, that is, $\omega_A/\omega_C = Cg/Ag$.

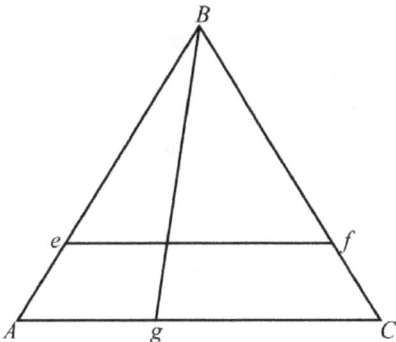

**Fig. 8.2:** The special lines in equilateral composition triangle.

## 3. Other composition presentations

### a. Isosceles composition triangle

When a component content of ternary system is less and the other two component contents are more, the alloy composition point will be near one side of an equilateral triangle. In order to make the phase diagrams clearly presented, two waists of the composition triangle are enlarged, so as to transform the equilateral triangle into an isosceles triangle. As shown in Fig. 8.3, as the component point $o$ is near the bottom, so just the isosceles trapezoid section is considered in practical application. The method for determination of composition of alloy point $o$ is the same as above equilateral triangle. Drawing the lines which parallel to two waists from point $o$, and the parallel lines and side $AC$ intersect at point $a$ and $c$, then we know that $w_A = Ca = 30\%$, $w_C = Ac = 60\%$. Drawing the line $ob$ which parallels to the side $AC$ from point $o$ and intersects the waist at point $b$, so $w_B = Ab = 10\%$.

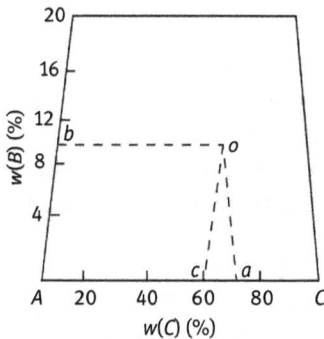

**Fig. 8.3:** Isosceles composition triangle.

### b. Orthogonal composition coordinate

When one component of a ternary system accounts for majority and other two components are minority, the alloy composition point will be close to the vertex angle of equilateral triangle. The phase diagram can be clearly expressed if the rectangular coordinates are used to represent the composition. Setting the origin of the rectangular coordinates as the component with high content, the two orthogonal axes represent the composition of the other two components. For example, the composition of the point $P$ is that $w(\text{Mn}) = 0.8\%$, $w(\text{Si}) = 0.6\%$, and the balance is ferrum, as shown in Fig. 8.4.

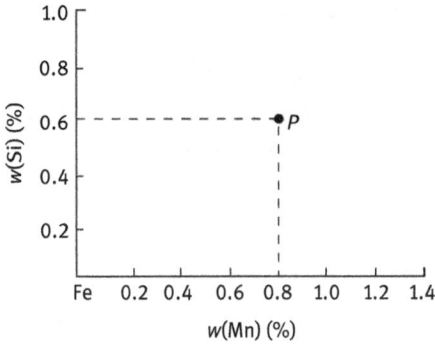

Fig. 8.4: Schematic illustration of local parts interested in complete composition triangle.

## c. Local presentation

If you only need to study the material within a certain composition range in ternary systems, it is more clear to take and enlarge the useful part of the composition triangle. The local ternary phase diagram is easier than the complete ternary phase diagram in determination, description, or analysis (see I, II, III in Fig. 8.5).

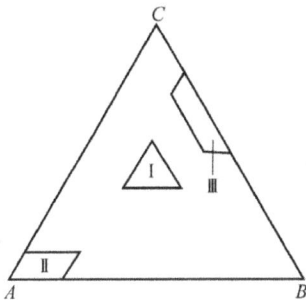

Fig. 8.5: Parts in composition triangle.

## 8.1.2 The space pattern of ternary phase diagrams

As previously mentioned, the ternary phase diagrams containing the component and temperature variables is three-dimensional graphics. The "phase diagram" of the ternary temperature composition under constant pressure can be drawn as a three-dimensional "space model," which is a right triangular prism with the equilateral composition triangle as the base and the temperature as the vertical axis. Because each side of the composition triangle represents a group of corresponding binary subsystem, three side faces of the triangle prism are, respectively, the three groups of binary phase diagrams. In the interior of the triangle prism, the phase regions are separated by a series of spatial curved surfaces.

Figure 8.6 is one of the simplest space patterns of ternary phase diagrams. The components $A$, $B$, $C$ form composition triangle and constitute the triangular prism frame. Three points $a$, $b$, $c$ are the melting points of components $A$, $B$, $C$, respectively. Since these three components are miscible with each other completely in both liquid and solid, the three sides are simple binary isomorphous phase diagrams. In the triangular prism body, the upward convex surface constituted by the liquidus lines of three binary systems are the liquidus surface of ternary system, which shows the solidification starting temperature of alloys with different compositions. The downward concave surface constituted by the solidus lines of three binary system is the solidus surface of ternary system, which shows the final temperature of solidification for alloys with different compositions. Above the liquidus surface is the liquid region and below the solidus surface is the solid region. The center region is liquid and solid two-phase equilibrium region, such as the temperature region between nodes 1 and 2 intersected by the liquidus and solidus surfaces of ternary system for composition $O$.

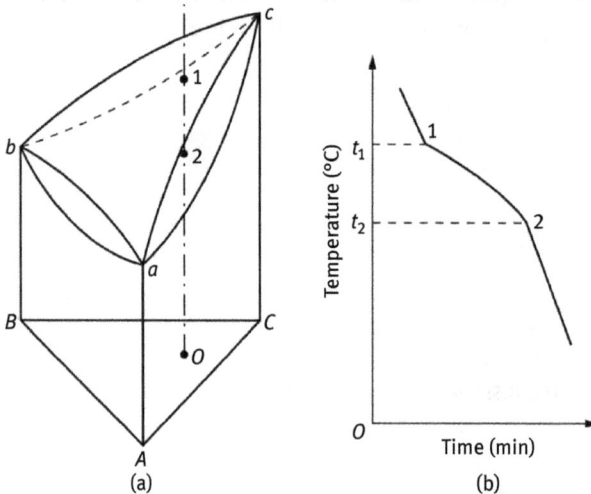

Fig. 8.6: Ternary isomorphous phase diagram and the alloy solidification: (a) ternary phase diagram and (b) cooling curve.

Apparently, even the simplest ternary phase diagrams are composed of a series of spatial curved surface, so it is difficult to describe the curvature change of the liquidus and solidus surfaces on paper clearly and accurately. And it is not so easy to determine the phase transition temperature of each alloy. It is more impossible to do these in the complex ternary phase diagrams.

Therefore, planarization makes the ternary phase diagram more practical.

### 8.1.3 Sections and projection drawings of ternary phase diagrams

To decompose the three-dimensional graphics into two-dimensional graphics, we must try to reduce a variable. For example, when the temperature variable is fixed at some value, there are only two component variables. The plane graph indicates the rule of ternary systems state with the variation of composition under a certain temperature. And it can fix one composition variable and remain the other composition variable and the temperature variable, the plane graph shows the variation rule of temperature and the composition variable. No matter what method is chosen, the graphics are sections of the three-dimensional phase diagram, so it is known as the section drawing.

#### 1. Horizontal section
The temperature axis of ternary phase diagrams is perpendicular to the composition triangle, so the sectional drawing at fixed temperature must be parallel to the composition triangle. The section is called horizontal section, also known as isothermal section.

The shape of the whole horizontal section should be consistent with the composition triangle. The curves of section are intersection lines obtained by the intersection of temperature section and phase interfaces in the space model. Figure 8.7 is the horizontal section of the ternary phase diagrams in two-phase equilibrium temperature range. The line $de$ and $fg$ are liquidus and solidus lines, respectively, which divide the horizontal section into liquid phase $L$, solid phase $\alpha$, and liquid and solid two-phase equilibrium region $L + \alpha$.

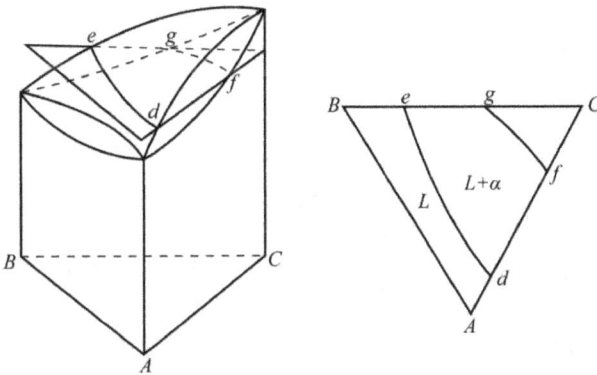

**Fig. 8.7:** Horizontal section of ternary phase diagrams.

## 2. Vertical section

The section fixing one composition variable and keeping temperature variable must be perpendicular to the composition triangle, so it is called vertical section. There are two commonly used vertical section. One is through the composition triangle vertex, fixing the ratio of the other two component contents, such as the vertical section $Ck$ in Fig. 8.8(a). Another one fixes one component composition and the content of the other two components is variable relatively, such as the vertical section $ab$ in Fig. 8.8(a). Every end of composition axis of section $ab$ does not mean the pure component, but corresponds to two components $A + B$ and $C + B$ in the two binary systems in which the component $B$ is fixed. For example, the composition of origin point $a$ in Fig. 8.8(b) is $w(B) = 10\%$, $w(A) = 90\%$, and $w(C) = 0\%$, and when the horizontal coordinate is at 50, the composition is $w(B) = 10\%$, $w(A) = 40\%$, $w(C) = 50\%$. It must be pointed out that although the vertical section of the ternary phase diagrams is similar to the shape of the binary phase diagrams, there are essential differences between them. The liquidus and solidus lines of binary phase diagrams can be used to express the law of liquid and solid concentration with the change of temperature in the process of alloy equilibrium solidification. However, vertical section of ternary phase diagrams cannot express the relationship of phase concentration with the change of temperature. It can only be used for understanding phase transformation temperature in the process of solidification. The tie-line cannot be applied to determine the mass fraction of two phases and the lever rule also cannot be used to calculate the two-phase relative amounts.

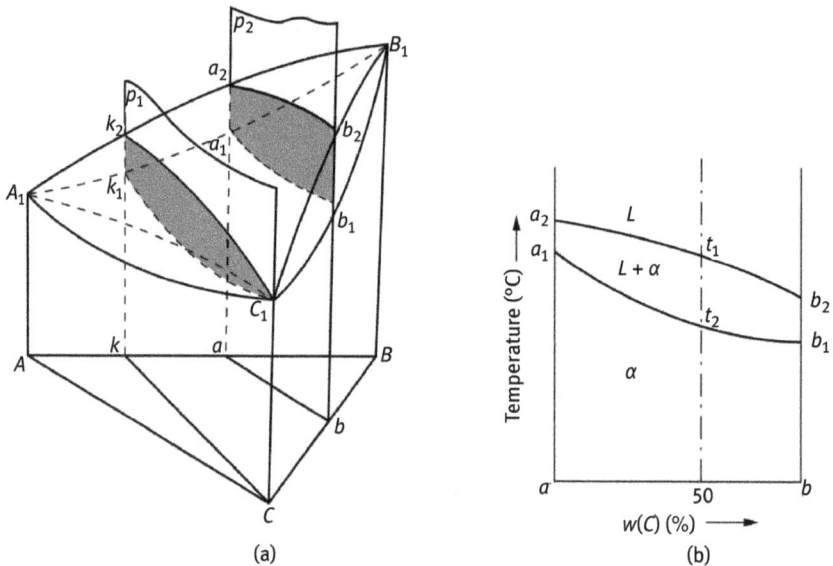

Fig. 8.8: Vertical section of ternary isomorphous phase diagrams.

### 3. Polythermal projections of liquidus and solidus surfaces

Similar to the contour map used in geography, a two-dimensional version of the ternary liquidus and solidus surfaces can be obtained through the vertical projection upon the base composition triangle. An example of polythermal projection of the liquidus and solidus is shown in Fig. 8.9. It is used to analyze the transformation of alloy during heating and cooling.

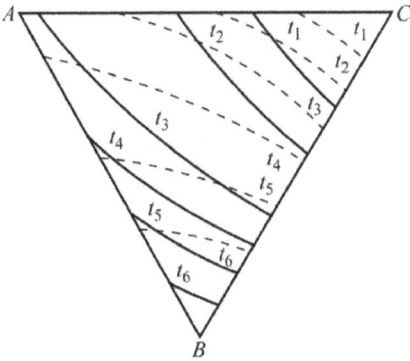

**Fig. 8.9:** The projection drawing example of ternary isomorphous phase diagrams.

If we project phase boundary lines of the horizontal sections at a series of different temperatures onto the composition triangle and label the corresponding temperatures in each projection, the constant temperature lines in Fig. 8.9 are named *liquidus isotherms* or *solidus isotherms* in *isothermal projection*. The isotherms can reflect the trend of the height of each phase interface with the change of temperature in the phase diagram, just as the contour lines in the map. If the temperature intervals of the adjacent isotherms are fixed, the denser the distance of isotherms is, the steeper the slope of phase interface is. On the contrary, the rarer the distance of isotherms is, the flatter the trend of the height of phase interface with the change of temperature is.

In order to make the complex projection of ternary phase diagram simpler and clearer, we only need to project part of the isothermal lines of phase interface down as needed. The liquidus projection or solidus projection is commonly used. Figure 8.9 shows the projection drawing of ternary isomorphous phase diagrams, in which the solid lines are the liquidus projection and the dotted lines are the solidus projection.

### 8.1.4 Lever rule and barycenter rule in ternary phase diagrams

In the study of three-component systems, we always tend to understand the composition and relative amount of constituent phases at different temperature when the composition of an alloy is known. In order to understand the above issues, the lever rule and barycenter rule are needed.

## 1. Linear rule

At a certain temperature, when one three-component alloy is in two-phase equilibrium, the composition point of the alloy and the composition points of its two equilibrium phases must be in a line in composition triangle, which is called linear rule. The proof is as follows.

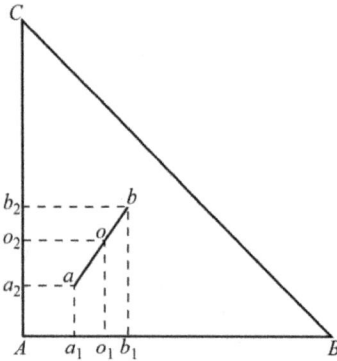

**Fig. 8.10:** The proof of linear law.

As shown in Fig. 8.10, it is supposed that the alloy with composition point $o$ is in the $\alpha + \beta$ two-phase equilibrium state, the composition point of phases $\alpha$ and $\beta$ is $a$ and $b$, respectively. We can read that the content of component $B$ of point $o$, phases $\alpha$ and $\beta$ is $Ao_1$, $Aa_1$, and $Ab_1$, respectively, from the figure. And the content of component $C$ is $Ao_2$, $Aa_2$, and $Ab_2$. If the mass fraction of phase $\alpha$ is $w_\alpha$, then the mass fraction of phase $\beta$ is $1 - w_\alpha$. The mass sum of components $B$ and $C$ in phases $\alpha$ and $\beta$ is equal to the mass of components $B$ and $C$ in the alloy. Thus, we can get that

$$Aa_1 \cdot w_\alpha + Ab_1(1 - w_\alpha) = Ao_1$$

$$Aa_2 \cdot w_\alpha + Ab_2(1 - w_\alpha) = Ao_2$$

After transpositions:

$$w_\alpha(Aa_1 - Ab_1) = Ao_1 - Ab_1$$

$$w_\alpha(Aa_2 - Ab_2) = Ao_2 - Ab_2$$

Then we can get

$$\frac{Aa_1 - Ab_1}{Aa_2 - Ab_2} = \frac{Ao_1 - Ab_1}{Ao_2 - Ab_2}$$

This relation is three collinear rule in analytic geometry. So it has proved that point $a$, $b$ and $c$ is in the same line. It can be also proved that the relation still holds when the composition triangle is equilateral triangle.

## 2. Lever rule

From the previous derivation, we can also infer that

$$\omega_\alpha = \frac{Ab_1 - Aa_1}{Ab_1 - Aa_1} = \frac{o_1 b_1}{a_1 b_1} = \frac{ob}{ab}$$

This is the lever rule of ternary systems.

The following conclusion is made by the linear law and lever rule. When a given material is in two-phase equilibrium state at a certain temperature, if one-phase composition is given, another phase composition point will be in the extension line of two known composition points; if the composition point of two-phase equilibrium is known, the component point of the material must be in the line of these two component points.

## 3. Barycenter rule

When one phase decomposes into three new phases or two new phases, the study of the relationship between the composition and the relative amount requires the barycenter rule.

According to the phase rule, when the ternary systems is in three-phase equilibrium, the degree of freedom is one. The composition of these three equilibrium phases at a given temperature should be determined. Alloy composition points should be located in the triangle made up of three equilibrium phase points. In the Fig. 8.11, the point $O$ is the alloy composition point and points $P$, $Q$, $S$ are the composition points of three equilibrium phases $\alpha$, $\beta$, $\gamma$, respectively. To calculate the relative amount of each phase in the alloy, we at first mix any two phases out of the three phases, such as phases $\alpha$ and $\gamma$. Then we mix the mixture of $\alpha$ and $\gamma$ and

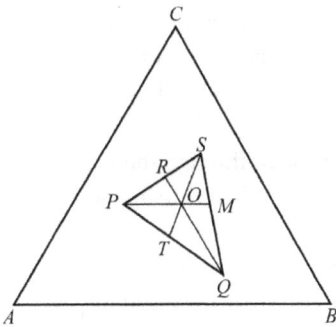

**Fig. 8.11:** The barycenter rule.

phase $\beta$ into alloy $O$. According to the linear law, the composition point of $\alpha - y$ mixture phase lies in the line $PS$; at the same time, the composition point must be in the extension line of line $QO$. So the intersection point of extension line $QO$ and line $PS$ is the composition point of the mixed phase $\alpha + y$. Then we can know the mass (mole) fraction of phase $\beta$:

$$\omega_\beta = \frac{OR}{QR}$$

By the same way, we can calculate the mass fraction of phases $\alpha$ and $y$:

$$\omega_\alpha = \frac{OM}{PM}$$

$$\omega_y = \frac{OT}{ST}$$

The results show that the point $O$ locates just in the mass center of the composition triangle $PQS$, that is the barycenter rule in ternary systems.

In addition to the geometric method, the relative amount of the three equilibrium phases can also be calculated directly by the algebraic method.

### 8.1.5 Analysis of equilibrium crystallization process of isomorphous phase diagrams

As shown in Fig. 8.12(a), the crystallization process is analyzed by taking an alloy with a composition at $O$ as an example. When temperature is over liquidus surface, the alloy is in liquid state. When temperature drops slowly to intersect with the liquidus surface temperature of $t_1$, the alloy begins to crystallize and the $\alpha$ solid solution with a composition of $S_1$ is crystallized from the liquid phase, in the meantime the liquid composition is almost equal to that of the alloy when a small supercoiling is needed. When temperature drops slowly to $t_2$, the amount of $\alpha$ phase gradually increases, accompanying with the decrease in the amount of liquid phase continuously. Meanwhile, the composition of solid phase changes from $S_1$ to $S_2$ along the solid surface, and the composition of the liquid varies from $L_1$ to $L_2$ along the liquid surface. During the cooling process, the liquid and solid phases are in equilibrium state at various temperatures. According to the linear law, the line connecting the component points of two equilibrium phases must go through the component point of original alloy. The tie-line is $L_1S_1$ at $t_1$ temperature, $L_2S_2$ at $t_2$ temperature, and $L_3S_3$ at $t_3$ temperature, respectively. When temperature drops down to intersect the solidus surface at $t_4$, the crystallization ends and the alloy just transforms into a single solid solution, at the moment the tie-line is $L_4S_4$ in which $L_4$ corresponds to the composition of the last drop of the liquid. At this time, the composition of the solid phase is

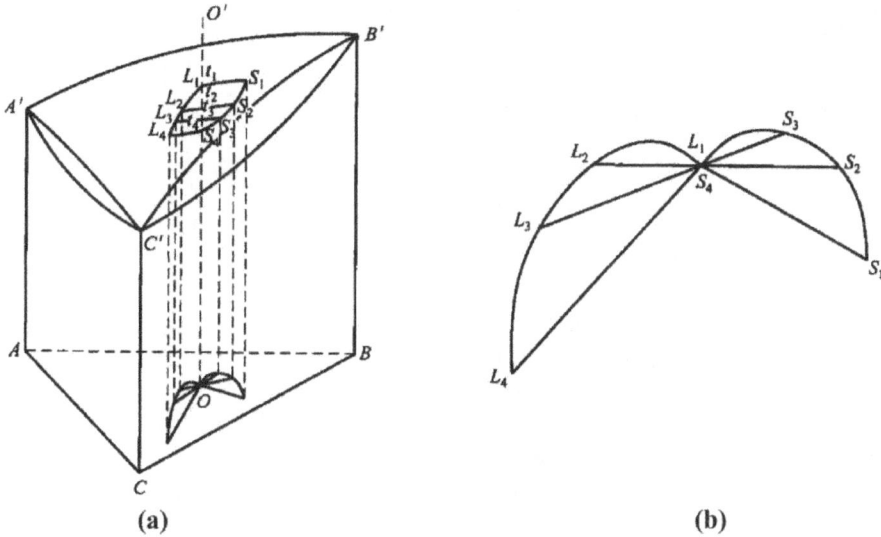

**Fig. 8.12:** Crystallization process of ternary isomorphous alloy: (a) composition changes of liquid and solid phases and (b) projection of tie-lines of two phases in the composition triangle.

equal to that of the original alloy. It is conceivable that during the process of crystalli-zation, the tie-lines rotate around the vertical axis$(OO')$ of original alloy composi-tion, that is, the liquid-phase composition points move along the spatial curve of $L_1L_2L_3L_4$ on liquidus surface, and the solid composition points move along the spatial curve of $S_1S_2S_3S_4$ on solid surface. The two spatial curves of $S_1S_2S_3S_4$ and $L_1L_2L_3L_4$ are neither on the same vertical plane nor on the same horizontal plane. But $L_1S_1$, $L_2S_2$, $L_3S_3$, and $L_4S_4$ as tie-lines are horizontal lines at different temper-atures, respectively, and they can be projected in the shape of a butterfly in the composition triangle, as shown in Fig. 8.12(b).

## 8.2 Ternary eutectic phase diagrams with insoluble components

### 1. Space pattern of phase diagrams

Figure 8.13 shows a simple eutectic ternary system, in which component $A$, $B$, and $C$ are insoluble with each other in the solid state and soluble completely in the liq-uid state. It is composed of three simple binary eutectic phase diagrams $A–B$, $B–C$, and $C–A$.

Points $a$, $b$, $c$ exhibit the three different melting points of the component $A$, $B$, $C$, respectively. Therefore, in eutectic ternary phase diagram, three downward con-verged liquidus surfaces are formed. Among them, surface $ae_1Ee_3a$ is the initial crystallization surface of component $A$. Surface $be_1Ee_2b$ is the initial crystallization

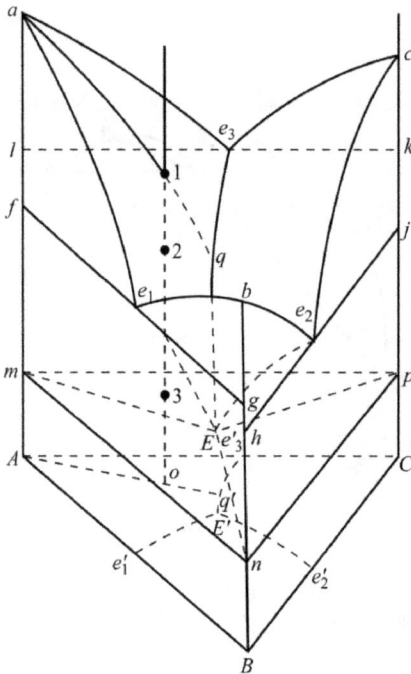

**Fig. 8.13:** The ternary eutectic diagram with no solid solubility.

surface of component B. Surface $ce_2Ee_3c$ is the initial crystallization surface of component C.

   Eutectic transition points $e_1$, $e_2$, $e_3$ of three binary eutectic systems stretch into
the eutectic transformation lines in ternary phase diagrams, that is, $e_1E$, $e_2E$, $e_3E$,
such "valleys" are called *univariant lines*, in which $P = 3$, so $f = 1$ (temperature).
When the liquid phase changes along the three univariant lines, the eutectic transition occurs:

$$e_1E \qquad L \to A + B$$
$$e_2E \qquad L \to B + C$$
$$e_3E \qquad L \to C + A$$

The three univariant lines meet at *ternary eutectic point E* where $f = 0$ and $P = 4$.
This point is the final solidification temperature in the alloy system, and is an invariant point since the compositions of all four phases in equilibrium and the temperature are fixed until all liquid is solidified. The liquid phase with composition $E$
takes place the eutectic transition at this temperature:

$$L_E \to A + B + C$$

The four-phase equilibrium surface is composed of point $E$ and the composition points $m$, $n$, $p$ of three solid phases at this temperature. Among them, the triangle $mEn$ is the bottom surface of three-phase equilibrium region where the eutectic reaction $L \rightarrow A + B$ occurs. The triangle $nEp$ is the bottom surface of three-phase equilibrium region where the eutectic reaction $L \rightarrow B + C$ occurs. The triangle $pEm$ is the bottom surface of three-phase equilibrium region where the eutectic reaction $L \rightarrow C + A$ occurs. The three-phase equilibrium region and the eutectic transformation initial surface are shown in Fig. 8.14. Below the temperature $E$, the alloy solidifies into a mixture composed of $A + B + C$ three pure component phases.

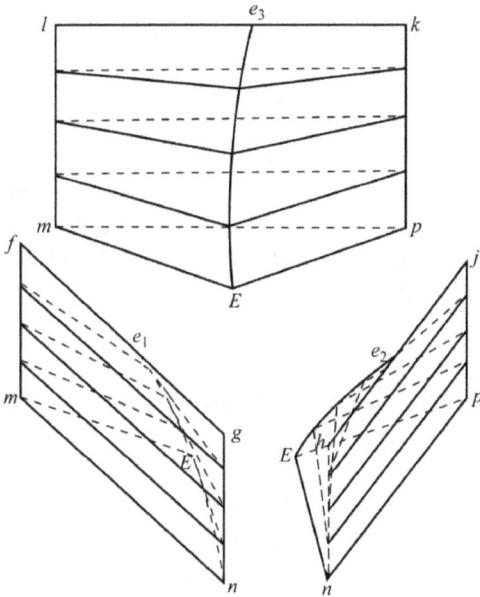

**Fig. 8.14:** Three-phase equilibrium region and two-phase eutectic surface.

**2. Section drawing**

The vertical sections $rs$ and $At$ in Fig. 8.15(a) are shown in Fig. 8.15(b) and (c), respectively. The component axis of the section $rs$ is parallel to the side $AC$ of the composition triangle. The liquidus lines $r'e'$ and $e's'$ in Fig. 8.15(b) are the intersection lines of the section $rs$ and liquidus surfaces $ae_1Ee_3a$ and $ce_2Ee_3c$ in Fig. 8.13. The curve $r_1d'$ in Fig. 8.15(b) is the intersection of the section $rs$ and transition surface $fe_1Emf$ in Fig. 8.13. Lines $d'e'$, $e'i'$, and $i's_1$ in Fig. 8.15(b) are the intersection of the section $rs$ and transition surfaces $le_3Eml$, $ke_3Epk$, and $je_2Epj$ in Fig. 8.13, respectively. The horizontal line $r_2s_2$ is the intersection of the section $rs$ and the eutectic plane of four-phase equilibrium.

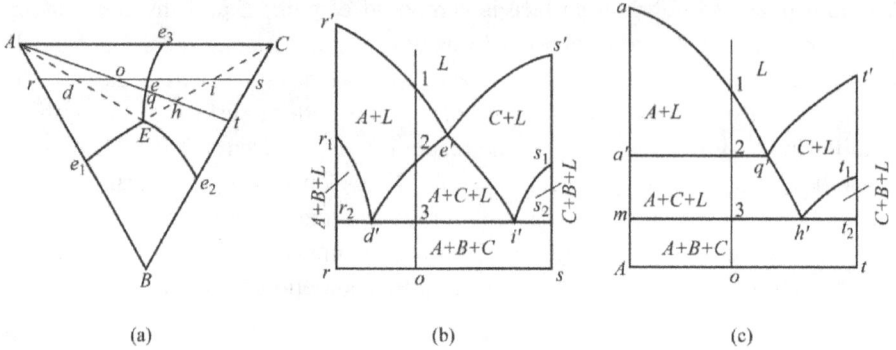

(a)  (b)  (c)

**Fig. 8.15:** The vertical section: (a) composition triangle, (b) section *rs*, and (c) section *At*.

We can use the vertical section to analyze the equilibrium solidification course of all alloy composition points on the section *rs* and determine the critical phase transition temperature. Consider the alloy *o* as an example. When it cools to point 1, the initial phase *A* begins to solidify. When it cools to point 2, it starts to enter the $A + C + L$ three-phase equilibrium region. The eutectic reaction $L \to C + A$ occurs and a binary eutectic alloy $A + C$ is formed. Point 3 is on the eutectic plane *mnp* (see Fig. 8.13). When it cools to point 3, the four-phase equilibrium eutectic reaction $L \to A + B + C$ occurs and a ternary eutectic alloy $A + B + C$ is formed. The microstructure of alloy does not change when it cools to room temperature, and its microstructure consists of initial phase *A* + two-phase eutectic alloy $(A + C)$ + three-phase eutectic alloy $(A + B + C)$ at room temperature.

The component axis of vertical section *At* passes the vertex *A* of composition triangle in Fig. 8.15(a). The intersection of this section *At* and transition surface *le₃Eml* in Fig. 8.13 is the tie-line of the solid equilibrium phase and liquid equilibrium phase, which is the horizontal line *a'q'* in the vertical section.

Figure 8.16 shows the horizontal sections of eutectic ternary phase diagrams at different temperatures. We can use these sections to know the equilibrium phase state of the alloy at different temperatures and analyze the solidification process of alloys with various composition during equilibrium cooling.

For example, alloy *R* in Fig. 8.16 is liquid when temperature *T* is above $T_b$. With the decrease in temperature during solidification, primary (initial) phase *B* forms in the liquid, then binary eutectic mixture of $B + C$ phases forms in the liquid, finally a ternary eutectic mixture of $B + C + A$ phases forms when temperature is lower than $T_E$. The microstructure of alloy *R* consists of primary *B* phase $+(B + C)$ binary eutectic mixture $+(B + C + A)$ ternary eutectic mixture.

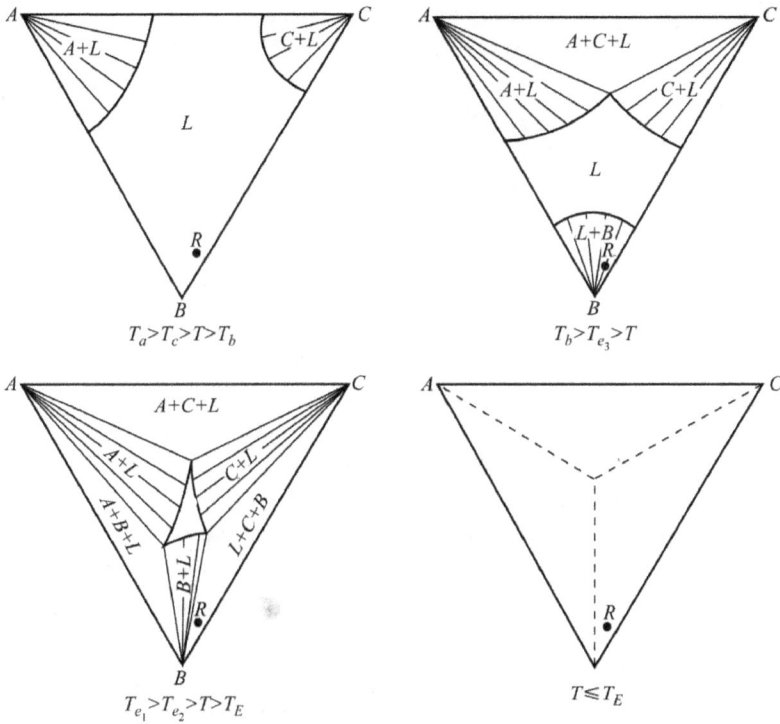

**Fig. 8.16:** The horizontal sections.

## 3. Projection drawing

In Fig. 8.17, the heavy lines $e_1E$, $e_2E$, and $e_3E$ are the projections of three *univariant lines*, in which point $E$ is the projection of the three-phase eutectic points. Three regions divided by these heavy lines are the projection of three liquidus surfaces. The fine lines marked letters $t_1$, $t_2$, and so on, are isotherms of liquidus surface.

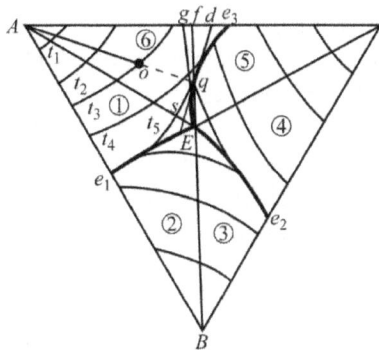

**Fig. 8.17:** The projection of the ternary eutectic phase diagrams with no solid solubility.

The solidification process of the alloy can be analyzed by using the projection drawing, which can not only determine the critical temperature of phase transition, but also the composition and relative amount of the phase. Still take the alloy $o$ as an example. When alloy cools to liquidus surface $ae_1Ee_3a$, the primary precipitation of $A$ occurs at temperature $t_3$ (corresponding to the point 1 in the space model in Fig. 8.13 and the vertical section in Fig. 8.15(c)). At the moment, the liquid composition is equal to that of the alloy, and the projection of tie-line in the two-phase equilibrium is line $Ao$. When alloy continues to cool down, the component $A$ precipitates constantly. The amount of component $A$ in liquid phase decreases and the amount of component $B$ and $C$ increases continuously. But the amount ratio of component $B$ and $C$ in the liquid phase keeps unchanged, so the composition of liquid phase should change along the extension direction of the tie-line $Ao$. At the temperature $t_5$ corresponding to the point 2 in the space model and the vertical section, the composition of liquid phase changes to the point $q$ on the univariant line $e_3E$ and the eutectic transition $L_q \rightarrow A + C$ occurs. Upon further cooling, the composition of liquid follows the univariant line $e_3E$ with coprecipitation of components $A$ and $C$, until the four-phase equilibrium eutectic reaction $L \rightarrow A + B + C$ occurs at point $E$ (corresponding to point 3 in the space model and the vertical section). When the temperature is slightly lower than $E$, after the finish of solidification no other changes will occur. So the equilibrium microstructure of the alloy at room temperature is the primary phase $A$ + two-phase eutectic mixture $(A + C)$ + three-phase eutectic mixture $(A + B + C)$.

The relative amount of constituent phases in the alloy can be calculated by using the lever rule. For example, when the binary eutectic reaction of alloy with composition $o$ just occurs, the mass fraction of the primary phase $A$ and the liquid phase $L$ with the liquid composition $o$ are, respectively:

$$\omega_A = \frac{oq}{Aq} \times 100\%$$

$$\omega_L = \frac{Ao}{Aq} \times 100\%$$

Since the two-phase eutectic transformation just begins when the composition of liquid is at point $q$, the amount of liquid is nearly 100% and the amount of eutectic mixture $(A + C)$ is close to zero, so the composition point of eutectic mixture $(A + C)$ should be the intersection point $d$ of the tangent at the point $q$ and the side $AC$. With the decrease in temperature, the compositions of both liquid-phase and two-phase eutectic mixtures $(A + C)$ will change. The composition of remaining liquid changes along the univariant line $e_3E$ and the composition of co-precipitated mixture $(A + C)$ can be determined by tangent line at the line $e_3E$ for every temperature. For instance, when the temperature of liquid phase decreases to the point $s$ along the univariant line $e_3E$, at the moment the composition of new two-phase eutectic mixture just formed is the intersection point $g$ of the tangent line at point $s$ and the

AC side. For liquid at the point $q$, when its composition reaches the point $E$ during cooling accompanying the maximum of $(A + C)$ co-precipitation before $(A + B + C)$ ternary eutectic reaction, the composition of the two-phase eutectic mixture $(A + C)$ is the point $f$, which is the intersection of line $Eq$ extension and side $AC$ according to linear rule and barycenter rule. Therefore, the mass fractions of the two-phase eutectic mixture $(A + C)$ and the three-phase eutectic mixture $(A + B + C)$ from the transition of liquid at the point $E$ in the alloy $o$ are respectively:

$$\omega_{(A+C)} = \frac{Eq}{Ef} \times \frac{Ao}{Aq} \times 100\%$$

$$\omega_{(A+B+C)} = \frac{qf}{Ef} \times \frac{Ao}{Aq} \times 100\%$$

The same method can be used to analyze the equilibrium cooling process and the microstructure of all alloy at room temperature. The microstructure of alloys at room temperature in various regions of the projection drawing (Fig. 8.17) is shown in Table 8.1.

**Table 8.1:** Microstructure of typical alloy in ternary alloy system with eutectic transformation and no solid solubility at room temperature.

| Region | Microstructure at room temperature |
|---|---|
| 1 | Initial crystal $A$ + two-phase eutectic crystal $(A + B)$ + three-phase eutectic crystal $(A + B + C)$ |
| 2 | Initial crystal $B$ + two-phase eutectic crystal $(A + B)$ + three-phase eutectic crystal $(A + B + C)$ |
| 3 | Initial crystal $B$ + two-phase eutectic crystal $(B + C)$ + three-phase eutectic crystal $(A + B + C)$ |
| 4 | Initial crystal $C$ + two-phase eutectic crystal $(B + C)$ + three-phase eutectic crystal $(A + B + C)$ |
| 5 | Initial crystal $C$ + two-phase eutectic crystal $(A + C)$ + three-phase eutectic crystal $(A + B + C)$ |
| 6 | Initial crystal $A$ + two-phase eutectic crystal $(A + C)$ + three-phase eutectic crystal $(A + B + C)$ |
| Line $AE$ | Initial crystal $A$ + three-phase eutectic crystal $(A + B + C)$ |
| Line $BE$ | Initial crystal $B$ + three-phase eutectic crystal $(A + B + C)$ |
| Line $CE$ | Initial crystal $C$ + three-phase eutectic crystal $(A + B + C)$ |
| Line $e_1E$ | Two-phase eutectic crystal $(A + B)$ + three-phase eutectic crystal $(A + B + C)$ |
| Line $e_2E$ | Two-phase eutectic crystal $(B + C)$ + three-phase eutectic crystal $(A + B + C)$ |
| Line $e_3E$ | Two-phase eutectic crystal $(A + C)$ + three-phase eutectic crystal $(A + B + C)$ |
| Point $E$ | Three-phase eutectic crystal $(A + B + C)$ |

**4. Phase region contacting rule**

The ternary phase diagram also follows the same phase region contacting rule of the binary phase diagram, that is, the phase number difference between adjacent phase regions is 1 (except for the point contact), and this rule is suitable for the spatial phase diagram, the horizontal section or the vertical section. Therefore, any single phase region is always adjacent to the two-phase region; the two-phase region is adjacent to the single-phase region or the three-phase region; the four phase region must be adjacent to the three-phase region, which can be found out in Figs. 8.13, 8.15, and 8.16 clearly. It is worthy to point out that when the phase region contacting rule is applied in the three-dimensional graphics, only the surface contacting with the phase region can be used to judge rather than the line or point contacting with the phase region. The section drawing can be judged by the line contacting with the phase region rather than the point contacting with the phase region. Furthermore, according to the phase contacting rule, except the composition point on the section with four phases (i.e., zero variable point), the intersection nodes of phase boundary lines on the sections must connect with four-phase boundary lines, which is one of the geometry rules to judge whether the section is correct or not.

## 8.3 Ternary eutectic phase diagrams with limited solid solubility

### 1. Analysis of phase diagrams

Figure 8.18 shows the ternary eutectic phase diagram of components with limited solid solubility. The only difference with simple ternary eutectic phase diagram with insoluble components shown in Fig. 8.13 is the additional surfaces of solid solubility and monophasic solid solution regions $\alpha$, $\beta$, and $y$ around pure component $A$, $B$, and $C$, respectively.

There is always a solidus surface that conjugates with a liquidus surface between liquid and solid two-phase equilibrium region, that is:

solidus surface *afmla* and liquidus surface $ae_1Ee_3a$ are conjugate;
solidus surface *bgnhb* and liquidus surface $be_1Ee_2b$ are conjugate;
solidus surface *cipkc* and liquidus surface $ce_2Ee_3c$ are conjugate.

Similar to simple ternary eutectic phase diagrams, a three-phase equilibrium region is adjacent to two two-phase region of liquid and solid, therefore, three three-phase equilibrium regions have six transitional surfaces with two-phase region. Also they converge to the three-phase eutectic surface *mnp* at $t_E$, indicating that the four-phase equilibrium eutectic transition of liquid phase with

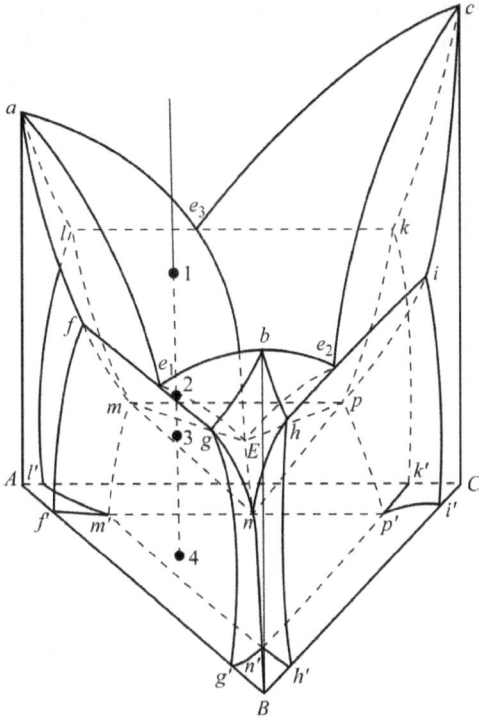

**Fig. 8.18:** Ternary eutectic phase diagrams with limited solid solubility.

composition $E$ takes places at this temperature. Two-phase eutectic transition and the three-phase eutectic transition are described in detail as follows:

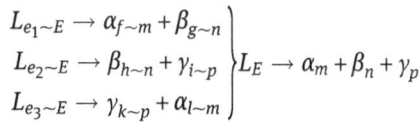

$$\left.\begin{array}{l} L_{e_1 \sim E} \rightarrow \alpha_{f \sim m} + \beta_{g \sim n} \\ L_{e_2 \sim E} \rightarrow \beta_{h \sim n} + \gamma_{i \sim p} \\ L_{e_3 \sim E} \rightarrow \gamma_{k \sim p} + \alpha_{l \sim m} \end{array}\right\} L_E \rightarrow \alpha_m + \beta_n + \gamma_p$$

The triangular prism beneath the four-phase equilibrium surface $mnp$ is three-phase equilibrium region of solid phase $\alpha$, $\beta$, and $\gamma$, whose tie triangle is $m'n'p'$ at room temperature. Two-phase regions between every two monophasic solid solution regions are surrounded by a pair of conjugate solubility surface: surface $fmm'f'f$ and $gnn'g'g$ for $\alpha + \beta$ region, surface $hnn'h'h$ and $ipp'i'i$ for $\beta + \gamma$ region, surface $kpp'k'k$ and $lmm'l'l$ for $\gamma + \alpha$ region.

Therefore, there are six kinds of phase interfaces in ternary eutectic phase diagrams with limited solid solubility: (1) three liquidus surfaces; (2) six initial surfaces of two-phase eutectic transition; (3) three single-phase solidus surfaces; (4) three terminal (finish) surfaces of two-phase eutectic transition; (5) one four-phase equilibrium eutectic surface; and (6) three pairs of conjugate solubility surfaces. The phase diagram is divided into six regions: a liquidus region, three monophasic solid solution regions, three two-phase equilibrium regions of liquid and solid, three two solid-phase equilibrium regions, three three-phase equilibrium regions where two-phase eutectic transition occurs, and a three solid-phase equilibrium region. Figure 8.19 shows the shape of three-phase equilibrium regions and two solid-phase equilibrium regions to help understanding.

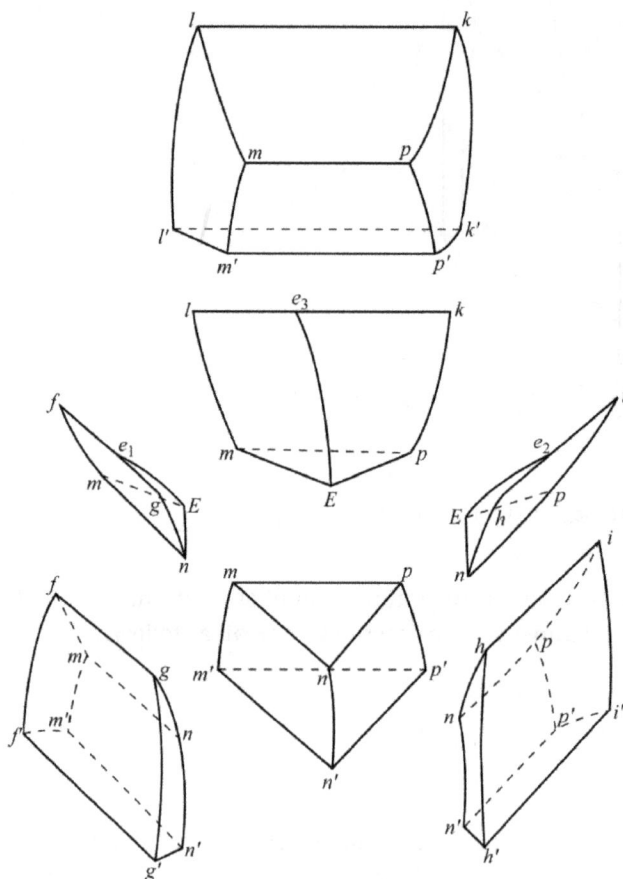

**Fig. 8.19:** Three-phase regions and two solid-phase regions in ternary eutectic phase diagrams.

## 2. Projection drawing
Figure 8.20 is a projection drawing of ternary eutectic phase diagrams. It is obvious that projections of three eutectic univariant lines $e_1E$, $e_2E$, and $e_3E$ divide the composition triangle into three regions: $Ae_1Ee_3A$, $Be_1Ee_2B$, and $Ce_2Ee_3C$. They are projections of three liquidus surfaces, and when temperature is lower than these liquidus surfaces, primary phase $\alpha$, $\beta$, and $y$ forms in corresponding region, respectively.

Three projections of solidus surfaces in two-phase equilibrium regions of liquid and solid conjugating with liquidus surfaces are $AfmlA$, $BgnhB$, and $CipkC$. Irregular regions being close to pure components $A$, $B$, $C$ on the outside of the solidus surfaces are single-phase regions of $\alpha$, $\beta$, and $\gamma$, respectively.

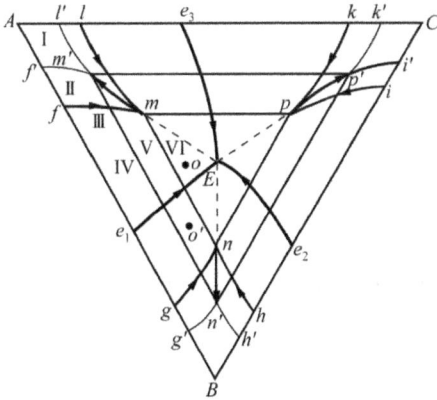

**Fig. 8.20:** Projection drawing of ternary eutectic phase diagrams.

In the projection drawing in Fig. 8.20, heavy lines are univariant lines, and the direction of arrow indicates the drop of temperature. We can see three univariant lines of three three-phase equilibrium eutectic regions: $e_1E(L)$, $fm(\alpha)$, and $gn(\beta)$ of $L + \alpha + \beta$ region, $e_2E(L)$, $hn(\beta)$, and $ip(\gamma)$ of $L + \beta + \gamma$ region, $e_3E(L)$, $kp(\gamma)$, and $lm(\alpha)$ of $L + \gamma + \alpha$ region. These regions start at eutectic transition lines $fg$, $hi$ and $kl$ of binary subsystems, and end at tie triangle $mEn$, $nEp$, and $pEm$ on four-phase equilibrium surface.

The triangle in the middle of projection drawing is four-phase equilibrium eutectic surface. After ternary eutectic transition of liquid with composition $E$, three-phase equilibrium region of $\alpha + \beta + \gamma$ forms. Obviously, the point $E$ is an invariant point with four-phase equilibrium eutectic transition and is an intersection of three univariant lines. This is a feature of ternary eutectic transformation projection drawings.

Figure 8.21 shows the three-phase composition triangle composed of tie-lines in ternary eutectic system before and after four-phase equilibrium. We can find that there are three two-phase eutectic transitions: $L \rightarrow \alpha + \beta$, $L \rightarrow \beta + \gamma$, and $L \rightarrow \gamma + \alpha$ before ternary eutectic transition. After ternary eutectic transition, there is a three-phase equilibrium of $\alpha + \beta + \gamma$. According to the phase rule, the freedom degree is zero during the solidification of four-phase equilibrium, which means the temperature and the composition of equilibrium phase are fixed, so the four-phase equilibrium surface is a plane triangle during solidification. The composition of alloy must be located within this triangle, whose three vertexes are composition points of formed solid phases, respectively.

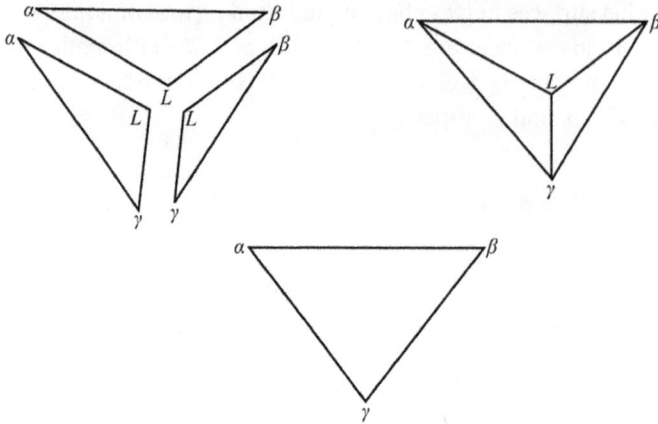

**Fig. 8.21:** Three-phase composition triangle of ternary eutectic system before and after four-phase equilibrium.

## 3. Section drawing

Figure 8.22 shows the horizontal sections of ternary eutectic phase diagram at different temperatures. They have several common features:

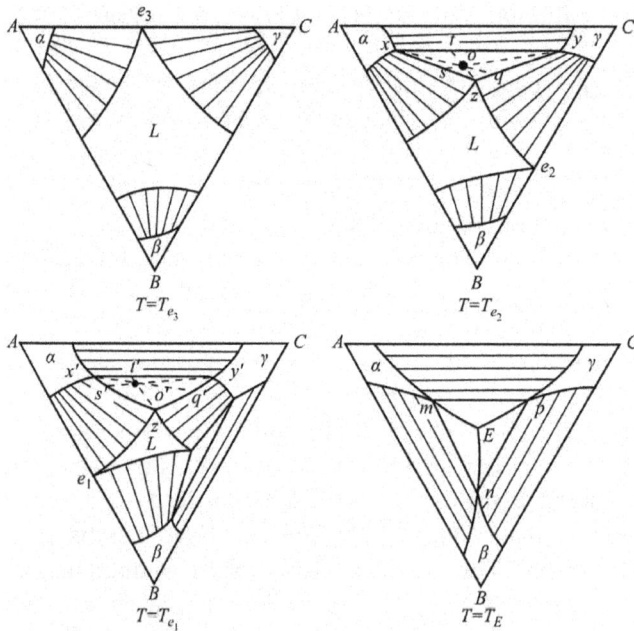

**Fig. 8.22:** The horizontal sections of ternary eutectic phase diagram.

(1) A three-phase region is known as a conjugate triangle, such as triangle composed of xyz. For example, alloy $O$ in the conjugate triangle will generate two-phase eutectic transition $(L \to \alpha + \gamma)$ at this temperature. Its three vertexes represent the compositions of three equilibrium phases ($\alpha$, $\gamma$, and $L$) at this temperature, respectively, and three vertexes are connected with three single-phase regions ($\alpha$, $\gamma$, and $L$).

(2) The boundary between three-phase region and two-phase region is the side of the triangle, such as xz. And it is also the conjugate line of adjacent two-phase regions composed of $\alpha$ and $L$ phases.

(3) Two-phase regions generally have two straight lines and two curves as boundaries. These straight lines are conjugated lines and are connected to three-phase regions or single-phase region, and a pair of conjugate curves separate the two single-phase region.

Figure 8.23 shows two typical vertical sections of ternary eutectic phase diagram. Figure 8.23(a) shows the corresponding positions of vertical sections in composition

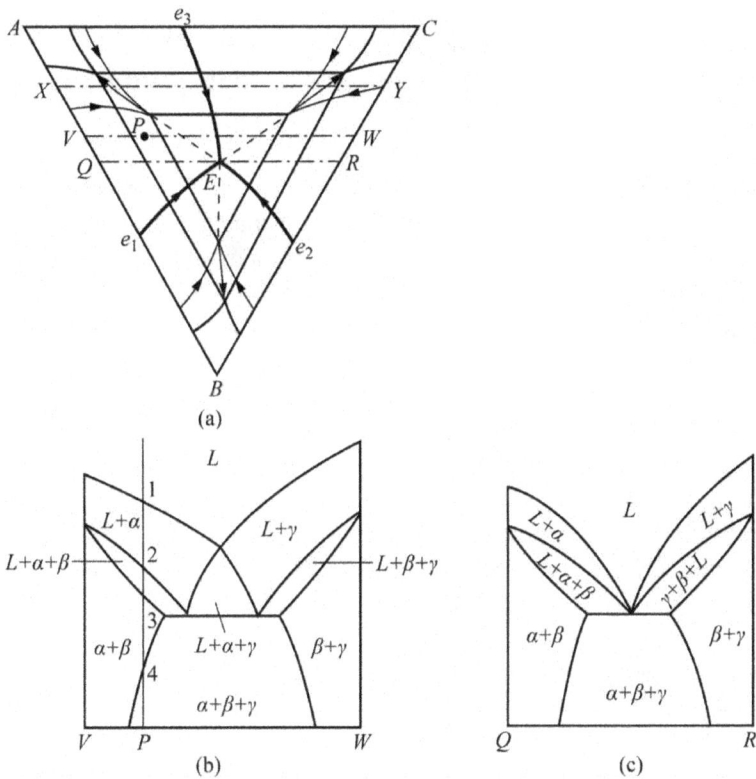

**Fig. 8.23:** Vertical sections of ternary eutectic phase diagram (a) projection drawing, (b) VW section, and (c) QR section.

triangle, and Fig. 8.23(b) is the *VW* vertical section. When cutting through the four-phase equilibrium eutectic surface, there will be a horizontal line connecting with curved triangle with upward vertex in vertical section. And the section will show common features of four-phase equilibrium eutectic region (horizontal line) and three-phase equilibrium region (curved triangle with upward vertex). According to 8.23(b), the solidification process of alloy *P* can be analyzed. When cooling to point 1, the primary crystal $\alpha$ begin to form. When cooling to point 2, it enters the three-phase equilibrium region and then the eutectic reaction $L\rightarrow\alpha+\beta$ occurs. Finally, the solidification ends at point 3. Because the $\alpha+\beta$ two-phase region between points 3 and 4, phase transition will not occur. When temperature is lower than point 4, phase $y$ will precipitate from $\alpha$ and $\beta$ because of the decrease of their solubility. The microstructure of alloy *P* at room temperature is $\alpha+(\alpha+\beta)+y$(a little). So it is obvious that the vertical section is more convenient than projection drawing if we only need to determine the critical phase transition temperature.

Figure 8.23(c) is the *QR* vertical section through point *E*, in which there is no three-phase equilibrium triangle of $L+\alpha+y$ in Fig. 8.23(b). But four-phase equilibrium eutectic transformation can be easily observed here.

## 8.4 Ternary phase diagrams consisting of two binary eutectic systems and one binary isomorphous system

Figure 8.24(a) shows the space pattern of this ternary phase diagrams. We can find that both *A–B* and *B–C* are binary eutectic subsystems of components with limited solubility in the solid state, while *A–C* is a binary isomorphous subsystem. $\alpha$ is ternary solid solution with solvent *A* or *C*, $\beta$ is solid solution with solvent *B*; $T_Aee_1T_CT_A$ and $T_Bee_1T_B$ are liquidus surfaces of $\alpha$ and $\beta$, respectively, and the curve $ee_1$ is the "valley" (univariant) line of these surfaces. When the composition point of liquid phase passes any point at curve $ee_1$, two-phase equilibrium eutectic transformation $L\rightarrow\alpha+\beta$ occurs at the corresponding temperature. Figure 8.24(b) shows the phase regions: $AT_Aacc_1a_1T_CCA$ is the single-phase region of $\alpha$; $BT_Bb_1d_1dbT_BB$ is the single-phase region of $\beta$; triangular prism $aa_1bb_1ee_1$ is the three-phase equilibrium region of $L+\alpha+\beta$; and the rest are two-phase regions $L+\alpha$, $L+\beta$, and $\alpha+\beta$. Figure 8.24(c) shows the projection drawing of freezing process of this ternary system on composition triangle, in which $bb_1$ is the trajectory of solubility when $\beta$ cools from eutectic temperature $T_e$ to $T_{e1}$; $aa_1$ is the trajectory of solubility when $\alpha$ cools from $T_e$ to $T_{e1}$; $dd_1$ is the projection of solubility curve when $\beta$ cools from eutectic temperature of *A–B* binary subsystem to that of *B–C* binary subsystem, $cc_1$ is projection of solubility curve when $\alpha$ cools from $e$ in *A–B* binary subsystem to $e_1$ in *B–C* binary subsystem. In the projection drawings, the movement direction of $L\rightarrow\alpha+\beta$ composition triangle with the decrease of temperature is also demonstrated. Figure 8.24(d)

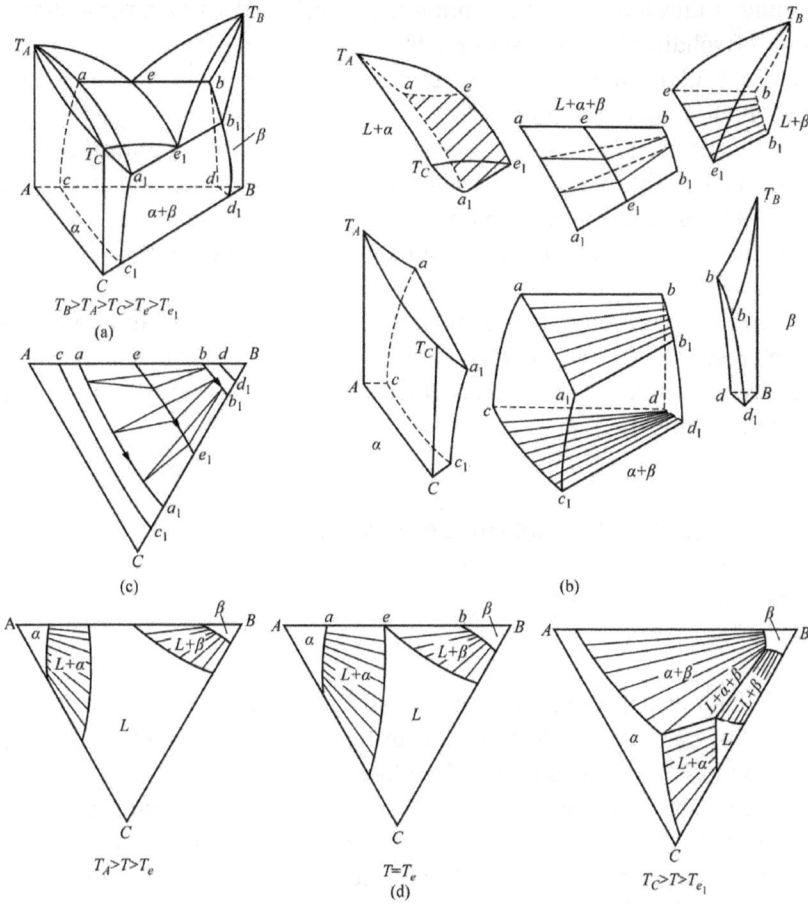

**Fig. 8.24:** Ternary phase diagrams consist of two binary eutectic systems and an binary isomorphous system. (a) Space pattern, (b) phase regions, (c) projection drawings, and (d) isothermal section.

is three horizontal section at different temperatures. It is convenient to analyze the equilibrium crystallization process of alloys with different compositions in this ternary system based on Fig. 8.24. Also, conclusions can be drawn as follows.

(1) Only alloys with composition points between the projection curves of univariant lines of $\alpha$ and $\beta$ will go through the three-phase equilibrium prism in space pattern, then two-phase eutectic transformation $L \rightarrow \alpha + \beta$ occurs.

(2) Among the alloys mentioned above, those with composition points between the projection curves of univariant line of $\alpha$ and liquidus univariant line have primary phase $\alpha$, its microstructure at the end of solidification is $\alpha + (\alpha + \beta)$; those with composition points between the projection curves of univariant line of $\beta$

and liquidus univariant line have primary phase $\beta$, and its microstructure at the end of solidification is $\beta + (\alpha + \beta)$; those with composition points on the liquidus univariant line do not have primary phase, its microstructure at the end of solidification is eutectic $(\alpha + \beta)$.

(3) Alloys with composition points between the projection curves of univariant line of $\alpha$ and curve $cc_1$ have equilibrium microstructure $\alpha + \beta_{II}$ at room temperature. Similarly, alloys with composition points between the projection curves of univariate line of $\beta$ and curve $dd_1$ have equilibrium microstructure $\beta + \alpha_{II}$ at room temperature.

(4) Alloys with composition points between curve $cc_1$ and side $AC$ have equilibrium microstructure $\alpha$ at room temperature, and alloys with composition point between curve $dd_1$ and vertex $B$ have equilibrium microstructure $\beta$ at room temperature.

## 8.5 Ternary quasi-peritectic phase diagrams

The quasi-peritectic reaction is

$$L + \alpha \leftrightarrow \beta + \gamma$$

Judging from the number of reactants, it has the features of peritectic transformation; meanwhile, it also has the features of eutectic transformation according to the number of products. So it was called quasi-peritectic transformation.

Figure 8.25(a) shows the space pattern of ternary system with four-phase quasi-peritectic transformation. There occurs peritectic transformation in $A$–$B$ and $A$–$C$ subsystems and eutectic transformation in $B$–$C$ subsystem. In Fig. 8.25(a) we know that $T_A > T_{p1} > T_{p1} > T_B > T_p > T_C > T_e$ ($T_p$ is the four-phase equilibrium temperature, and quadrilateral $abpc$ is the quasi-peritectic transformation plane.

It is obvious that two three-phase equilibrium prisms above plane $abpc$ belong to the peritectic transformation $L + \alpha \rightarrow \beta$ and $L + \alpha \rightarrow \gamma$; after four-phase equilibrium quasi-peritectic transformation, there will be a three-phase equilibrium eutectic transformation $L \rightarrow \beta + \gamma$ and a three-phase equilibrium region $\alpha + \beta + \gamma$. To illustrate this, Fig. 8.25(b) shows isothermal sections before and after quasi-peritectic transformation, and Fig. 8.25(c) shows that quasi-peritectic transformation plane is quadrilateral, and tie-lines of the composition points of reactants and products are its diagonal lines.

The projection drawing in Fig. 8.26(a) shows the cooling process of this ternary system. And Fig. 8.26(b) shows the $a_2$-2 vertical section, from which the microstructure change during cooling can easily be analyzed.

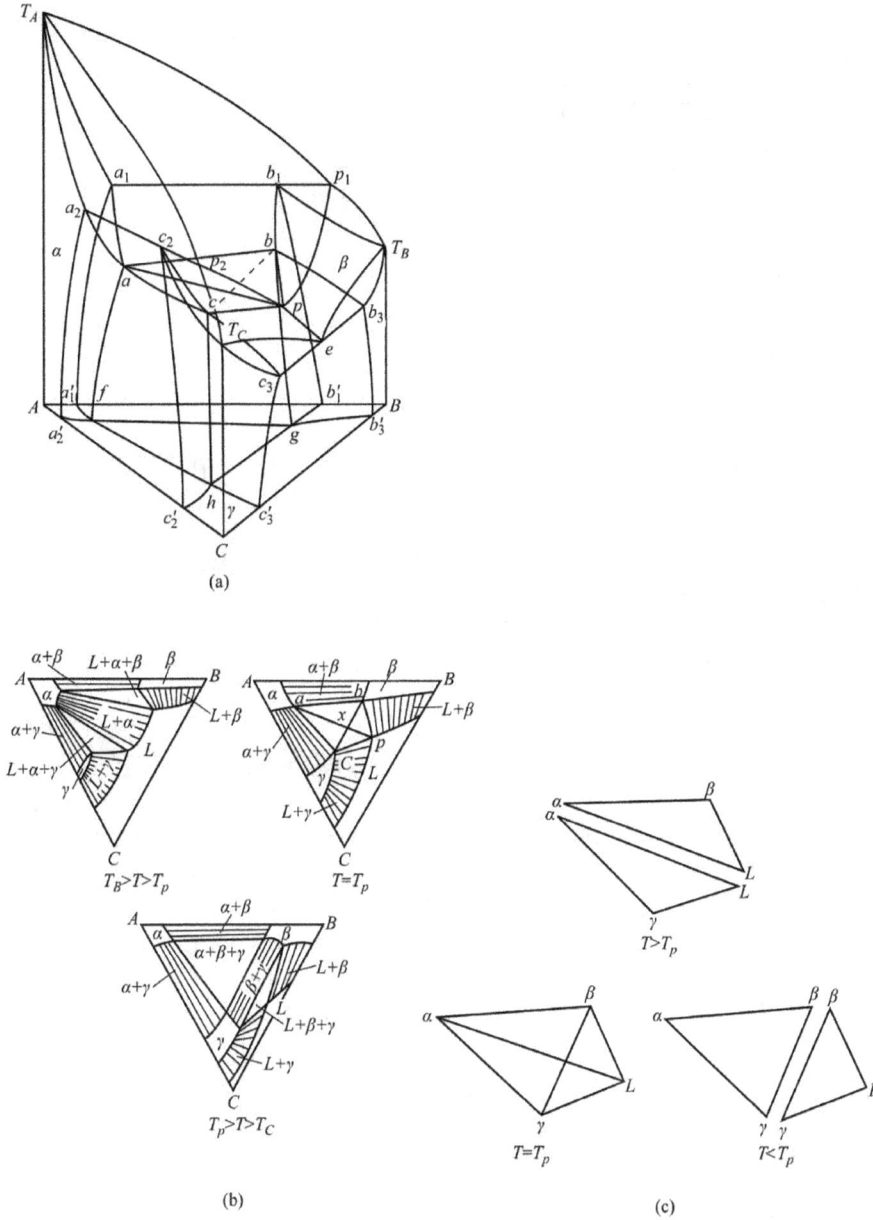

**Fig. 8.25:** Ternary system with four-phase quasi-peritectic transformation. (a) space pattern (b) isothermal section (c) three-phase composition triangle before and after four-phase quasi-peritectic equilibrium.

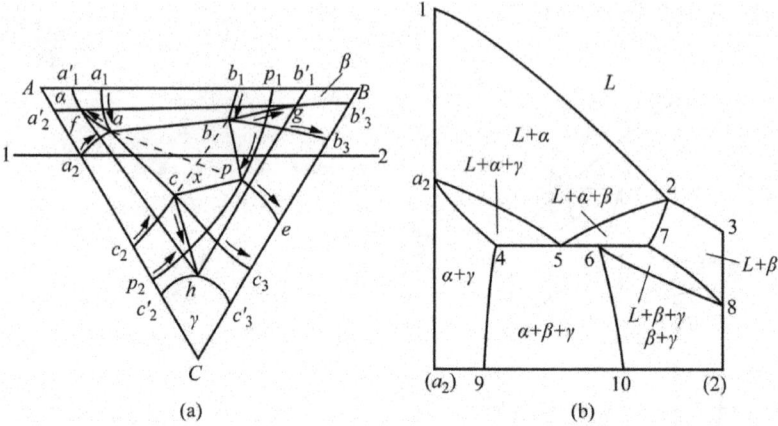

**Fig. 8.26:** Projection drawing of cooling process (a) and $a_2$-2 vertical section (b).

## 8.6 Ternary phase diagrams with four-phase equilibrium peritectic transformation

Four-phase equilibrium peritectic reaction is

$$L + \alpha + \beta \rightarrow \gamma$$

It shows that there is $L + \alpha + \beta$ three-phase equilibrium before the reaction occurs. Also three reactants cannot completely disappear or all be kept at the same time. Usually, only one reactant will disappear, the other two will have residual to form a new three-phase equilibrium with $\gamma$.

Figure 8.27(a) shows the space pattern of ternary phase diagram with ternary peritectic four-phase equilibrium transformation. There is eutectic transformation in $A–B$ subsystem and peritectic transformation in $A–C$ and $B–C$ subsystems. We can also find that $T_A > T_B > T_{e_1} > T_p > T_{p_2} > T_{p_3} > T_C$, where $T_p$ is the four-phase equilibrium temperature and the following peritectic transformation occurs:

$$L + \alpha + \beta \rightleftharpoons \gamma$$

The four-phase equilibrium region in space pattern is a plane triangle $abp$, it was called the four-phase equilibrium peritectic transformation plane. There is a three-phase equilibrium prism (eutectic $L \rightarrow \alpha + \beta$) above the plane and three three-phase equilibrium prism: $\alpha + \beta + y$ three-phase region, a $L + \alpha \rightleftharpoons y$ peritectic transformation region and another $L + \beta \rightleftharpoons y$ peritectic transformation region beneath it (Fig. 8.27(b)). Figure 8.27(c) shows the projection drawing of the cooling process of this ternary system.

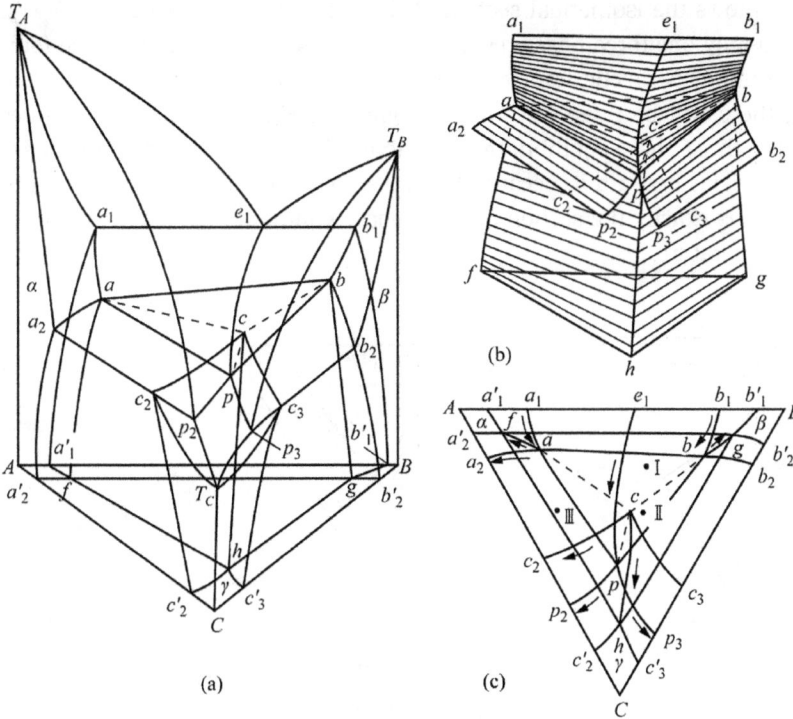

**Fig. 8.27:** Space pattern of ternary phase diagram with ternary peritectic four-phase equilibrium transformation.

Figure 8.28 shows the composition triangles of three phases before and after ternary peritectic four-phase equilibrium, we can also find that the composition point which forms $y$ in ternary peritectic transformation is in the tie-triangle formed by composition points of three reaction phases (reactants), respectively.

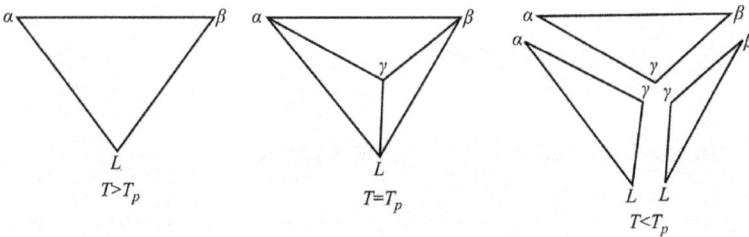

**Fig. 8.28:** The composition triangles of three phases before and after ternary peritectic four-phase equilibrium.

Figure 8.29 shows the isothermal section of this ternary system. When $T_{e_1} > T > T_p$, there will be only one three-phase region in Fig. 8.29(a); and it is exactly the four-phase equilibrium peritectic transformation plane in Fig. 8.29(b) if $T = T_p$; when $T_p > T > T_{p_2}$, there will be three three-phase regions in Fig. 8.29(c). This illustrates that there is a three-phase equilibrium prism above the plane and three three-phase equilibrium prisms beneath it, because the three-phase equilibrium region in horizontal section is exactly the section of three-phase equilibrium prism at the corresponding temperature.

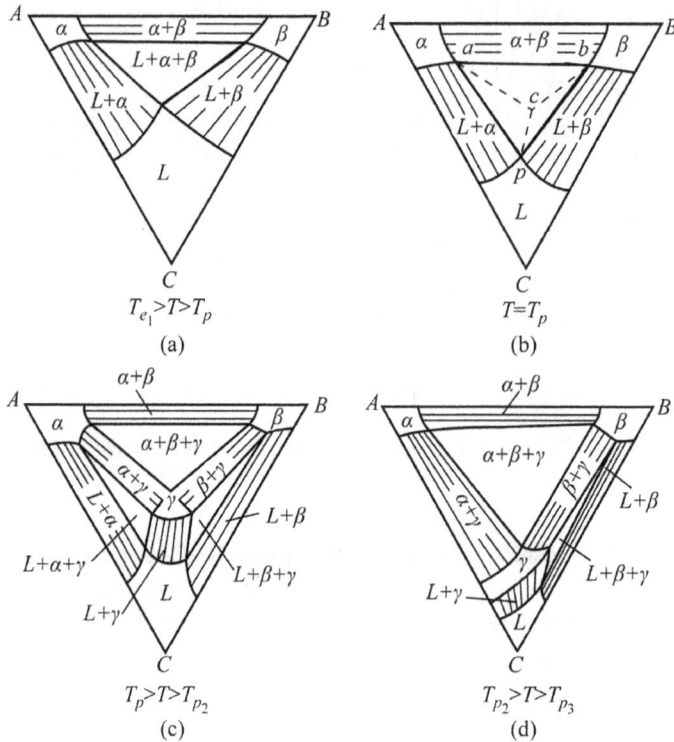

**Fig. 8.29:** Isothermal sections of a ternary peritectic system.

## 8.7 Ternary phase diagrams with stable compound

In the ternary system, if there is one or several binary subsystems which can form one or several stable binary compounds that would not decompose below the melting point or change its microstructure, or a stable ternary compound formed by the three components, we can regard these compounds as independent components during analyzing the phase diagrams. Compounds can form pseudobinary system by compound(s) or/and pure component(s) each other, and the phase diagrams will be divided into several independent regions so that each becomes a simple ternary phase diagram.

Figure 8.30 shows a ternary phase diagram with stable compound formed in a binary subsystem. We can find that there is a stable compound $D$ formed in $A-B$ subsystem, compound $D$ and another component $C$ forms a pseudobinary system. The $D-C$ system divides the whole phase diagram into two simple ternary eutectic phase diagrams. In $A-D-C$ system, four-phase equilibrium eutectic transformation $L_{E1} \rightarrow A + D + C$ occurs; and in $B-D-C$ system, $L_{E2} \rightarrow B + D + C$ occurs. The vertical section through component axis $D-C$ in Fig. 8.31 is similar to binary eutectic phase diagram. We should know that one end of this vertical section is compound rather than pure metal, so it was called pseudobinary phase diagram. And the vertical section $X-X$ paralleling to the side $A-B$ of composition triangle consist of two vertical sections of ternary eutectic phase diagrams.

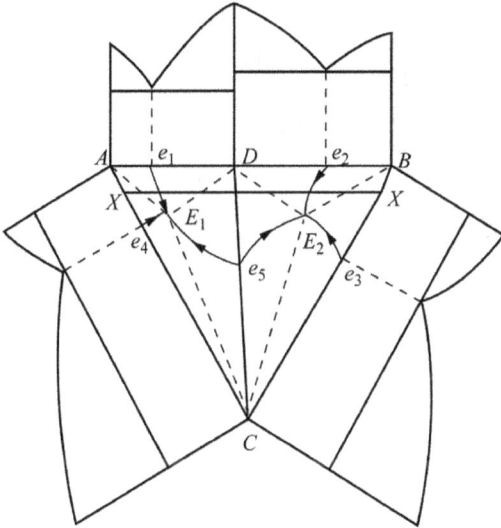

**Fig. 8.30:** Ternary phase diagram with stable compound formed in a binary subsystem.

So we can refer to knowledge in Section 8.2 to analyze this kind of ternary phase diagrams and study the transformation of microstructure.

However, if $A-B$ can form stable compound $\delta$, but $C-\delta$ does not have the feature of pseudobinary system as shown in Fig. 8.32, we cannot divide $A-B-C$ ternary system into two ternary systems to discuss the alloys or cooling process. Here, $T_P > T_E$, and quasi-peritectic transformation occurs at $T_P$:

$$L_P + \beta \rightleftharpoons \delta + \gamma$$

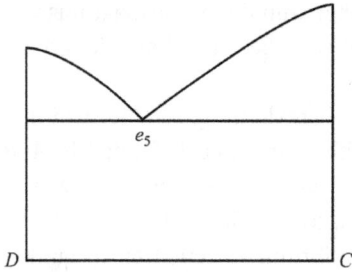

**Fig. 8.31:** *D–C* vertical section.

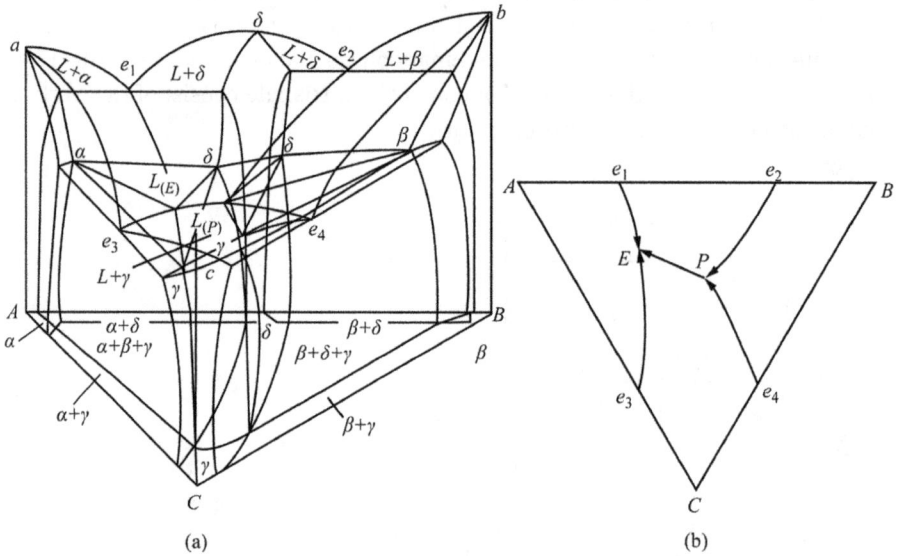

**Fig. 8.32:** Ternary phase diagram where δ–C system does not have the feature of pseudobinary system (a) space pattern and (b) polythermal projection.

And ternary eutectic transformation occurs at $T_E$:

$$L_E \rightleftharpoons \alpha + \beta + \delta$$

Figure 8.32(b) shows the polythermal projection of this ternary system, in which univariant lines meet at a ternary quasi-peritectic $P$ (two arrows entering and one arrow leaving), and a ternary eutectic $E$ (three arrows converging).

## 8.8 Examples of ternary phase diagrams

### 1. Vertical sections of Fe–C–Si ternary system
Figure 8.33 illustrates the two vertical sections of Fe–C–Si ternary system with mass fraction of Si containing 2.4% and 4.8%, respectively. Both of them are parallel to

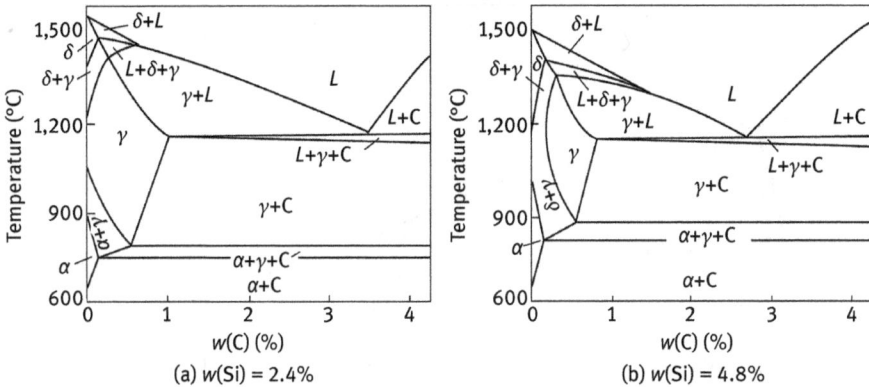

**Fig. 8.33:** Vertical sections of Fe–C–Si ternary system.

the Fe–C side in the composition triangle of Fe–C–Si system. These vertical sections are important evidences to analysis the contents of components and the variations of microstructure of grey cast iron.

There are four single-phase regions in these two vertical sections: liquid phase $L$, ferrite $\alpha$(BCC), high temperature ferrite $\delta$(BCC), and austenite $\gamma$(FCC). Besides, seven two-phase regions and three three-phase regions are also included in these two vertical sections. Figure 8.33 reveals the similarity between vertical sections and the Fe–C binary phase diagram. The only discrepancies are as follows: three-phase equilibrium regions, like peritectic reaction $(L + \delta \to \gamma)$, eutectic reaction $(L \to \gamma + C)$, and eutectoid reaction $(\gamma \to \alpha + C)$ are phase regions limited by several boundaries rather than horizontal lines. On the other hand, due to the addition of element Si, the positions of peritectic point, eutectic point, and eutectoid point move slightly. With the increase in Si, the peritectic reaction temperature drops and both the eutectic reaction temperature and eutectoid reaction temperature increase, and the $\gamma$ phase region shrinks gradually.

## 2. Phase diagram of Fe–Cr–C ternary system

The Fe–Cr–C ternary-type alloys, such as chromium stainless steels 0Cr13, 1Cr13, 2Cr13, and the high carbon and high chromium die steel Cr12, are widely used in the industries. Additionally, Fe–Cr–C alloys also play a significant role in most of the other popular steels and casting irons which belong to multicomponent alloys. Figure 8.34 illustrates the vertical section of Fe–Cr–C ternary system with the mass fraction of Cr containing 13%. The shape of the section is slightly more complicated than the one in the Fe–C–Si ternary system. There are three horizontal lines of four-phase equilibrium in addition to four single-phase regions, eight two-phase regions, and eight three-phase regions.

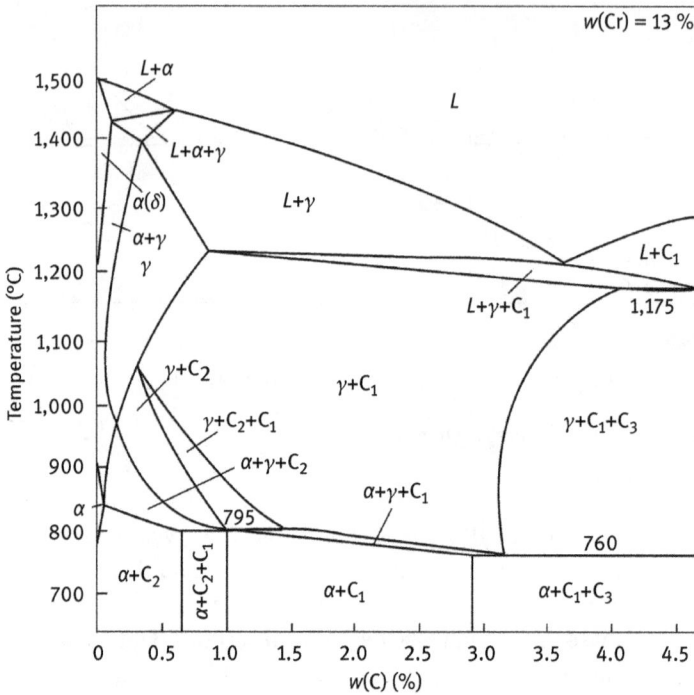

**Fig. 8.34:** Vertical section of Fe–Cr–C ternary system with mass fraction of Cr containing 13%.

The four single-phase regions contain liquid phase $L$, ferrite $\alpha$, high-temperature ferrite $\delta$, and austenite $\gamma$. $C_1$ and $C_2$ are $Cr_7C_3$- and $Cr_{23}C_6$-based carbides with Fe solutions. $C_3$ is $Fe_3C$-based alloyed cementite with Cr solutions. Every reaction occurring in the two-phase region, three-phase region, and four-phase region is listed in Table 8.2.

**Table 8.2:** Reactions occurring in every phase region during cooling of alloys in the vertical sections of Fe–Cr–C ternary system (mass fraction of $w_{Cr} = 13\%$).

| Two-phase equilibrium region | Three-phase equilibrium region | Four-phase equilibrium region |
|---|---|---|
| $L \rightarrow \alpha$ | $L + \alpha \rightarrow \gamma$ | $L + C_1 \xrightarrow{1,175\ ^{\circ}C} \gamma + C_3$ |
| $L \rightarrow \gamma$ | $L \rightarrow \gamma + C_1$ | $\gamma + C_2 \xrightarrow{795\ ^{\circ}C} \alpha + C_1$ |
| $L \rightarrow C_1$ | $\gamma \rightarrow \alpha + C_1$ | $\gamma + C_1 \xrightarrow{760\ ^{\circ}C} \alpha + C_3$ |
| $\alpha \rightarrow \gamma$ | $\gamma + C_1 \rightarrow C_2$ | |
| $\gamma \rightarrow \alpha$ | $\gamma \rightarrow \alpha + C_1$ | |
| $\gamma \rightarrow C_1$ | $\gamma + C_1 \rightarrow C_3$ | |
| $\gamma \rightarrow C_2$ | $\alpha + C_1 + C_2$ | |
| $\alpha \rightarrow C_2$ | | |
| $\alpha \rightarrow C_1$ | $\alpha + C_1 + C_3$ | |

Figure 8.35 illustrates the horizontal sections of Fe–Cr–C ternary system at temperature 1,150 °C and 850 °C, respectively. The contents of Cr and C are represented by rectangular coordinates in different proportions in both of these two sections. Figure 8.35 shows the existence of the single-phase regions of $\alpha$, $y$, $C_1$, and $C_3$ in both sections. However, the section at 1,150 °C contains liquid region, indicating that some alloys has melted at this temperature. The three-phase regions in both sections are triangles whose vertexes are all connected with the single-phase regions and three-phase equilibrium regions are separated by two-phase equilibrium regions.

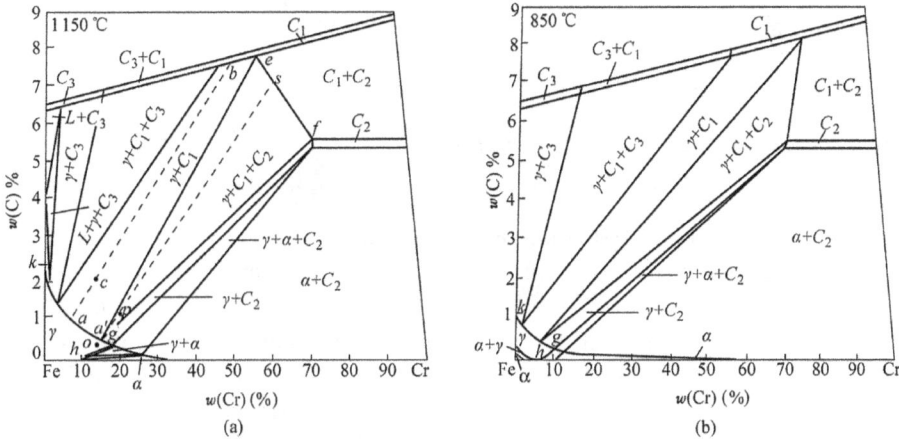

**Fig. 8.35:** Horizontal sections of Fe–Cr–C ternary system.

## 3. Horizontal sections of Fe–C–N ternary system

Figure 8.36 illustrates the horizontal sections of Fe–C–N ternary system at temperature 565 and 600 °C, respectively. Horizontal sections are usually used to help analyze the microstructure of penetration layers of carbon steels after carbonization or nitridation treatment. In Fig. 8.36, $\alpha$ is ferrite, $y$ is austenite, $C$ is cementite; $\varepsilon$ represents phase $Fe_{2-3}(N, C)$; $y'$ represents phase $Fe_4(N, C)$; $\chi$ represents carbide. Figure 8.36(a) shows a big triangle whose vertexes are connected with single-phase region $\alpha$, $y'$, and $C$ and whose three sides are connected with two-phase regions. This is the four-phase equilibrium eutectoid reaction plane: $y \rightarrow \alpha + y' + C$. When mass fraction of $w(C)$ is 0.45% (dotted lines in Fig. 8.35) and the nitrogen content on surface of 45 steel is high enough, the constituent phases in every layer of the component from the surface to core are: $\varepsilon$, $y' + \varepsilon$, $C + y'$, $\alpha + C$ after nitridation treatment at slightly lower than 565 °C, respectively, and the constituent phases are: $\varepsilon$, $\varepsilon + y'$, $y + \varepsilon$, $y$, $\alpha + y$, $\alpha + C$ after nitridation treatment at 600 °C, respectively.

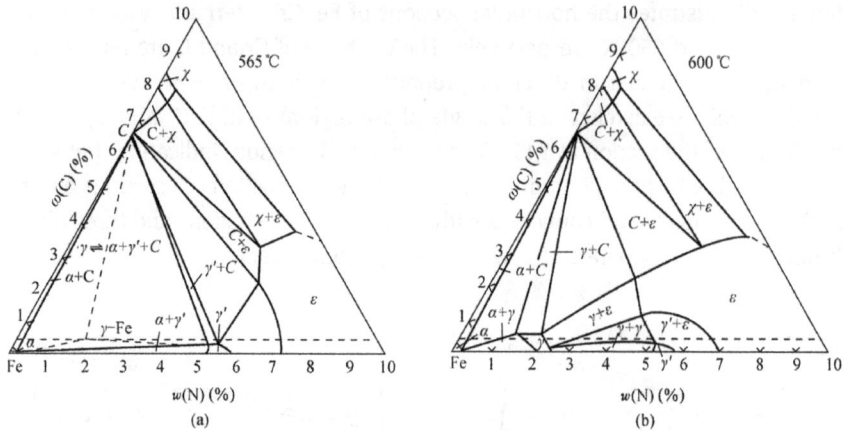

Fig. 8.36: Horizontal sections of Fe–C–N ternary system.

## 4. Projection drawing of Al–Cu–Mg ternary system

Figure 8.37 illustrates the polythermal projection of Al-rich part in the Al–Cu–Mg ternary system. The fine solid lines are isotherms ($x$ °C). The heavy solid lines with arrows are not only the projections of intersections of liquidus surface, but also the representation of projections of liquidus univariant lines of three-phase equilibrium reactions. There is a saddle point at 518 °C, and this is maximum temperature, which lowers in opposite direction along univariant $E_T P_2$. In Fig. 8.37, every liquidus surface is marked by letter which stands for the primary phase. The meanings of those letters are listed as follows:

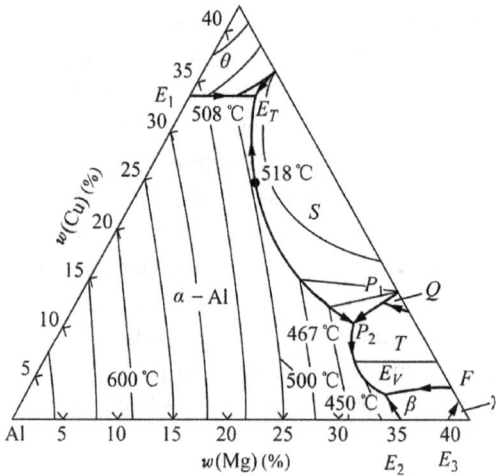

Fig. 8.37: Projection drawing of liquidus surface of Al–Cu–Mg ternary system.

α-Al (the solid solution with Al solvent)

$\theta$ $CuAl_2$    $\beta$ $Mg_2Al_3$        $y$ $Mg_{17}Al_{12}$
$S$ $CuMgAl_2$  $T$ $Mg_{32}(Al, Cu)_{19}$  $Q$ $Cu_3Mg_6Al_7$

According to the features of the four-phase equilibrium reaction surface, the following four-phase equilibrium reactions are included in this ternary system:

$L \rightarrow \alpha + \theta + S$        $(E_T)$
$L + Q \rightarrow S + T$        $(P_1)$
$L \rightarrow \alpha + \beta + T$        $(E_V)$
$L + S \rightarrow \alpha + T$        $(P_2)$

Figure 8.38 illustrates the projection drawing of Al-rich part of solidus surface in
the Al–Cu–Mg ternary system. It includes several parts as described in the following
sections.

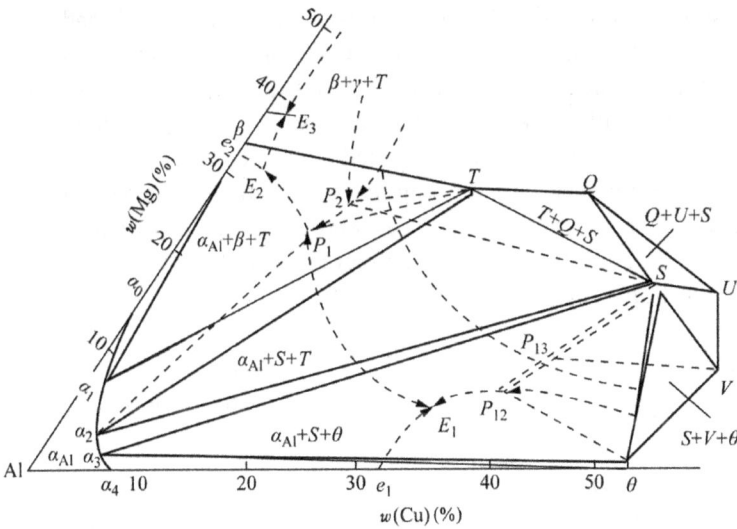

**Fig. 8.38:** Projection drawing of Al-rich part of solidus surface of Al-Cu-Mg ternary system.

### a. Seven four-phase equilibrium horizontal surfaces

In Fig. 8.38, quadrangle $P_{13}SUV$ represents the projection drawing of quasi-peritectic
four-phase equilibrium reaction $L + U \rightleftharpoons S + V$ and the triangle $SUV$ stands for solidus
surface; quadrangle $P_{12}SV\theta$ represents the projection drawing of quasi-peritectic
four-phase equilibrium reaction $L + V \rightleftharpoons S + \theta$ and the triangle $S\theta V$ stands for solidus surface; triangle $P_{13}QU$ represents the peritectic four-phase equilibrium reac-
tion $L + U + Q \rightleftharpoons S$ and the triangle $QUS$ stands for solidus surface; quadrangle

$P_2TQS$ represents the projection drawing of quasi-peritectic four-phase equilibrium reaction $L + Q \rightleftharpoons S + T$ and the triangle $TQS$ stands for solidus surface; triangle $\alpha_3 S\theta$ represents the eutectic four-phase equilibrium reaction $L \rightleftharpoons \alpha_{Al} + S + \theta$; quadrangle $P_1 TS\alpha_2$ represents the projection drawing of quasi-peritectic four-phase equilibrium reaction $L + S \rightleftharpoons \alpha_{Al} + T$ and the triangle $\alpha_2 TS$ stands for solidus surface; triangle $\alpha_1 T\beta$ represents the eutectic four-phase equilibrium reaction $L \rightleftharpoons \alpha_{Al} + \beta + T$.

### b. Four three-phase equilibrium reaction finale surface

In Fig. 8.38, the eutectic three-phase equilibrium reaction $L \rightleftharpoons \alpha_{Al} + \theta$, the composition of every phase varies according to $e_1 E_1$ and $\alpha_4 \alpha_3$ while the composition of $\theta$ is unchanged, when the temperature drops from 548 to 508 °C and the curve connecting $\alpha_3 \alpha_4$ and $\theta$ is the reaction finale surface whose projection is $\alpha_3 \alpha_4 \theta$. In the eutectic three-phase equilibrium reaction $L \rightleftharpoons \alpha_{Al} + S$, the composition of every phase varies along $P_1 E_1$ and $\alpha_2 \alpha_3$ curves, respectively, when the temperature drops from the highest temperature 518 °C along the $P_1 E_1$ univariant line to 508 and 467 °C in opposite direction, and the curve consisting of $\alpha_2 \alpha_3$ (combined with Fig. 8.37) and $S$ is a final reaction surface whose projection is $\alpha_2 \alpha_3 S$. In the eutectic three-phase equilibrium reaction $L \rightleftharpoons \alpha_{Al} + T$, the composition of every phase varies, respectively, along $P_1 E_2$ and $\alpha_2 \alpha_1$ when the temperature drops from 467 to 450 °C, the curve consisting of $\alpha_2 \alpha_1$ and $T$ is a final reaction surface whose projection is $\alpha_1 \alpha_2 T$. In the eutectic three-phase equilibrium reaction $L \rightleftharpoons \alpha_{Al} + \beta$, the composition of every phase varies, respectively, along $e_2 E_2$ and $\alpha_0 \alpha_1$ when the temperature drops from 451 to 450 °C and the curve consisting of $\alpha_0 \alpha_1$ and $\beta$ is a final reaction surface whose projection is $\alpha_0 \alpha_1 \beta$.

### c. One solidification finale surface of primary phase

The projection of final solidification surface of primary phase $\alpha_{Al}$ is $Al\alpha_0 \alpha_1 \alpha_2 \alpha_3 \alpha_4 Al$.

### 5. Projection drawing of Al–Mg–Si ternary system

Figure 8.39 illustrates the projection drawing of Al-rich part in Al–Mg–Si ternary system. After the formation of $Mg_2Si$ by element Mg and Si, pseudobinary system of Al and $Mg_2Si$ is formed. After eutectic reaction at temperature 595 °C, eutectic alloy $E_0$ with $\omega(Mg)$ containing 8.15% and $\omega(Si)$ containing 4.75% is formed:

$$L \rightleftharpoons a + Mg_2Si$$

where $a$ is the Al-rich solid solution. Al and Mg can form $Al_3Mg_2$, and Al and $Al_3Mg_2$ ($Al_8Mg_5$) can form eutectic $E_2$ ($L \rightarrow \alpha + Al_3Mg_2$). While Al and Si can form eutectic

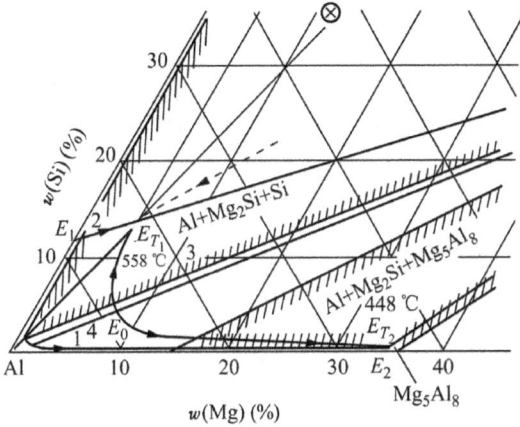

**Fig. 8.39:** Projection drawing of Al–Mg–Si ternary system.

$E_1$ ($L \rightarrow \alpha + $ Si). In the Al–Si–Mg$_2$Si, four-phase equilibrium reaction can occur at temperature 558 °C:

$$L_{ET1} \rightleftharpoons a + Mg_2Si + Si$$

In the Al–Mg$_2$Si–Al$_3$Mg$_2$, four-phase equilibrium reaction can occur at temperature 448 °C:

$$L_{ET2} \rightleftharpoons a + Mg_2Si + Al_3Mg_2$$

As for alloy-2, three-phase eutectic reaction occurs first: $L \rightarrow \alpha + $ Si, forming eutectic ($\alpha + $ Si). Then, four-phase eutectic reaction occurs, forming eutectic ($\alpha + Mg_2Si + $ Si). As for alloy-1, the solidification of solution $\alpha$ occurs first. Then three-phase eutectic reaction occurs: $L \rightarrow \alpha + Mg_2Si$, forming eutectic ($\alpha + Mg_2Si$). Finally, four-phase eutectic reaction occurs, forming eutectic ($\alpha + Mg_2Si + Al_3Mg_2$).

### 6. Ternary phase diagram of ceramics
Figure 8.40 illustrates the projection drawing of liquidus surface of MgO–Al$_2$O$_3$–SiO$_2$, and the four-phase equilibrium reactions of points marked in this figure are listed in Table 8.3. The names of binary and ternary compounds are as follows:

| | |
|---|---|
| MgO · SiO$_2$ (clinoenstatite) | Decomposing at 1,830 K |
| 2MgO · SiO$_2$ (forsterite) | Decomposing at 2,173 K |
| 3Al$_2$O$_3$ · 2SiO$_2$ (mullite) | Decomposing at 2,123 K |
| 3MgO · Al$_2$O$_3$ (spinelle) | Decomposing at 2,408 K |
| 2MgO · 2Al$_2$O$_3$ · 5SiO$_2$ (cordierite) | Decomposing at 1,813 K |
| 4MgO · 5Al$_2$O$_3$ · 2SiO$_2$ (sapphirine) | Decomposing at 1,478 K |

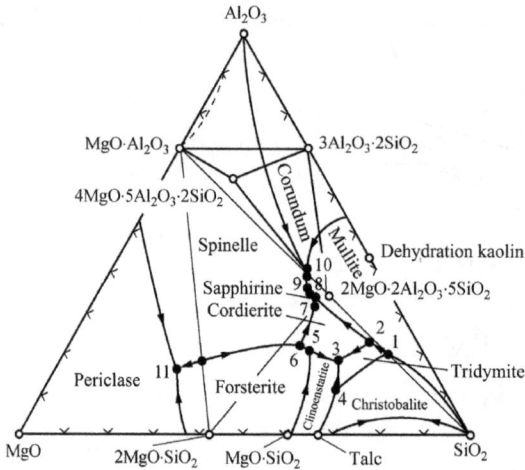

**Fig. 8.40:** Projection drawing of liquidus surface of $MgO–Al_2O_3–SiO_2$.

**Table 8.3:** Equilibrium reaction temperature and the constitute of liquid in $MgO–Al_2O_3–SiO_2$ ternary system.

| Marks on Fig. 8.40 | Equilibrium of phases | Equilibrium temperature (K) | Mass fraction of component (%) | | |
|---|---|---|---|---|---|
| | | | MgO | Al$_2$O$_3$ | SiO$_2$ |
| 1 | Cristobalite + solution⇌tridymite + mullite | 1,743 ± 5 | 5.5 | 18 | 76.5 |
| 2 | 3Al$_2$O$_3$ · 2SiO$_2$ + solution + α-tridymite ⇌cordierite | 1,713 ± 5 | 9.5 | 22.5 | 68 |
| 3 | MgO · SiO$_2$ + α-tridymite + cordierite⇌solution | 1,708 ± 5 | 20.5 | 17.5 | 62 |
| 4 | Cristobalite + solution + clinoenstatite⇌tridymite | 1,743 ± 5 | 26.5 | 8.5 | 65 |
| 5 | 2MgO · SiO$_2$ + MgO · SiO$_2$ + cordierite⇌solution | 1,633 ± 5 | 25 | 21 | 54 |
| 6 | MgO · Al$_2$O$_3$ + solution⇌2MgO · SiO$_2$ + cordierite | 1,643 ± 5 | 25.5 | 23 | 51.5 |
| 7 | Sapphirine + solution⇌cordierite + spinelle | 1,726 ± 5 | 17.5 | 33.5 | 49 |
| 8 | Mullite + solution⇌cordierite + sapphirine | 1,733 ± 5 | 16.5 | 34.5 | 49 |
| 9 | MgO · Al$_2$O$_3$ + mullite + solution⇌sapphirine | 1,755 ± 5 | 17 | 37 | 46 |
| 10 | Corundum + solution⇌mullite + spinelle | 1,851 ± 5 | 15 | 42 | 43 |
| 11 | MgO + MgO · Al$_2$O$_3$ + 2MgO · SiO$_2$⇌solution | 1,973 ± 5 | 51 | 20 | 29 |

Besides, Figs. 8.41–8.44 illustrate phase diagrams of ternary system of $Na_2O–CaO–SiO_2$, $K_2O–SiO_2–Al_2O_3$, $CaO–SiO_2–Al_2O_3$, and $Li_2O–SiO_2–Al_2O_3$ for reference and comparison, respectively.

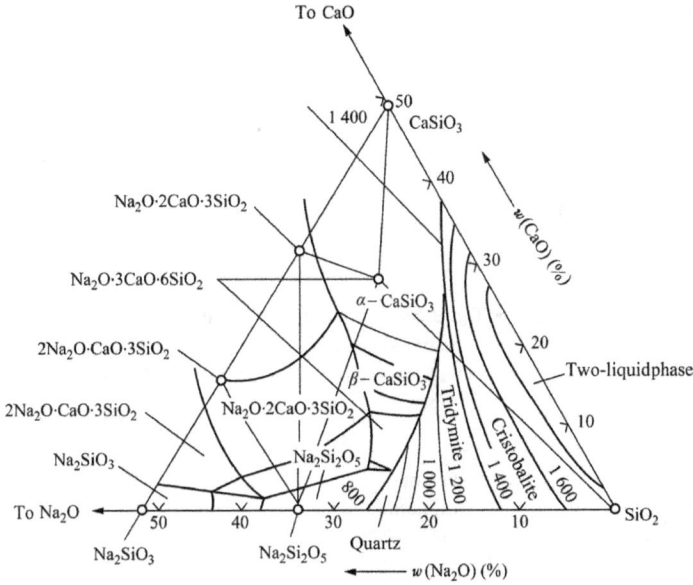

**Fig. 8.41:** Phase diagram of $Na_2O-CaO-SiO_2$.

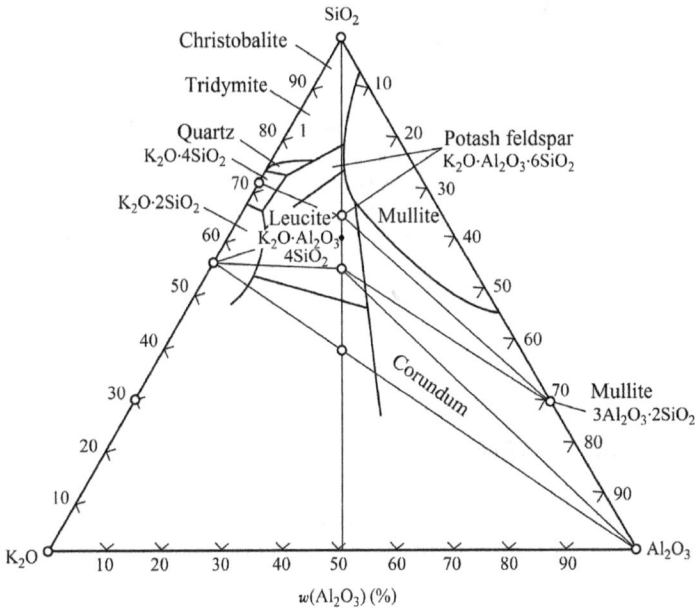

**Fig. 8.42:** Phase diagram of $K_2O-SiO_2-Al_2O_3$.

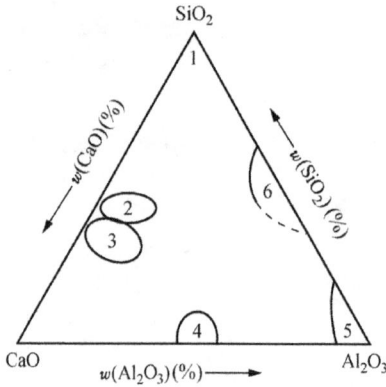

**Fig. 8.43:** Phase diagram of CaO–SiO$_2$–Al$_2$O$_3$: 1, quartz glass; 2, cinder; 3, cement; 4, low-silicon clay; 5, high-alumina refractory; 6, refractory.

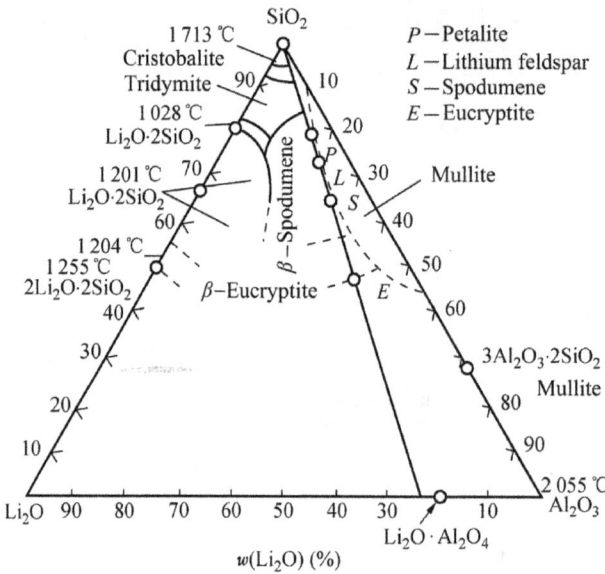

P—Petalite
L—Lithium feldspar
S—Spodumene
E—Eucryptite

**Fig. 8.44:** Phase diagram of Li$_2$O–SiO$_2$–Al$_2$O$_3$..

## 8.9 Summary of ternary diagram

Comparing with binary diagram, the ternary diagram is much more complex since an additional component variable is considered, which means there are two component variables.

According to the phase rule, the equilibrium phase number in the ternary system can range from single phase to four phase under different conditions. The features of phase equilibriums and phase regions are summarized as follows:

## 1. Single phase

When the ternary system is under the condition of single phase, the number of degree of freedom can be calculated as $f = 4-1 = 3$ according to the Gibbs phase rule. It contains one temperature variable and two independent variables of phase components. Therefore, in a ternary phase diagram, the single-phase region covers certain region of temperatures and compositions. In this region, temperature and composition can vary independently, and thus the section can be a plane figure in any possible shape.

## 2. Two-phase equilibrium

The number of freedom degree of two-phase equilibrium in the ternary system is two, indicating one temperature variable and one independent variable of phase components. That is, one of three components in a phase is independently variable, and other two components are not independent variables. In ternary system, there is a conjugate relation between two equilibrium phases. In both vertical sections and horizontal sections, a pair of curves acts as the boundaries separating them from two single phases.

The boundaries between two-phase regions and three-phase regions are comprised of the conjugate curves of two equilibrium phases at different temperatures. Therefore, in the horizontal sections, the two-phase region is separated from the three-phase region by straight line which is also the conjugate curve at this temperature.

## 3. Three-phase equilibrium

The number of degree of freedom of three-phase equilibrium region in the ternary system is one, indicating there is only one independent variable among temperature and the composition of phases. This system is called univariate system and the three-phase equilibrium transformation is called univariate transformation.

The three-phase equilibrium transformations in ternary system are as follows:

(1)  The type of the eutectic transformation $I \rightleftharpoons II + III$, including:

| | |
|---|---|
| Eutectic transformation | $L \rightleftharpoons \alpha + \beta$ |
| Eutectoid transformation | $y \rightleftharpoons \alpha + \beta$ |
| Monotectic transformation | $L_1 \rightleftharpoons L_2 + \alpha$ |
| Melt crystal transformation | $y \rightleftharpoons L + \alpha$ |

(2)  The type of the peritectic transformation $I + II \rightleftharpoons III$, including:

| | |
|---|---|
| Peritectic transformation | $L + \alpha \rightleftharpoons \beta$ |
| Peritectoid transformation | $\alpha + y \rightleftharpoons \beta$ |
| Waferwork transformation | $L_1 + L_2 \rightleftharpoons \alpha$ |

In the space pattern, the composition points of three equilibrium phases can form three space curves called univariate lines as the temperature changes. Every two univariate lines can form a space surface and three univariate lines can form an irregular space triprism volume whose edges are connected to the single-phase region and whose cylinders are connected to two-phase region. The tri-prism volume can begin or end at the three-phase equilibrium lines of binary subsystem and it also can begin or end at the four-phase equilibrium surfaces. Both the three-phase regions including liquid phase in Figs. 8.13 and 8.18 begin with the three-phase equilibrium lines and end with the four-phase planes.

The horizontal section of every three-phase space is conjugate triangle whose vertexes connect to three single-phase regions, respectively. And the conjugate lines which connect two vertexes are the boundaries of phase region between two-phase region and three-phase region. However, all the vertical sections of triangle space are curved triangles.

Taking the reaction which occurs during alloy cooling as an example. No matter what kinds of three-phase equilibrium reactions occur, the positions of univariate lines of reaction phases are higher than the ones of produced phases. Therefore, the movement of conjugate triangles is led by the composition points of reaction phase. On the other hand, in the vertical sections, the positions of phase regions of reaction phases are higher than the ones of three-phase regions and the positions of phase regions of produced phases are lower than the ones of three-phase regions. Specifically, as for eutectic reaction ($L \rightarrow \alpha + \beta$), the movement of conjugate triangle is led by one vertex ($L$) as only one reaction phase is included, shown in Fig. 8.45(a). In eutectic reactions, the movement paths of the compositions of three phases are achieved by intersecting the tangent of the composition of liquid phase with the $\alpha\beta$ side. The vertical sections of three-phase regions are curved triangles with upward vertex (shown in Figs. 8.15 and 8.23); As for peritectic reaction ($L + \beta \rightarrow \alpha$), the movement of conjugate triangle is led by one side ($L\beta$) because there are two reaction phases and one produced phase, shown in Fig. 8.45(b). The movement paths of the compositions of three phases are achieved by intersecting the tangent of the composition of the liquid phase with the extension cord of the $\alpha\beta$ side or by intersecting the tangent of the composition of the $\alpha$ phase with the $L\beta$ side. The vertical sections of three-phase regions are curved triangles with the bottom side upward (shown in Fig. 8.26(b)).

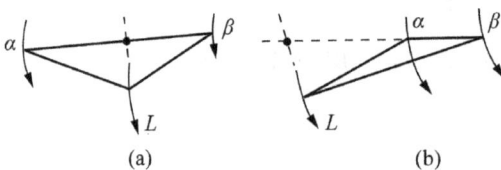

Fig. 8.45: The movement paths of eutectic triangle (a) and peritectic triangle (b).

## 4. Four-phase equilibrium

According to the Gibbs phase rule, the number of freedom degree of the four-phase equilibrium in the ternary system is zero, indicating that the equilibrium temperatures and the compositions of equilibrium phase are fixed.

The four-phase equilibrium transformations in ternary system can be classified into three types:

(1) The type of eutectic transformation I$\rightleftharpoons$II + III + IV, including:

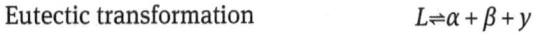

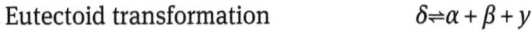

    Eutectic transformation                              $L \rightleftharpoons \alpha + \beta + y$

    Eutectoid transformation                           $\delta \rightleftharpoons \alpha + \beta + y$

(2) The type of quasi-peritectic transformation I + II$\rightleftharpoons$III + IV, including:

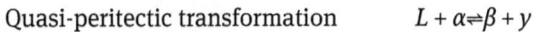

    Quasi-peritectic transformation        $L + \alpha \rightleftharpoons \beta + y$

    Peritectoid–eutectoid transformation   $\delta + \alpha \rightleftharpoons \beta + y$

(3) The type of peritectic transformation I + II + III$\rightleftharpoons$IV, including:

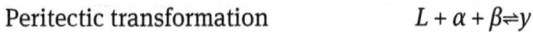

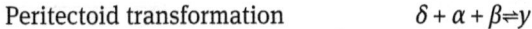

    Peritectic transformation                         $L + \alpha + \beta \rightleftharpoons y$

    Peritectoid transformation                      $\delta + \alpha + \beta \rightleftharpoons y$

The four-phase equilibrium regions are horizontal surfaces in the phase diagram of ternary systems and are horizontal lines in the vertical sections.

The four-phase surfaces are featured in connecting the compositions of four equilibrium phases with four single-phase regions, and the conjugate lines of two equilibrium phases are the boundaries between two-phase regions and single-phase region, and these boundaries connect with six two-phase regions and four three-phase regions, respectively. The relationships between all kinds of surfaces of four-phase equilibrium transformations and surrounding phase regions in space pattern are listed in Fig. 8.46.

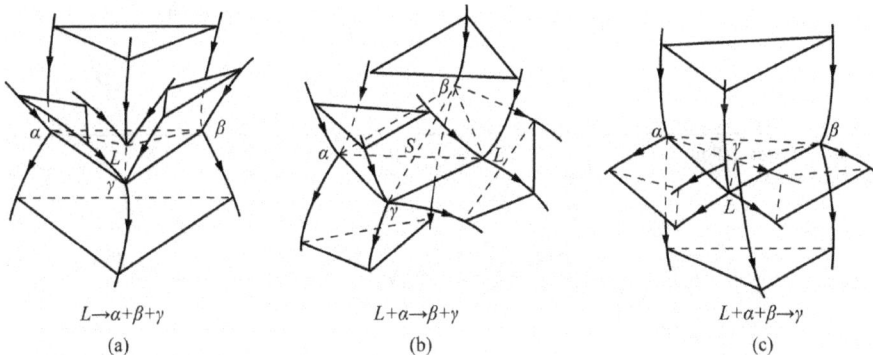

$L \rightarrow \alpha + \beta + y$                 $L + \alpha \rightarrow \beta + y$               $L + \alpha + \beta \rightarrow y$

(a)                           (b)                         (c)

**Fig. 8.46:** Three types of space patterns of four-phase equilibrium region.

The space patterns of different kinds of four-phase surfaces are diverse. In other words, all the possible existing three-phase equilibriums during the process of four-phase reactions of alloys are different, meanwhile, the space movements of univariate lines are dissimilar as well.

Therefore, the type of four-phase equilibrium transformations can be determined by three-phase space during four-phase reaction or movement trend of the univariate lines. Table 8.4 lists all the features of different types of four-phase equilibrium transformations before and after and the univariant lines, in which for ternary eutectic reaction, the arrows of three univariant lines converge, and for ternary quasi-peritectic transformation, two arrows enter and one arrow leaves, and for ternary peritectic transformation, one arrows enters and two arrows leave.

**Table 8.4:** Four-phase equilibrium reactions of ternary system.

| Type of reactions | $L \rightarrow \alpha+\beta+\gamma$ | $L+\alpha \rightarrow \beta+\gamma$ | $L+\alpha+\beta \rightarrow \gamma$ |
|---|---|---|---|
| Three-phase equilibrium before transformation | | | |
| Four-phase equilibrium | | | |
| Three-phase equilibrium after transformation | | | |
| Projection of intersection of liquidus surface | | | |

Although the main content of this chapter is the phase diagrams of ternary systems, the components of many materials are actually over three. If the components are four or more, it is impossible to show the variation of constituent phases with temperature and composition in three-dimensional space. In general, only one or two component variables exist by fixing the content of other components, then based on experiments and calculations, it is applicable to get two dimensional or three dimensional plotting with temperature axis. As a result, such a method can be used to analyze the constituent phase and phase transformation in quaternary system, quinary system, and so on by the same way as the binary or ternary phase diagram. And this kind of phase diagram is called pseudobinary or pseudoternary phase diagram.

# Chapter 9
# Foundation of solid phase transformation

In a broad sense, the process of changing the aggregate state (phase state) of atoms (or molecules) of a substance is called phase transition. For example, the solidification process from the liquid phase to solid phase and the evaporation process from liquid phase to gas phase. As the pressure and temperature change, the internal structures of metals and ceramics and other solids will change. The change occurs from one phase to another, which is called solid-state phase transformation. The state before phase transformation is called the old phase or the parent phase and after phase transformation, the state of phase is called the new phase. After the occurrence of solid-state phase transformation, there must be some differences between the new phase and the parent phase. These differences are either manifested in the crystal structure (such as allotropic transformation); either in the chemical composition (such as spinodal decomposition), or in the surface energy (such as powder sintering), or in the strain energy (such as deformation recrystallization), or in the interface energy (such as grain growth), or all possession (such as dissolution of supersaturated solid solution).

There are many kinds of solid-state phase transitions in metals and many materials undergo several different types of phase transitions under different conditions. Mastering the law of the solid-state phase transition of metals is helpful to take measures (such as the specific heating and cooling processes) to control the solid phase transition process and obtain the expected structures. This can obtain the desired properties and maximize the potential of existing metals, and in turn, develop new materials according to property requirements.

## 9.1 Introduction of solid-state phase transformation in metal

### 9.1.1 Main classification

#### 1. Classification by equilibrium diagram
According to the equilibrium state diagram of the metal materials, the solid phase transition can be divided into equilibrium phase transition and nonequilibrium phase transition.

https://doi.org/10.1515/9783110495379-009

## (1) Equilibrium phase transition

A balanced phase transition refers to a transition of a balanced microstructure that can be obtained in a slow heating or cooling rate. The equilibrium phase transitions of metal materials in solid state are mainly as follows:

### a) Allotropic transformations and polymorphic transformation

When temperature and pressure change, the process of converting pure metal from one crystal structure to another is called allotropic transformation. It is called a polymorphic transformation when it occurs in a solid solution. For example, the transition from ferrite to austenite or austenite to ferrite when steel is heated or cooled belongs to polymorphic transformation.

### b) Equilibrium dissolution precipitation

In the case of slow cooling, the process of precipitating the excess phase from the supersaturated solid solution is called equilibrium dissolution precipitation. As is reflected by the schematic $A-B$ binary alloy phase in Fig. 9.1, when the alloy of $b$ component is heated to $T_1$, the $\beta$ phase will completely dissolve into $\alpha$ phase to form a single solid solution. If the temperature is cooled slowly from $T_1$, $\beta$ phase will be continuously precipitated along the solution curve $MN$, which is called equilibrium dissolution precipitation. The characteristic is that the parent $\alpha$ phase does not disappear, but as the new $\beta$ phase precipitates, the component and volume fracture of the parent phase change constantly. The structure and composition of the new phase are different from that of the parent phase, and the composition of the new phase is generally changed.

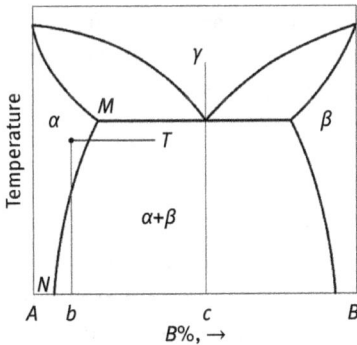

Fig. 9.1: The binary equilibrium phase diagram containing dissolution precipitation.

### c) Eutectoid transformation

When the alloy is cooled, the transformation from one solid phase to two different solid phases is known as eutectoid transformation. As shown in Fig. 9.1, when the alloy of the $c$ component is cooled slowly from the $y$ state, a eutectoid transition occurs when the temperature is lower than the critical temperature: $y{\rightarrow}\alpha+\beta$. The eutectoid phase is similar to the eutectic reaction in eutectic crystallization, and the structure and component of the two phases are different from those of the parent phase. $\alpha + \beta \rightarrow y$ can also occurs during heating, which is called inverse eutectoid phase change. For example, the transition of austenite($y$) to pearlite($\alpha + Fe_3C$) during cooling ($y \rightarrow \alpha + Fe_3C$) and the transition from pearlite to austenite during heating ($\alpha + Fe_3C \rightarrow y$) belong to eutectoid and inverse eutectoid transition.

### d) Spinodal decomposition

Some alloys have homogenous single-phase solid solution at high temperature, but when cooled to a range of certain temperatures they can be decomposed into two micro zones with the same structure but different composition, like $\alpha \rightarrow \alpha_1 + \alpha_2$. This change is known as spinodal decomposition. The characteristic of spinodal decomposition is that there is no obvious interface and component mutation between the two microzones formed at the initial stage of change, but the homogeneous solid solution becomes inhomogeneous through uphill diffusion finally.

### e) Order transformation

In the solid solution (including the solid solution based on mesophase), the relative position of the atoms in the crystal lattice is changed from disorder to order (refers to long-range order), which is called order transformation. This order transformation can occur in many alloy systems, such as Cu–Zn, Cu–Au, Mn–Ni, Fe–Ni, and Ti–Ni.

### (2) Nonequilibrium phase transition

If heating or cooling rate is fast, the equilibrium phase transitions above will be suppressed. Solid materials may undergo transformation that cannot be reflected in the equilibrium diagram and obtain an metastable structure, which is called the nonequilibrium phase transition. The nonequilibrium phase transition is mainly listed as the following:

### a) Pseudoeutectoid phase transition

Figure 9.2 is the bottom left part of the Fe–C alloy equilibrium diagram. When the austenite ($y$) is cooled slowly below the *GES* line from the high temperature, the ferrite ($\alpha$) or cementite ($Fe_3C$) will be precipitated, while the carbon content of the austenite is close to the *S* point. When the carbon content reaches the *S* point, it will be transformed into pearlite ($\alpha + Fe_3C$) through eutectoid transformation. But if the cooling rate is faster, those changes cannot be carried out. The austenite of noneutectoid component will be supercooled to the temperature below the extension of GS and ES lines (the shadow area in Fig. 9.2), and ferrite and cementite will be precipitated simultaneously. The transition process and transition product are similar to eutectoid phase transition, but the ratio of ferrite to cementite (or the average composition of the product) is not constant which varies according to the carbon content of austenite. This is called pseudoeutectoid phase transition.

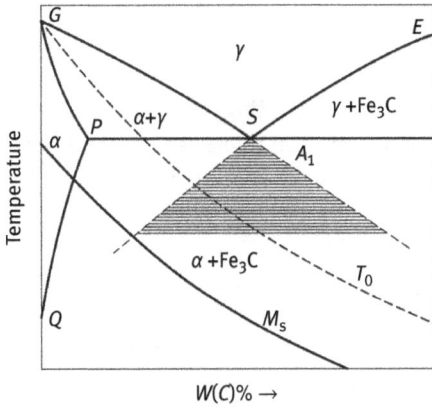

**Fig. 9.2:** A part of the equilibrium phase diagram of Fe–C alloy.

### b) Martensite phase transition

Also, taking Fe–C alloy as an example, if the cooling rate is further improved, the pseudoeutectoid phase transition cannot be carried out, and the austenite will be supercooled to lower temperature. The austenite can only be transformed into the $\alpha$ lattice by shearing from the $y$ lattice in the absence of atomic diffusion and composition change, because iron atoms and carbon atoms are not able or easy to diffuse at low temperature. This is called martensite phase transition, and the transition product is called martensite (in order to distinguish the $\alpha$ phase formed in the equilibrium phase change, it is called $\alpha'$ phase) which has the same composition as the parent's austenite. In Fig. 9.2, $T_0$ is the temperature at which the $\alpha'$ phase and $y$ phase of the same composition have equal free energy. The free energy of the $\alpha'$ phase is lower than that of the $y$ phase below $T_0$, and the $y$ phase will change to the $\alpha'$ phase, that is, the martensitic transition occurs. But in fact, for various reasons, martensitic transformation in steel occurs at the $M_S$ point (onset temperature of martensitic transition), which is about 250 °C cooler than $T_0$ instead of occurring near $T_0$.

In addition to Fe–C alloy, martensitic transformations can also occur in many other alloys and ceramics.

Not only in the cooling, martensitic phase transformation can also occur when heating, customarily known as reverse martensitic phase transformation.

### c) Bainite phase transition

In steel, for example, when the austenite is cooled to the temperature range between the pearlite transition and the martensite transition, the iron atoms are no
longer diffused as the result of lower temperature, but the carbon atoms still have
a certain diffusion capacity. Thus, there is a unique nonequilibrium phase transition, where the carbon atoms diffuse while the iron atoms do not. This phase
change is called bainite phase transition (or known as medium-temperature transformation). The transformation product is also a mixture of $\alpha$ phase and carbide,
however, the carbide content and morphology of $\alpha$ phase and the morphology and
distribution of carbides are different from that of the pearlite, which is known as
the bainite.

### d) Nonequilibrium dissolution precipitation

The equilibrium diagram of the alloy is shown in Fig. 9.1, if the alloy of the $b$ component is rapidly cooling at the $T_1$ temperature, the $\beta$ phase cannot be precipitated
in the cooling process, and then the supersaturated $\alpha$ solid solution will be obtained
when the temperature decreases to room temperature. If the solute atoms still have
a certain diffusion capacity at room temperature or below the temperature of the
solid solubility curve $MN$, the supersaturated solid solution $\alpha$ may decompose and
gradually precipitate new phase at the above isothermal temperature. However, in
the initial stage of precipitation, the composition and structure of the new phase
are different from the equilibrium dissolution precipitation. This process is called
nonequilibrium dissolution precipitation (or aging).

### 2. Classification by atomic migration

According to the migration of atoms in phase change process, the metal solid-
state transition can be divided into diffusion transformation and nondiffusion
transformation.

### (1) Diffusion transformation

During phase transition, the phase interface moves through atomic short-range or
long-range diffusion, which is called diffusion transformation, also known as non-
cooperative phase change. The diffusion transition occurs only when the temperature is high enough and the atomic activity is strong enough. The higher the
temperature is, the stronger the atomic activity is, and the farther the diffusion distance
is. Allotropic transformation, polymorphic transformation, dissolution transformation,
eutectoid phase transition, spinodal decomposition, and order transformation all belong to the diffusion transformation.

The characteristics of the diffusion transformation are:
i)   There is atom diffusion in the process of phase transition, and the phase transition rate is controlled by the atom diffusion velocity.
ii)  The composition of the new phase and the parent phase are generally different.
iii) There is no macroscopic shape change except the volume change caused by the different specific volume between the new phase and the parent phase.

## (2) Nondiffusion transformation

During the phase transition, the atoms do not undergo diffusion, and the motion of all the atoms involved in the transition is a coherent phase transition called the nondiffusion phase transition, also known as the cooperative phase transition. In the nondiffusion transformation, the atoms only migrate regularly so that the crystal lattice is reorganized. During migration, the relative moving distances of neighboring atoms do not exceed one atom spacing while the relative positions of adjacent atoms remain unchanged. Martensite phase transition and allotropic transformation of some pure metals at low temperatures are nondiffusion transition, and the atoms are unable (or difficult) to diffuse during the process.

The characteristics of the nondiffusion transformation are:
i)   Due to the macroscopic shape change caused by the uniform shear, a relief phenomenon occurs on the surface of the polished sample.
ii)  The chemical composition of the new phase and the parent phase are the same because the phase transition does not need atomic diffusion.
iii) There is a certain degree of crystallographic orientation relationship between the new phase and the parent phase.
iv)  When some materials undergo a nondiffusion transition, the travelling speed of phase interface is very fast which is close to the sound velocity.

## 3. Classification by phase transformation

According to the method of phase change, metal solid-state phase change can be divided into nuclear phase change and nonnuclear phase change.

## (1) Nuclear phase transition

Nuclear phase transition is carried out by nucleation and growth. The new phase nuclei can be formed uniformly in the parent phase, and can also be preferentially formed in some favorable position in the matrix. The phase transition process is completed by the constant growth of the new phase after nucleation. There are phase interfaces between the new phase and the parent phase. Most of the metal solid-state phase transitions belong to the nuclear phase transition.

### (2) Nonnuclear phase transition

Nonnuclear phase transition has no nucleation stage. The nonnuclear phase transition starts at the beginning of the composition fluctuation and forms a high concentration region and a low concentration region through this process. However, there is no obvious boundary between the two groups, the composition is continuously increased from the low concentration to the high concentration. Then, the concentration difference is gradually increased by the uphill diffusion, and finally a single-phase solid solution will be decomposed into two phases associated with a coherent interface, in which the compositions are different while the lattice structures are the same. Such as the spinodal decomposition in alloy is the nonnuclear phase transition.

In summary, although there are many types of the solid-state phase transition in metals, the changes occurred in the phase transition process are essentially divided into three aspects: structure, composition, and degree of ordering. Some phase transitions have only one change, while others have two or more changes simultaneously. The same metal can undergo different phase transition under different conditions, so as to obtain different structures and properties. For example, the hardness of eutectoid steel with pearlite structure after equilibrium transformation is about HRC 23; if it is converted into martensite by cooling rapidly, the hardness can be over HRC 60. The tensile strength of Al–4%Cu alloy with equilibrium microstructure is only 150 MPa; the tensile strength can be up to 350 MPa if the alloy undergoes nonequilibrium dissolution precipitation. From here, we can see that by changing the heating and cooling condition, a certain kind of transformation can be occurred to obtain a certain kind of microstructure, which can improve the performance of the materials to a great extent.

### 9.1.2 The main characteristics of phase transformation in solids

Most solid-state phase transitions (except spinodal decomposition) are accomplished through nucleation and growth. Therefore, the liquid crystal theory of metal and its basic concepts still apply to solid-state phase change in principle. However, because the phase transition is carried out under the specific conditions of the solid state, the atoms of the solid crystal are regularly arranged and have many crystal defects. Therefore, the solid-state phase transition has many characteristics which are different from the liquid crystalline process of metals.

### 1. Phase interface

When a solid phase changes, both old and new phases are solid. Depending on the crystallographic matching between the old and the new atoms phases, the interface can be divided into coherent boundary, semicoherent boundary, and noncoherent boundary, as shown in Fig. 9.3. The interface structure between the new phase and the old phase has a great influence on the nucleation and growth process of the solid-state phase change and the morphology of the structure after change.

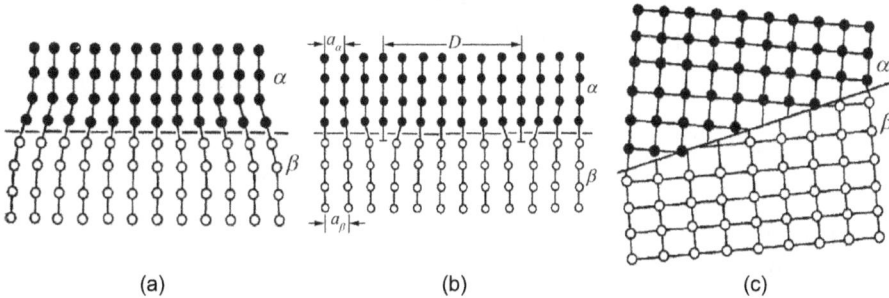

**Fig. 9.3:** Schematic diagram of the boundary structure in solid phase transformation: (a) coherent boundary, (b) semicoherent boundary, and (c) noncoherent boundary.

### (1) Coherent boundary

If the crystal structure is the same and the lattice constant is equal, or the crystal structure and the lattice constant of the two phases are different, there is a set of specific crystallographic planes which can make the perfect matching between the two atoms. At this point, the position of the atoms in the interface happens to be the common position of the two-phase lattices, and the atoms in the interface are shared by the two lattices. This interface is called a coherent interface (Fig. 9.3(a)). The elastic strain energy and the interfacial energy are close to zero under the ideal coherent boundary (such as the twin boundary). In fact, there are always some differences between the two lattices, such as the lattice types and the lattice parameters, thus, when the two phases interface is perfectly coherent, elastic strain will be generated near the phase interface. When the coherent relationship between two phases depends on the positive strain, it is called the first type of coherent lattice while the shear strain is used to maintain it, it is called the second type of coherent lattice, and both sides of the grain boundary have a certain distortion of lattice, shown in Fig. 9.4. Figure 9.4(a) is the first type of the coherent lattice, which is compressed on the one side of the grain boundary and stretched on the other side. Figure 9.4(b) is the second type of the coherent lattice, and there is lattice bending near the grain boundary.

Generally speaking, the coherent boundary is characterized by small interfacial energy and large elastic strain energy due to the distortion near the interface. Coherent interface must rely on elastic energy to maintain. The elastic strain of the coherent interface can be increased when the new phase grows up, and when it exceeds the yield limit of the parent phase, the plastic deformation will occur; as a result, the common lattice relationship will be ruined.

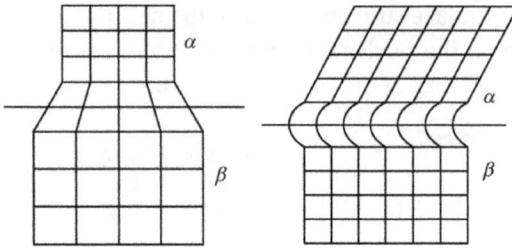

**Fig. 9.4:** The first type (a) and second type (b) of coherent boundary.

### (2) Semicoherent boundary

The magnitude of elastic strain energy in the coherent boundary depends on the relative difference of the atomic distance between the two atoms (referred to as degree of mismatch). If $a_\alpha$ and $a_\beta$, respectively, represent the atomic spacing of the two phases along the crystal direction paralleling to the interface, the difference between the spacing of the two atoms in this direction is indicated by $\Delta a = |a_\alpha - a_\beta|$ , thus the degree of the mismatch $\delta$ is

$$\delta = \frac{|a_\alpha - a_\beta|}{a_\alpha} = \frac{\Delta a}{a_\alpha} \tag{9.1}$$

Obviously, the greater the mismatch $\delta$ is, the greater the elastic strain energy is; when $\delta$ increases to a certain degree, it is difficult to maintain total coherent relationship. Therefore, some edge dislocations will be generated at the interface to compensate the huge difference between the atomic spacing, so that the elastic strain energy of the interface can be decreased. At this point, the two atoms in the interface become partially mismatch (Fig. 9.3(b)), so called the semicoherent interface. It can be seen that the mismatch of one-dimensional lattice can be compensated by a set of edge dislocations without producing a long-range strain field. The spacing $D$ of this set of dislocations shall be

$$D = \frac{a_\beta}{\delta} \tag{9.2}$$

The interface is almost perfectly matched except for the core part of dislocation. The structure of the core part of dislocation is seriously distorted, and the lattice plane is discontinuous.

### (3) Incoherent boundary

When the arrangement of the two atoms in the boundaries is very different, that is when the degree of mismatch $\delta$ is large, the mismatch relationship between the two phases is no longer maintained. This interface is called incoherent boundary (Fig. 9.3(c)). The structure of the incoherent boundary is similar to that of the large-angle grain boundary, which is composed of very thin transition layers arranged by atoms irregularly.

It is generally believed that two phases can form a perfectly coherent boundary when the mismatch is less than 0.05. When the mismatch is greater than 0.25, it is easy to form a incoherent boundary. When the misfit is between 0.05 and 0.25, it is easy to form a semicoherent boundary.

When the solid-state phase transition is occurred, the interfacial energy of two phases is related to the structure and the composition of the interface. The irregular arrangement of atoms on the two phase interface will lead to the increase of the interfacial energy and the absorption of solution atoms meanwhile. The lattice strain energy can be caused by the lattice distortion due to the presence of solution atoms in the lattice, and the interface strain energy can be reduced when the solute atoms are distributed at the interface. Therefore, the solute atoms tend to segregate at the interface to reduce the total energy.

## 2. Orientation relationship and habit plane

In many cases, there is a certain orientation relationship between the new phase and the parent phase in solid-state phase change of metals, and the new phase is often formed on a certain lattice plane of the parent phase, which is called the habit plane represented by the crystal index of the parent phase. The existence of the habit plane means that the atomic arrangements of the new phase and the parent phase are similar in this plane, and the matching is better, which helps to reduce the interface energy. It also means that there must be a certain orientation relationship between the new phase and the parent phase. Since the orientation relationship between the two-phase crystals relative to their habit plane is determined, respectively, the orientation relationship between the two phases is determined, and the result is the existence of a parallel relationship between the grain surface and the crystal orientation in two phases. For example, when the transition from austenite to martensite occurs in the steel, the close-packed plane $\{111\}_{\gamma}$ of the austenite is parallel to the close-packed plane $\{110\}_{\alpha'}$ of the martensite; the close-packed orientation $<110>_{\gamma}$ of the austenite is parallel to the close-packed orientation $<111>_{\alpha'}$, which is known as K-S relationship, and is remembered as

$$\{111\}_{\gamma}//\{110\}_{\alpha'}; <110>_{\gamma}//<111>_{\alpha'}$$

Generally speaking, there must be some orientation relationship between the new phase and the parent phase while the two-phase interface is coherent or semicoherent boundary; if there is no certain orientation relationship, the two-phase interface must be incoherent boundary. On the other hand, they may not have a coherent or semicoherent boundary though there is a certain relationship between the two phases. This is due to the destruction of the coherent and semicoherent boundary with the growth of the new phase.

### 3. Elastic strain energy

When the solid-state phase transition occurs in the metal, the volume may change due to the different specific volume between the new phase and the parent phase. The new phase cannot expand and shrink freely as a result of the constrains from the surrounding mother phase, so the elastic strain and stress must be generated between the new phase and the surrounding mother phase, which will add an additional elastic strain energy to the system. Studies have shown that the elastic strain energy due to phase change in the complete crystal is not only related to the different specific volume and the modulus of elasticity between the new phase and the mother phase, but also to the shape of the new phase. Assuming that the new phases of the different shapes are treated as rotational ellipsoids, the ratio $c/a$ reflects the specific shape of the rotational ellipsoid, a represents the equatorial diameter of the rotational ellipsoid while $c$ represents the distance between the poles of the rotating axis. When $c < a$ it represents a disk; when $c = a$ it is a sphere; and when $c > a$ it is a bar (needle). Figure 9.5 shows the influence of geometry of the new phase on the strain energy generated due to the different specific volume (relative value), which can be seen that the strain energy is the maximum when the new phase is spherical, and that is the smallest in the disk shape while that is centered in the bar (needle) shape.

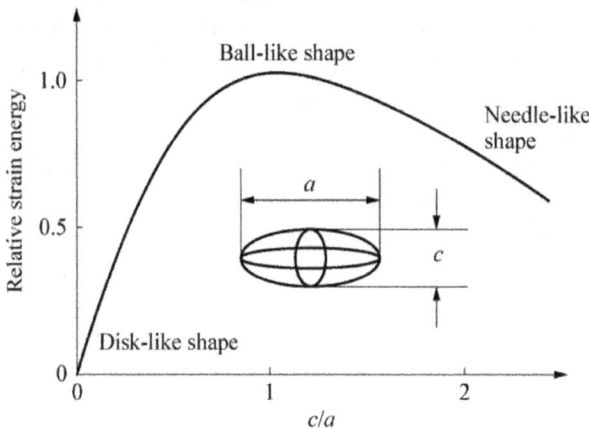

**Fig. 9.5:** The relationship between the shape of new phase and relative strain energy.

The elastic stain energy is produced by the mismatch between the two-phase interface, in addition to the different specific volume. The elastic strain energy due to mismatch is the largest in the coherent boundary, the semicoherent boundary is next (the elastic strain energy decreases due to the generation of interfacial dislocations), and the incoherent boundary is zero.

From the above, the phase transition resistance includes both the interfacial energy and the elastic strain energy. The interface types between the new phase and the mother phase have different effects on the interfacial energy and the elastic strain energy. The interfacial energy can be reduced when the interface is coherent, but the elastic stain energy can be increased. When the interface is incoherent, the elastic strain energy of disk shape is the lowest while the interfacial energy is higher. However, the spherical shape has the lowest interfacial energy, but the highest elastic strain energy. Whether the interfacial energy or the elastic strain energy plays the leading role during the phase transition depends on the specific conditions. If the undercooling is very deep, the radius of critical nucleus is small and the unit volume of new phase interface is very big, the leading role is played by the interfacial energy because the huge of energy increases nucleation energy to become the main resistance of the phase transition. The interface is easy to be coherent to reduce the interfacial energy, and the reduction of the interfacial energy exceeds the increase of the elastic strain energy caused by the coherent boundary, which can reduce the total nucleation energy and easy to nucleation. In the case of very low degree of undercooling, the radius of critical nucleus is big, the incoherent interface is easy to form because the interfacial energy is no longer dominant. At this situation, the elastic strain energy plays the leading role on account of huge difference between the specific volume of two phases, and the new phase in shape of disk will form to reduce the elastic strain energy; if the volume difference between two phases is small and the elastic strain energy has little effect, a spherical phase is formed to decrease the interfacial energy.

**4. The formation of the transition phase**
According to the phase transition thermodynamics, the phase transition is caused by the negative free energy between the new phase and the mother phase, and it is also aimed to change from the unstable mother phase with high energy to the stable new phase with the lowest energy. However, when crystal structures of the stable new phase and the parent phase are different, only a incoherent boundary of high energy can be formed between the two phases. The radius of critical nucleus of the new phase is small, and the unit volume of the new phase interface is very big. Furthermore, the interfacial energy which can prevent nucleation and the nucleation energy of the incoherent interface are both large, which helps the phase transition difficult to happen. In this case, instead of being directly transformed into the stable phase with the lowest free energy from the parent phase, the metastable transition phase is formed firstly, whose crystal structure or composition is close to the mother phase while the free energy is lower than the mother phase. In order to reduce the nucleation energy and make it easy to happen, the transition phase generally has a coherent or semicoherent interface with lower interfacial energy.

Although the transition phase can exist under certain condition, it has the tendency to change continuously to the equilibrium phase because the free energy of the transition phase is still higher than that of the equilibrium phase, and this tendency will increase with the increase of temperature. If the transition phase obtained by proper heat treatment is used in the room temperature, the tendency of changing into equilibrium phase is often slow to be negligible.

**5. The influence of crystal defects**

There are many crystal defects such as grain boundary, sub-grain boundary, vacancy, and dislocation in solid crystals, thus the lattice distortion occurs in the surrounding which is stored the distortion energy. Generally speaking, the nucleation of the new phase in solid-state phase transition is always preferentially formed at the crystal defects. This is because crystal defects are the largest region of energy fluctuations, structure fluctuations, and component fluctuations. When the nuclei are formed in these areas, the activation energy of atom diffusion is low, the diffusion speed is fast, and the phase transition stress is easily relaxed. For example, the grain boundaries which has a lower nucleation energy is the origin of the nonspontaneous nucleation, thus it catalyzes the phase transition.

Dislocation also has a significant catalytic effect on phase change. It is generally believed that the dislocation line will disappear when nucleation is on that dislocation, then the distortion energy in the dislocation center can be released and the free energy of the system will be decreased. The energy released can be used to overcome the formation of the new phase interface and the energy of the transition strain, thus accelerating the phase change. There is another way to catalyze the phase transition. The dislocation adheres to the new phase interface and forms a part of the dislocations in the semicoherent boundary instead of disappearing. As a result, the free energy of the system will also be reduced. In conclusion, from the energy point of view, the nucleation energy of homogeneous nucleation is the largest, the vacancy nucleation is the second, the dislocation nucleation is the third, and the grain boundary inhomogeneous nucleation is the smallest.

**6. The diffusion of atoms**

In many cases, the solid-state phase transition must be carried out through the diffusion of some components because of the different composition between the new phase and the mother phase, then the diffusion becomes the governing factor in the phase transition. Thus, the diffusion rate of the atom has a significant influence on the solid-state phase transition because the diffusion rate of the atom in solid metal is much lower than that in liquid metal. A solid-state phase transition controlled by atomic diffusion can produce a large degree of undercooling. The phase transition drive force increases with the increase of undercooling, and velocity of the phase change also increases. However, when the degree of undercooling increases to a certain degree, the phase change decreases with the increase of the degree of undercooling because the capacity of the atomic diffusion is declining. If the degree of undercooling is further increased, the phase transition in diffusion type can be inhibited, and the nondiffusion phase transition will occur to form a metastable phase at a low temperature. For example, when steel is rapidly cooling from austenite, the diffusion phase transition can be suppressed, while in the low temperature, the nondiffused martensite phase change occurs in the sheer mode to generate the metastable martensite.

## 9.2 Thermodynamics of solid phase transformation

### 9.2.1 Thermodynamic condition of solid phase transformation

#### 1. Driving forces of phase transformation

The thermodynamics of phase transition points out that the stability of the system state is determined by the free energy, and the system is the most stable when the free energy reaches the lowest. All systems have a spontaneous tendency to reduce the free energy to achieve a steady state. If the system has the conditions for lowering the system free energy, it will spontaneously change from a high-energy state to a low-energy state. This change is called spontaneous transformation. The solid phase transformation is also spontaneous transition. So, only when the free energy of new phase is lower than the old phase, the phase transformation can occur. The free energy difference of the new and the old phases and the lower free energy of the new phase are the driving forces for the spontaneous transformation of the old phase into the new phase. The above is the thermodynamic condition of phase change. Thus, for a solid phase transformation to occur, it is necessary to create a negative free energy difference between the new and the old phases. Otherwise, the phase change is impossible.

Free energy $G$ is a feature parameter of the system. Set $H$ as enthalpy, $S$ as entropy, and $T$ is absolute temperature, and

$$G = H - TS \tag{9.3}$$

The free energy of any phase is a function of temperature. By changing the temperature, it is possible to obtain thermodynamic conditions of phase transformation. To investigate the characteristics of this relation, the first and second derivative of $G$ with respect to $T$ can be done:

$$dG = dH - TdS - SdT \tag{9.4}$$

For reversible processes, the general equations of the first and second laws of thermodynamics can be written as,

$$TdS = dH + dW$$

In solid materials, there are only small volume changes caused by phase change, so these volume changes can be neglected. Suppose that W is expansion work, so in the constant volume process, the volume ($V$) is constant, $dW = 0$, and $TdS = dH$. Put it into eq. (9.4), we obtain $dG = -SdT$, and the first derivative is

$$\left(\frac{\partial G}{\partial T}\right)_V = -S \tag{9.5}$$

Since $S$ is always positive, $(\partial G/\partial T)_V$ should always be negative, that is, $G$ always decreases as $T$ increases. And the second derivative is

$$\left(\frac{\partial^2 G}{\partial T^2}\right)_V = -\left(\frac{\partial S}{\partial T}\right)_V \tag{9.6}$$

Also, since entropy $S$ always increases with $T$ increasing, so $(\partial S/\partial T)_V$ is positive and $(\partial^2 G/\partial T^2)_V$ is negative. This means that the characteristic curve of free energy $G$ and temperature $T$ is always concave downward. Figure 9.6 shows the change of the free energy $G$ with temperature $T$ in the $\alpha$ and $y$ phases in a material. Both of the free energy decrease with the increase in temperature, but because of the different of the entropy value between two phases and the degree of entropy varying with the temperature, the free energy curve of the two phases may intersect at one point $T_0$. At $T_0$, $G_\alpha = G_y$, the two phases are in equilibrium and $T_0$ is called the theoretical transition temperature. Since the system tends to minimize the free energy, when the temperature is lower than $T_0$, $G_\alpha$ is lower than $G_y$, and phase $y$ should change into phase $\alpha$;

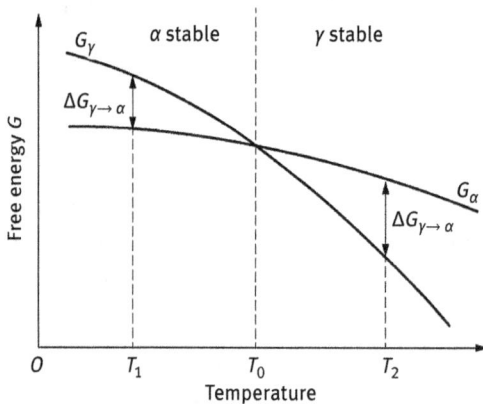

Fig. 9.6: The relationship between free energy of each phase and temperature.

conversely, when temperature is higher than $T_0$, the phase $\alpha$ changes to phase $y$. However, this change does not happen in $T_0$. Only by cooling or heating, get the necessary undercooling ($\Delta T = T_0 - T_1$) or superheating ($\Delta T = T_2 - T_0$) to obtain the sufficient free energy difference for phase change ($\Delta G_{y \to a}$ or $\Delta G_{a \to y}$). In other word, the phase change $\alpha \to y$ or $y \to \alpha$ can occur only when the energy condition of the thermodynamics of phase transformation is satisfied. Obviously, as the undercooling or superheat increases, the driving force of phase transformation also increases, which is beneficial to the phase transformation. And the phase transformation is always in the direction of decreasing the free energy.

**2. Phase transformation barrier**

To make the system change from the old to the new phase, in addition to the driving force of phase transformation, the phase transformation barrier also need to be overcome. The phase transformation barrier refers to the interatomic attraction of reorganization the crystal lattice overcome during phase transformation. In Fig. 9.7, the state I represents the unstable old phase $y$, and the free energy is higher; the state II represents the stable new phase $\alpha$, the free energy is lower. According to thermodynamic conditions, the free energy of phase $\alpha$ is lower than that of phase $y$. So, there is a free energy difference $\Delta G_{y \to a} = G_a - G_y$, and $\Delta G_{y \to a} < 0$. Phase $y$ has a spontaneous tendency to change to phase $\alpha$. But to make the phase change possible, not only the free energy difference $\Delta G_{y \to a}$ is needed, also need the additional energy $\Delta g$ to overcome the phase transformation barrier caused by interatomic attraction.

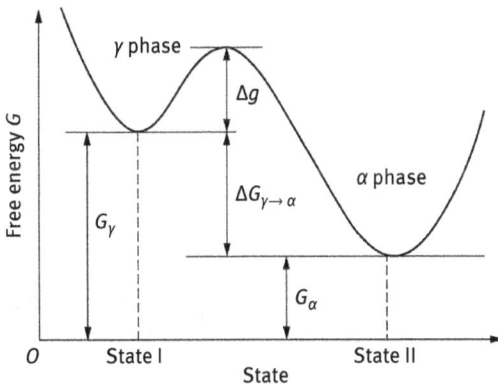

**Fig. 9.7:** Schematic diagram of the phase transformation barrier.

In crystals, atoms can acquire this additional energy in two ways. One is the heterogeneity of atomic thermal vibration. It makes some atoms to have very high thermal vibrational energy to overcome the interatomic attraction and leave the equilibrium position to obtain additional energy. The two is mechanical stress. For example,

elastic deformation or plastic deformation can destroy the regularity of the arrangement of atoms in crystals, cause an internal stress in crystal and force some atoms
out of equilibrium position to get the additional energy.

The barrier can be approximately expressed by activation energy $Q$. The activation energy is the energy required to move the atoms in crystal away from the equilibrium positions to another new equilibrium or nonequilibrium position. Obviously,
the greater the activation energy, the higher the phase transformation barrier. The
activation energy is related to temperature. The higher the temperature is, the smaller
the activation energy is. This is because the increase of the distance between atoms,
the decrease of attraction. Therefore, the higher the temperature, the easier the phase
transformation. However, in more cases, the barrier is represented by the self-
diffusion coefficient $D$ of the crystal atom, and the self-diffusion coefficient $D$ decreases exponentially with the decrease of temperature.

$$D = D_0 \exp\left(-\frac{Q}{RT}\right) \tag{9.7}$$

In the formula, $D_0$ is the coefficient (frequency factor), $R$ is the gas constant, $T$ is
the absolute temperature, and $Q$ is the activation energy. Thus, the greater the self-
diffusion coefficient, the stronger the ability to overcome the barrier, and the easier
the phase transformation.

## 9.2.2 Nucleation of solid phase transformation

Most of the solid phase transformation is done through the process of nucleation and
growth. The nucleation process is often firstly forming the essential component and
structure of the new phase in some small areas of the parent phase, called the nuclear embryo. If the size of a nuclear embryo exceeds a critical value, it will be stable
and grow spontaneously then become crystal nucleus of new phase. If the crystal nucleus is uniformly distributed not preferentially in the parent phase, it is called homogeneous nucleation; if the crystal nucleus is preferentially distributed unevenly in
some regions of the parent phase, it is called heterogeneous nucleation.

Researches point out that the solid phase transformation is similar to the crystallization process of liquid metal, which rarely occurs homogeneous nucleation.
And the new phase mainly heterogeneously nucleates at the grain boundaries of
the parent phase, stacking faults, dislocations, and other crystal defects. To facilitate the analysis, first discuss the condition of homogeneous nucleation.

### 1. Homogeneous nucleation

Compared with the crystallization process of liquid metal, the driving force of homogeneous nucleation in solid phase transformation is also the difference of the

free energy between the old and new phases. But in addition to the interfacial energy, the resistance to nucleation also has the elastic strain energy. The interfacial energy of the crystal nucleus is proportional to the surface area of the crystal nucleus, and the elastic strain energy is proportional to the quality of crystal nucleus. According to the classical nucleation theory, during the homogeneous nucleation in solid phase transformation, the total change of the free energy $\Delta G$ of the system is equal to:

$$\Delta G = -V \cdot \Delta G_V + S\sigma + V\varepsilon \tag{9.8}$$

In the formula, $V$ is the volume of new phase; $\Delta G_V$ is the free energy difference per unit volume between the new phase and the parent phase; $S$ is surface area of new phase; $\sigma$ is the interfacial energy per unit area between the new phase and the parent phase; $\varepsilon$ is elastic strain energy per unit volume of new phase. In formula (9.8), $V \cdot \Delta G_V$ is volume free energy difference, that is, the driving force of phase change. And $S\sigma$ is the interfacial energy, $V\varepsilon$ is elastic strain energy, both of which are phase transformation resistance. It is easily seen that only when $V \cdot \Delta G_V > S\sigma + V\varepsilon$, the formula can be negative, $\Delta G < 0$, and the new phase nucleation is possible. It can only be achieved at a certain undercooling when a new phase nucleus larger than the critical size is formed in the high-energy micro region.

If the crystal nucleus of the new phase is spherical, formula (9.8) can be written as

$$\Delta G = -\frac{4}{3}\pi r^3 \Delta G_V + 4\pi r^2 \sigma + \frac{4}{3}\pi r^3 \varepsilon \tag{9.9}$$

For $d\Delta G/dr = 0$, the critical nucleation radius of the new phase $r_c$ is

$$r_c = \frac{2\sigma}{\Delta G_V - \varepsilon} \tag{9.10}$$

The nucleation energy for the critical nucleation is

$$W = \Delta G_{max} = \frac{16\pi\sigma^3}{3(\Delta G_V - \varepsilon)^2} \tag{9.11}$$

From formulas (9.10) and (9.11), it is found that the critical nucleation radius $r_c$ increases and the nucleation energy $W$ increases as the surface energy $\sigma$ and elastic strain energy $\varepsilon$ increase. Therefore, the coherent new phase nucleus having a low interfacial energy and high elastic strain energy tends to form disk or flake shape; the incoherent new phase nucleus having a high interfacial energy and low elastic strain energy is easy to form equiaxed shape. But, if the interfacial energy of new phase nucleus is highly anisotropic, the latter may be lamellar or acicular.

The critical nucleation radius and nucleation energy are both functions of the free energy difference, so they also vary with undercooling (superheat). With the increase of undercooling (superheat), the critical nucleation radius and nucleation energy decrease, the probability of new phase nucleation increases, and the number of new phase crystal nucleus increases, that is, phase transformation is easy to occur. The same as the additional energy needed to overcome the phase transition barrier, the energy required for nucleation energy comes from two aspects: one is the energy fluctuations in the parent phase; the other is the internal stress caused by deformation and other factors.

Similar to the crystallization of liquid metals, the nucleation rate $I$ of homogeneous nucleation in solid phase transformation can be expressed by the following formula:

$$I = nv \exp\left(-\frac{Q+W}{kT}\right) \tag{9.12}$$

In the formula, $n$ is the number of atoms in a unit volume parent phase, $v$ is atomic vibration frequency, $Q$ is atomic diffusion activation energy, $k$ is the Boltzmann constant, and $T$ is temperature of phase transformation. The diffusion activation energy of solid metal atoms is relatively large, and the nucleation rate of metal solid phase transition is much lower. At the same time, a large number of crystal defects in solid materials can provide energy to promote nucleation. Therefore, heterogeneous nucleation becomes the main nucleation mode of the solid phase transformation.

### 2. Heterogeneous nucleation

All kinds of crystal defects existing in the parent phase can be used as nucleation sites, and the energy stored in the crystal defects can reduce nucleation energy and make nucleation easy. When the new phase nucleus is formed at the crystal defects of the parent phase, the total change of the system free energy is

$$\Delta G = -V \cdot \Delta G_V + S\sigma + V\varepsilon - \Delta G_d \tag{9.13}$$

Compared with formula (9.8), the last item is added, which is the energy caused by the disappearance or reduction of crystal defects. Therefore, the crystal defects can promote nucleation. The effects of crystal defects in nucleation are explained below.

### (1) Nucleation at grain boundary

In the polycrystal, the boundary between two adjacent grains is called the interface; the common junction line of the three grains is called the boundary edge; the common point of the four grains is called the boundary corner. Interfaces, boundary edge and boundary corner are not plane, line and point in geometric sense, and they all occupy a certain volume. If $\delta$ represents the boundary thickness, and $L$ represents the mean diameter of the grain. It can be approximately estimated that the volume fraction of the interface, boundary edge, and boundary corner in the polycrystal is $(\delta/L)$, $(\delta/L)^2$, and $(\delta/L)^3$, respectively. The interface, boundary edges, and boundary corner, all can provide the stored distortion energy to promote nucleation. At the interface nucleation, there is only one interface for the crystal nucleus to swallow, at the boundary edge nucleation, there are three interfaces, and the boundary corner is six. So, from the energy point of view, the boundary corner provides the largest energy, the boundary edge is the second, and the interface is minimum. However, from the volume fraction of the three nucleation position, the interface occupies the most and the boundary corner is the smallest. Considering the two factors comprehensively, in different interface position the heterogeneous nucleation rate $N$ can be expressed as

$$N = nv\left(\frac{\delta}{L}\right)^{3-i}\exp\left(-\frac{Q}{kT}\right)\exp\left(-\frac{A_i W}{kT}\right) \qquad (9.14)$$

In the formula, $i = 0, 1, 2, 3$, respectively, represent the boundary corner nucleation, boundary edge nucleation, interface nucleation and homogeneous nucleation. $A_i$ is the ratio of the nucleation energy of nucleation at different positions in grain boundaries to the nucleation energy of homogeneous nucleation. $A_0 < A_1 < A_2 < 1$, $A_3 = 1$.

In order to reduce the surface area of nucleation and the interfacial energy, the interface is always spherical in incoherent nucleation. The shape of incoherent crystal nucleus of the interface, boundary edge and boundary corner are biconvex lens, curved triangular prism with both ends of the tip and spherical tetrahedron, respectively, as shown in Fig. 9.8. While the coherent and semicoherent interfaces are usually planes. It has been said before that there is a certain the crystallographic relationships between the new phase and the parent phase on the interfaces. When gain boundary nucleation at large angles, it cannot have the crystallographic relationships with grains on both sides of the grain boundary. Therefore, the new phase crystal nucleus can only be coherent or semicoherent with parent phase crystal of one side, and the other side is incoherent. It results in the change in the shape of crystal nucleus, one side is a spherical cap shape, and the other side is a plane, as shown in Fig. 9.9.

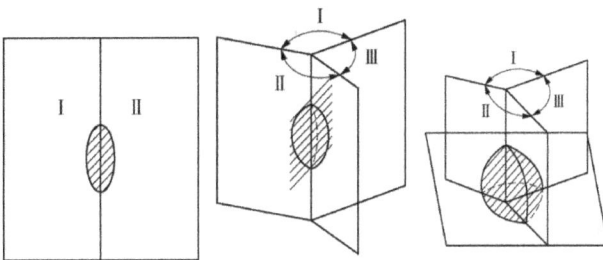

**Fig. 9.8:** The shape of the incoherent nucleus on boundaries: (a) Interface nucleation, (b) edge nucleation, and (c) corner nucleation.

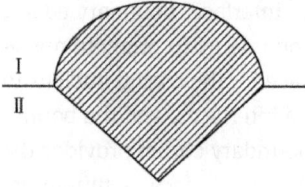

**Fig. 9.9:** Interface nucleus with coherent boundary on one side.

When $\alpha$ is the parent phase and $\beta$ is the new phase, in grain boundary nucleation in the total change of the free energy of the system is expressed as

$$\Delta G = -V \cdot \Delta G_V + S_{\alpha\beta}\sigma_{\alpha\beta} + V\varepsilon - S_{\alpha\alpha}\sigma_{\alpha\alpha} \tag{9.15}$$

In formula (9.19), $S_{\alpha\beta}$ stands for the surface area of phase $\beta$; $\sigma_{\alpha\beta}$ is the interfacial energy per unit area of phase $\alpha$ and $\beta$; $S_{\alpha\alpha}$ is $\alpha$ phase grain boundary area swallowed by $\beta$; $\sigma_{\alpha\alpha}$ is the interfacial energy per unit area of phase $\alpha$. Rewrite formula (9.15):

$$\Delta G = -V \cdot \Delta G_V + V\varepsilon + S_{\alpha\beta}\sigma_{\alpha\beta}\left(1 - \frac{S_{\alpha\alpha}}{S_{\alpha\beta}} \cdot \frac{\sigma_{\alpha\alpha}}{\sigma_{\alpha\beta}}\right) \tag{9.16}$$

Set $\chi = \frac{\sigma_{\alpha\alpha}}{\sigma_{\alpha\beta}}$, and the nucleation energy of grain boundary nucleation can be derived:

$$W = \Delta G_{max} = \frac{16}{3} \cdot \frac{\pi\sigma_{\alpha\beta}{}^3 \left(1 - \frac{S_{\alpha\alpha}}{S_{\alpha\beta}} \cdot \chi\right)^3}{(\Delta G_V - \varepsilon)^2} \tag{9.17}$$

For the interfacial nucleation, from the balance of the interfacial tension (Fig. 9.10a), the relationship between the interfacial energy is shown as follows:

$$2\sigma_{\alpha\beta} \cos\theta = \sigma_{\alpha\alpha}$$

$$\chi = \frac{\sigma_{\alpha\alpha}}{\sigma_{\alpha\beta}} = 2\cos\theta \tag{9.18}$$

If the crystal nucleus is double spherical cap shape, $R$ is the radius of curvature

$$S_{\alpha\alpha} = \pi R^2 \sin^2\theta = \pi R^2 (1 - \cos^2\theta)$$

$$S_{\alpha\beta} = 4\pi R^2 (1 - \cos\theta)$$

$$\frac{S_{\alpha\alpha}}{S_{\alpha\beta}} = \frac{1}{4}(1 + \cos\theta) = \frac{1}{4}\left(1 + \frac{1}{2}\chi\right) \tag{9.19}$$

Depending on the formula (9.17), when $1 - \frac{S_{\alpha\alpha}}{S_{\alpha\beta}} = 01 - \frac{S_{\alpha\alpha}}{S_{\alpha\beta}} = 0$, $W = 0$, formula (9.19) can be rewritten as

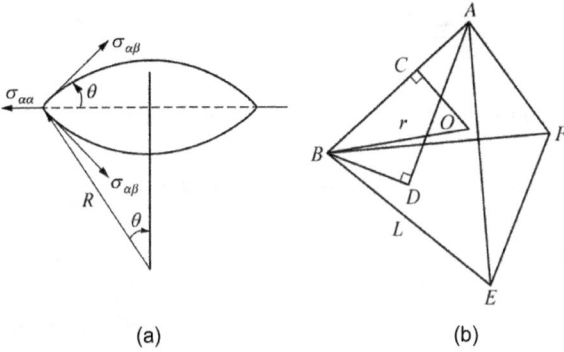

**Fig. 9.10:** The surface area and the absorbed boundary area of the interface and corner nucleus: (a) interface nucleation and (b) corner nucleation.

$$\frac{1}{2}\chi^2 + \chi - 4 = 0 \qquad (9.20)$$

The solution of the quadratic equation $\chi = 2$, $\chi = -4$. Thus, in the interface nucleation, as long as $\chi = \frac{\sigma_{\alpha\alpha}}{\sigma_{\alpha\beta}} \geq 2$, nucleation will no longer require additional energy.

In the boundary corner nucleation, the crystal nucleus can be approximately regarded as a tetrahedron (Fig. 9.10b) for the convenience. Set the length of a tetrahedron as $L$, the distance from the center of the tetrahedron $O$ to the vertex is $r$. In the Fig. (9.10b), the extension line of the $OA$ intersects the $BEF$ plane at point $D$. And the point $D$ should be the center of the triangle $BEF$, $BD \perp AD$. Make the vertical line $OC$ of line $AB$, and as $\triangle AOC \sim \triangle ABD$, so,

$$\frac{OC}{BD} = \frac{AC}{AD}$$

and

$$\therefore AC = \frac{1}{2}L, BD = \frac{2}{3} \cdot \frac{\sqrt{3}}{2}L = \frac{1}{\sqrt{3}L}, AD = \sqrt{L^2 - \frac{1}{3}L^2} = \sqrt{\frac{2}{3}}L$$

$$\therefore OC = \frac{AC \cdot BD}{AD} = \frac{1}{2\sqrt{2}}L$$

The triangle $OAB$ is one of the six swallowed interfaces, the area of which is $S_1 = \frac{1}{2}L \cdot OC = \frac{1}{4\sqrt{2}}L^2$. So the total area swallowed is $S_{\alpha\alpha} = 6S_1 = \frac{3}{2\sqrt{2}}L^2$. The surface area of tetrahedral crystal nucleus is $S_{\alpha\beta} = 4 \cdot \frac{1}{2} \cdot \frac{\sqrt{3}}{2}L^2 = \sqrt{3}L^2$. Therefore,

$$\frac{S_{\alpha\alpha}}{S_{\alpha\beta}} = \frac{\sqrt{3}}{2\sqrt{2}} \qquad (9.21)$$

Put this into the formula (9.17), when $1 - \frac{S_{\alpha\alpha}}{S_{\alpha\beta}} = 0$ and $W = 0$, $1 - \frac{\sqrt{3}}{2\sqrt{2}}\chi = 0$, So

$$\chi = \frac{2\sqrt{2}}{\sqrt{3}} \tag{9.22}$$

That is, when $\chi = \frac{\sigma_{aa}}{\sigma_{a\beta}} \geq \frac{2\sqrt{2}}{\sqrt{3}}$, the boundary corner nucleation does not have energy barrier.

For boundary edge nucleation, calculation results show that when $\chi = \frac{\sigma_{aa}}{\sigma_{a\beta}} \geq \sqrt{3}$, there is no energy barrier.

The above analysis indicates that boundary corner nucleation has the smallest energy barrier. However, whether the boundary corner can become the preferred nucleation site, it depends on undercooling and values of $\sigma_{aa}/\sigma_{a\beta}$. When undercooling is rather large, the nucleation driving force increases, and the nucleation energy decreases. There is little energy barrier in either position. At this time, the interface nucleation with a larger volume fraction has the greater contribution to the nucleation. And when $\chi = \frac{\sigma_{aa}}{\sigma_{a\beta}} \geq 2$, these positions all have no energy barrier. Of course, the interface nucleation contributes most.

### (2) Dislocation nucleation

There are three forms of dislocation nucleation.

The first form: the new phase nucleates at dislocation line and this dislocation line disappears. The released distortion energy reduces the nucleation energy and promotes nucleation. If the new phase nucleus around the dislocation is considered approximately as a cylinder with a radius of $r$, the distortion energy of the unit length due to the disappearance of the dislocation line is

$$A \ln \frac{r}{r_0} = A(\ln r - \ln r_0) = A \ln r \tag{9.23}$$

For edge dislocation: $A = Gb^2/4\pi(1-v)$; for screw dislocation: $A = Gb^2/4\pi$. Here, $r_0$ is the radius of the central hole of the dislocation supposed, $G$ is the shear modulus, $B$ is Burgers vector, and $v$ is Poisson's ratio. It is obvious that the distortion energy of dislocations is related to Burgers vector $B$. The larger the value of Burgers vector, the greater the role of dislocations in nucleation. At this time, the change of the free energy of a unit length crystal nucleus should be

$$\Delta G = -A \ln r - \pi r^2 (\Delta G_V - \varepsilon) + 2\pi r \sigma \tag{9.24}$$

And the critical radius of crystal nucleus $r_c$ can be deduced from eq. (9.24):

$$r_c = \frac{2\pi\sigma \pm \sqrt{4\pi^2\sigma^2 - 8\pi A(\Delta G_V - \varepsilon)}}{4\pi(\Delta G_V - \varepsilon)} \tag{9.25}$$

When $\Delta G_V$ and $A$ is too large, $4\pi^2\sigma^2 < 8\pi A(\Delta G_V - \varepsilon)$, the $r_c$ has no real root. In this situation, dislocation nucleation has no energy barrier.

The second form: the dislocation line does not disappear and is attached to the new phase interface. Then it becomes the dislocation part in the semicoherent interface, compensating the mismatch and reducing the interfacial energy. So, the nuclear energy of the new phase decreases.

The third form: when the composition of the new phase is different from that of the matrix, the solute atoms segregate on dislocation line (form atmosphere). It is favorable for the formation of precipitate nucleation, and therefore promotes the phase transition.

It is estimated that when the driving force of phase transformation is very small, and the interfacial energy between the new phase and the parent phase is about $2 \times 10^{-5} J/cm^2$, the nucleation rate of homogeneous nucleation is only $10^{-70}/(cm^3 \cdot s)$. If the density of dislocation is $10^8/cm$, the nucleation rate of heterogeneous nucleation caused by dislocation can reach $10^8/(cm^3 \cdot s)$. Therefore, when there is a higher density of dislocation in the crystal, it is difficult for the solid phase transformation to be homogeneous nucleation.

### (3) Vacancy nucleation

Vacancies promote nucleation by influencing diffusion or by providing their energy to the driving force of phase transformation. For example, in the case of desolvation decomposition of supersaturated solid solution, when the solid solution cools rapidly from the high temperature, the supersaturated solute atoms are retained in the solid solution. At the same time, a large number of supersaturated vacancies are also retained. On the one hand, they promote the diffusion of solute atoms. On the other hand, they also promotes nucleation of heterogeneous nucleation as the nucleation site of precipitation phase, and distribute the precipitate in the whole matrix. But near the grain boundary there is always the precipitation free zone, in which no precipitate is found. This is because the supersaturated vacancies near the grain boundary diffuse to the grain boundary and disappear, so there is no inhomogeneous nucleation here. More vacancies remain away from the grain boundaries, and precipitates are easy to nucleate and grow here.

### 9.2.3 Nucleus growth of solid phase transformation

#### 1. Growth mechanism of new phase

The growth of the new phase crystal nucleus is essentially the migration of the interface towards the parent phase. Different types of the solid phase transformation have different growth mechanism. For eutectoid phase transformation and dissolution transformation and so on, due to the different composition between new phase and the parent phase, the growth of the new phase crystal nucleus must depend on the long-range diffusion of solute atoms in the parent phase. Until the composition near the interface meets the new requirements, the new phase crystal nucleus can grow up. When this type of phase transformation occurs, there must be a mass transfer process. On the contrary, for allotropic transformation and martensitic transformation and so on, the composition of the new phase and the parent phase are the same, and crystal growth does not need the mass transfer process. In the growth of the new phase crystal nucleus, the atoms near the interface only do the short-range diffusion or even do not diffuse at all.

If the new phase nucleus and parent phase have a certain crystallographic relationships, this relationships will remain during growing. The growth mechanism of new phase is also related to the interface structure of the crystal nucleus (like coherent, semicoherent, or incoherent interface). In fact, the situation that the new phase nucleus is completely matched with the parent phase and forms the perfectly coherent interfaces is very rare. Usually, it forms the semicoherent or incoherent interfaces. These two interfaces have different migration mechanism.

### (1) Migration of the semicoherent interface

Because the semicoherent interface has a lower interfacial energy, the interface tends to remain plane during growing process. Taking the martensitic transformation as an example, the growth of crystal nucleus is accomplished by the shear of atoms in one side of the parent phase on the semicoherent interface. Its characteristic is that a large number of atoms move regularly in one direction with the distance less than one interatomic spacing, and maintain the original crystallographic relationships, as shown in Fig. 9.11. This type of growth of crystal nucleus is called synergistic growth or displacement growth. Because the migration of atoms during phase transformation is less than one interatomic spacing, it is also called diffusionless phase transformation. This synergistic growth using a homogeneous shear manner results in tilting on the surface of the specimen during polishing, as Fig. 9.12.

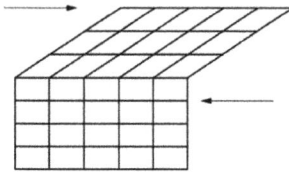

Fig. 9.11: Displacement growth caused by shear deformation.

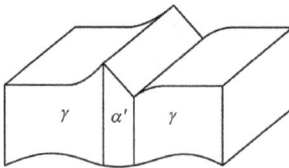

Fig. 9.12: Schematic diagram of surface tilting caused by martensite transformation.

In addition to the above shear mechanism, it can also use the motion of interfacial dislocation on the semicoherent interface, make the interfaces migrate along the normal direction, thus promote the growth of new phase crystal nucleus. The possible structure of the semicoherent interface including interfacial dislocations is shown in Fig. 9.13. Figure 9.13(a) is a flat interface. The interface dislocations are on the same plane, and the Burgers vectors of edge dislocations are parallel to the interface. Now, if the interface migrates along the normal direction, the interface dislocation must climb to move with the interface. It is difficult to achieve without the external force or enough high temperature, so it hinders the migration of the interface and interferes the growth of crystal nucleus. However, if as shown in Fig. 9.13(b), the interface dislocations are distributed on the stepped interface, the Burgers vector of edge dislocation and interface are at a certain angle. Thus, the slip motion of the dislocation can make the steps migrate laterally across the interface, causing the interface to move toward the normal direction and promoting the growth of new phase as shown in Fig. 9.14. This type of growth of crystal nucleus is called stepped growth.

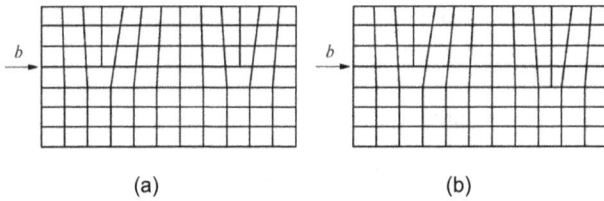

(a)                                    (b)

**Fig. 9.13:** The possible structure of the semicoherent interface: (a) straight interface and (b) step interface.

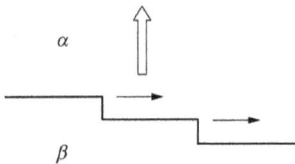

**Fig. 9.14:** Schematic diagram of the nucleus growth by step mode.

## (2) Migration of incoherent interface

In many cases, the interface between the new phase crystal nucleus and the parent phase is incoherent interface. The arrangement of atoms at the interface is disordered, forming a thin transition layer with irregular arrangement, and the possible structure is shown in Fig. 9.15(a). The movement of the atoms on the interface is not cooperative, without a certain order, and the relative displacement of the distance is different and the adjacent relationships may also change. This interface can accept or output atoms in any position. As the parent atoms move toward the new phase, the interface itself moves along its normal direction, thus make the new phase grow up gradually. However, some people think that the micro region in the

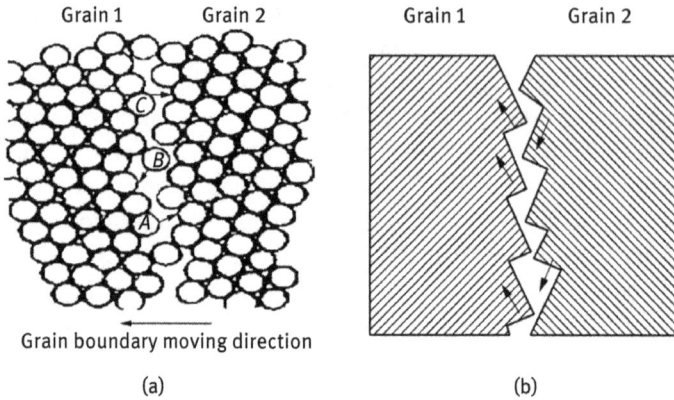

**Fig. 9.15:** The possible structure of the incoherent interface: (a) transition thin layer of irregular arrangement of atoms and (b) step shape incoherent interface.

incoherent interface may also have a step-like structure, as shown in Fig. 9.15(b). The plane of this step is the most close-packed plane of atoms, and the height of step is approximately equal to an atomic layer. The atoms transfer from the end of the parent phase step to the step of new phase, so that the step of new phase occurs the lateral movement. Therefore, the interface develops toward the normal direction and the new phase grow up. Because the migration of this incoherent interface is through interfacial diffusion, this phase transformation is also called diffusion-type phase transformation.

It should be noted that the solid phase transformation does not necessarily belong to either a pure diffusion type or a diffusion free type. For example, the bainitic transformation in steel has both the characteristics of diffusion-type transformation and diffusion free phase transformation. In other words, it is not only consistent with the migration mechanism of the semicoherent interface, but also has the diffusion behavior of solute atoms.

**2. Nuclei growth speed**

New phase growth speed is determined by the migration speed of phase interface. For the interface migration realized by lattice shear mechanism, atomic diffusion is not needed and the growth activation energy is zero, thus the growth speed is usually very high. If the interface migration relies on atomic diffusion, the new phase growth speed is relatively lower since diffusion needs time. Diffusion can be divided into short-range diffusion and long-range diffusion. When atoms only diffuse on the short-range level, new phase growth will not lead to composition change. On the contrary, new phase growth relying on atomic long-range diffusion is accompanied by composition change.

**(1) New phase growth speed when diffusion is on short-range level without composition change**

In this case, new phase growth speed is controlled by interface diffusion (short-range diffusion). Assuming the parent phase is $\gamma$, the new phase is $\alpha$, and the phase transformation happens during the cooling process. Atoms in the parent phase vibrate around their equilibrium position at a vibration frequency of $v_0$. Because of the existence of energy fluctuation, some atoms get additional energy surpassing phase transformation energy barrier $\Delta G$ and leave their equilibrium position migrating into the new phase. The probability for vibrating atoms surpassing energy barrier is $\exp((-\Delta g)/kT)$, thus the frequency for atoms in the $\gamma$ phase to enter into the $\alpha$ phase can be expressed as follows:

$$v_{\gamma \to \alpha} = v_0 \exp\left(\frac{-\Delta g}{kT}\right) \tag{9.26}$$

In the same way, atoms in the $\alpha$ phase could also move across the phase interface into the $\gamma$ phase, but these atoms need more energy to surpass the energy barrier $(\Delta g + \Delta G_{\alpha \to \gamma})$, and their vibration frequency is

$$v_{\alpha \to \gamma} = v_0 \exp\left[\frac{-(\Delta g + \Delta G_{\alpha \to \gamma})}{kT}\right] \tag{9.27}$$

Apparently, interface migration speed or new phase growth speed is proportional to the net frequency for atoms migrating from the parent phase to the new phase $\Delta v = v_{\gamma \to \alpha} - v_{\alpha \to \gamma}$, that is:

$$v \propto \exp\left(\frac{-\Delta g}{kT}\right)\left[1 - \exp\left(\frac{-\Delta G_{\alpha \to \gamma}}{kT}\right)\right] \tag{9.28}$$

Two cases are discussed in the following:
a) The supercooling is very small. In this case, $\Delta G_{\alpha \to \gamma} \to 0$. Since $e^x \approx 1 + x$ (when $|x|$ is very small), formula (9.28) can be simplified as

$$v \propto \frac{\Delta G_{\alpha \to \gamma}}{kT} \exp\left(\frac{-\Delta g}{kT}\right) \tag{9.29}$$

This shows that when the supercooling is very small, the new phase growth speed is proportional to the free energy difference between the parent phase and the new phase, indicating that the new phase growth speed increases when temperature drops.
b) The supercooling is very large. In this case, $\Delta G_{\alpha \to \gamma} \ll kT$, $\exp((-\Delta G_{\alpha \to \gamma})/kT) \to 0$, and formula (9.28) can be simplified as

$$v \propto \exp\left(\frac{-\Delta g}{kT}\right) \tag{9.30}$$

This shows that when the supercooling is very large, the new phase growth speed decreases exponentially when temperature drops.

In summary, during the whole phase transformation process, new phase growth speed exhibits increase first and then decrease when the temperature drops, as shown in Fig. 9.16.

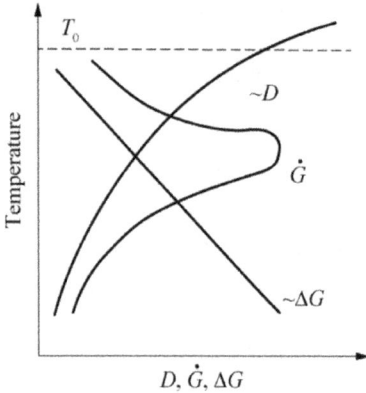

**Fig. 9.16:** Relationship between the nuclei growth speed and temperature.

### (2) New phase growth speed when diffusion is on long-range level with composition change

When the new phase growth is realized by atomic diffusion, redistribution of solute atoms will be inevitably caused, and the growth speed is controlled by long-range atomic diffusion. When phase transformation takes place at a certain temperature, two phases have fixed compositions at the interface, which can be derived from phase diagrams. There are only two possible situations, solute atom concentration in the new phase is higher or lower than that of the parent phase, as shown in Fig. 9.17. There exists concentration gradient between the interface (at the parent phase side) and the parent phase interior, and diffusion of the solute atoms from high-concentration region to low-concentration region will inevitably happen. The result of the diffusion is that the solute concentration near the interface (at the parent phase side) will increase or decrease, broking the phase equilibrium at that temperature. Only if the interface moves towards the parent phase (new phase grows), the new phase in Fig. 9.17(a) excludes solute atoms, or the new phase in Fig. 9.17(b) absorbs solute atoms, the phase interface can be sustained until the concentration gradient inside the phase is absent and a dynamic equilibrium is achieved. When the temperature drops, the above process will repeat until the whole phase transformation is over.

The phase interface migration speed can be calculated according to the law of diffusion. Assuming the unit area phase interface moves by $dx$ in the time of $dt$, the amount of solute atoms absorbed or excluded for the volume increase of the new

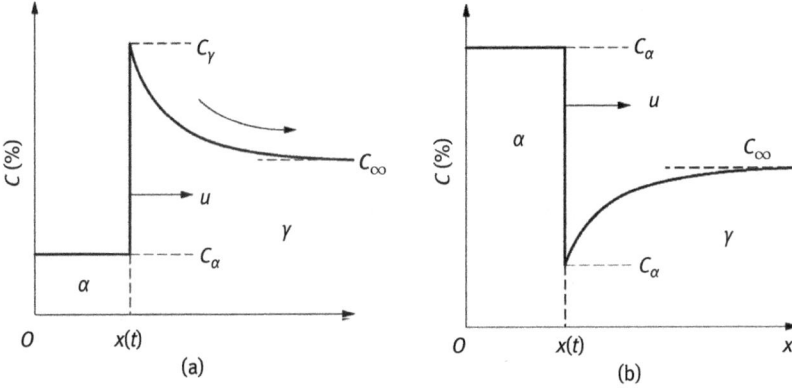

**Fig. 9.17:** Concentration distribution of solute atoms during the nuclei growth process. (a) The concentration solute atoms in the new phase is lower than that in the parent phase and (b) the concentration solute atoms in the new phase is higher than that in the parent phase.

phase is $|\Delta G_{\alpha/\gamma}|dx$, where $\Delta G_{\alpha/\gamma}$ is the concentration difference at the interface of the two phases. The absorption and exclusion of solute atoms for the new phase are both diffusion process inside the parent phase. Assuming the diffusion process is steady-state diffusion, the amount of atoms absorbed or excluded in unit area and unit time is $D(\partial C_r/\partial x)_{x_0} dt$, and there exists:

$$|\Delta C_{\alpha/\gamma}|dx = D(\partial C_r/\partial x)_{x_0} dt \tag{9.31}$$

Then the new phase growth speed can be expressed as

$$v = dx/dt = D/|\Delta C_{\alpha/\gamma}| \times (\partial C_r/\partial x)_{x_0} \tag{9.32}$$

This shows that the new phase growth speed $v$ is proportional to diffusion coefficient $D$, also proportional to the concentration gradient of the parent phase near the interface $\partial C_y/\partial x$, and inversely proportional to equilibrium concentration difference of the two phases on the interface $\Delta C_{\alpha/\gamma}$.

### (3) Relationship between temperature and the new phase growth speed of the diffusion-type phase transformation

New phase growth speed of the diffusion-type phase transformation is controlled by driving force and diffusion coefficient. For the cooling transformation process, the relationship between driving force as well as diffusion coefficient and the transformation temperature is contradictory. The greater the degree of undercooling, the greater the driving force, and the harder the atomic diffusion. As a result, there is a maximum value of the phase transformation speed during the cooling process. For the heating transformation process, the relationship between driving force as well as diffusion coefficient and the transformation temperature is consistent. The greater superheating, the greater the driving force, and the stronger the atomic diffusion capacity. As a result, the phase transformation rate increases monotonically with temperature.

## 9.3 Phase transformation macro kinetic equation

Solid-state phase transformation speed is determined by nucleation rate and growth speed of the new phase, both of which change with temperature and time. As a result, it is very hard to get a precise formula of the phase transformation speed. Assuming that nucleation rate and growth speed do not change with time at a certain temperature, Johnson–Mehl derived the relationship between the volume fraction of the new phase $f$ and the time $\tau$, which is the famous Johnson–Mehl formula:

$$f = 1 - \exp\left(-\frac{\pi}{3}NG^3\tau^4\right) \tag{9.33}$$

where $N$ is the nucleation rate; $G$ is the linear growth speed of nucleus.

Johnson–Mehl formula is suitable for diffusion-type phase transformation with constant nucleation rate and linear growth speed.

In fact, nucleation rate and linear growth speed change with time, and Avrami empirical formula can be used in this case:

$$f = 1 - \exp(-k\tau^n) \tag{9.34}$$

where $k$ is the constant, which is determined by phase transformation temperature, parent phase composition, and grain size; $n$ is the constant, which is determined by the type of phase transformation.

Most of experimental data of solid-state phase transformation is in good agreement with Avrami empirical formula (9.34).

## 9.4 Phase transformation kinetic curve

Aiming at different values of $N$ and $G$ in formula (9.33) (i.e., different phase transformation temperatures), the relation curve between the new phase volume fraction and time can be derived. These curves all exhibit an "s" shape, as shown in Fig. 9.18(a). Their characteristic is that phase transformation speed is relative low at the beginning and ending but highest in the middle. In real production, it is more convenient to characterize phase transformation by temperature–time coordinate. Connecting the beginning points and ending points at different temperatures in Fig. 9.18(a) by smooth curves respectively, Fig. 9.18(b) is obtained. Because this figure exhibits "C" shape, it is called C-curve, or time–temperature–transformation curve.

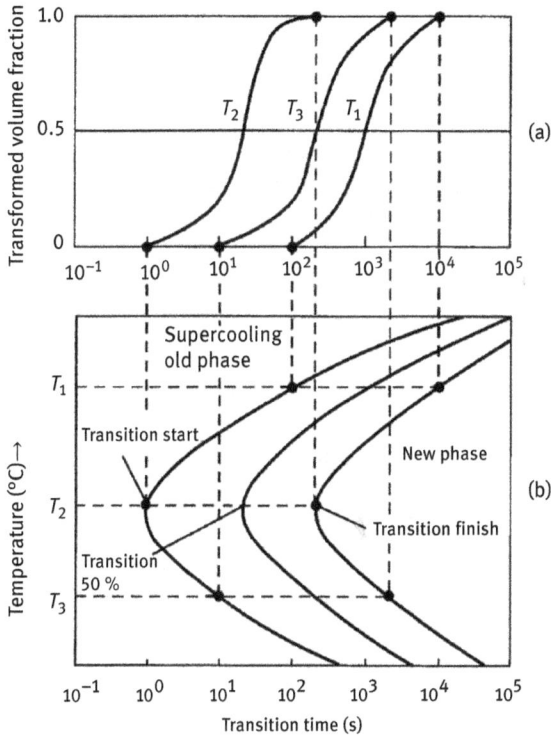

**Fig. 9.18:** Kinetic curves of phase transformation (a) and the corresponding isothermal transformation diagram (b).

In the left of transformation start line is the parent phase supercooling region. Although the thermodynamic phase transformation condition is satisfied, phase transformation does not happen in this region since a nurture period is needed. The temperature corresponding to the shortest nurture period is called "the nose temperature." When supercooling is too large or too small, the nurture period will be elongated and the supercooled parent phase will be stabilized, which is caused by the inconsistence between relations of driving force as well as atomic diffusion capacity with the temperature. In the right of transformation end line is the new phase region. The horizontal distance between start and end lines represents the transformation quantity of the new phase at different temperatures.

Kinetic curves for isothermal (cooling) diffusion-type phase transformation all exhibit "C" shape. C-curves of metallic materials are all obtained by experiments.

# Chapter 10
# Functional properties of materials

Solid materials are generally divided into two categories in terms of performance: structural materials and functional materials. Strength and toughness are the primary properties of structural materials. Whereas a particular function, such as electrical property, thermal property, magnetic property, or optical property, is a primary property of the functional materials. The performance of functional materials depends on the electron structure and electron motion of atoms (revolving, scattering, inspiring, jumping, and so on), which is different from structural materials. The performance of structural materials depends on the bonding between atoms (e.g., metallic bonds, ionic bonds, covalent bonds, and hydrogen bonds) and microstructure (e.g., crystal structure, grain size, phase morphology, dislocation substructure, and second-phase characteristics), but is irrelevant to the electron movement. This chapter reviews the physical basis of the functional properties. Emphasis is put on the description of the behaviors and mechanisms of the functional materials with electricity, heat, magnetism, light behavior, and other functional properties.

## 10.1 Introduction to physical basis of functional materials

### 10.1.1 Energy band

Energy band theory is the main theoretical basis for studying the electron motions in solids. It is the starting point of energy band theory that the shared electrons are no longer restricted to individual atoms, but move in the whole solid. The electrons are on the different discrete energy levels for the single atom. When a large number of atoms form crystals, energy levels of every atom are separated because of the overlap of electron clouds. In each level, the electron energy is approximately continuous because of very small energy difference among the electrons in the same energy level. The energy that is continuously distributed in the energy level forms an energy band. The relationship of the energy level and the energy band is shown in Fig. 10.1. The energy band of the lower energy level is narrower, while that of higher energy level is wider. The lowest energy band corresponds to the innermost layer electrons which have the small electron orbit and rarely overlap with each other. This leads to the narrow energy band. On the other hand, the electrons in the outer orbit have higher energy, and thus are easily disturbed. This causes more overlap between different atoms, forming wide energy band. Ignoring the interaction between atomic states, there is a simple corresponding relation between the energy band and the energy level. According to the subshell electron energy levels

https://doi.org/10.1515/9783110495379-010

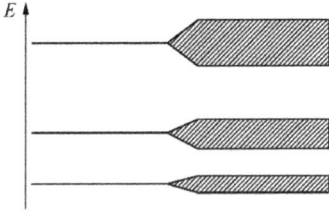

**Fig. 10.1:** Energy levels and energy bands.

of angular momentum quantum number, $s$, $p$, $d$, . . ., the corresponding energy bands are named as $ns$ band, $np$ band, $nd$ band, and more. Some regions between the energy bands, called forbidden band or band gap, are energy levels that electrons do not possess. Considering a solid consisting of $N$ atoms, each energy band contains $N$ divisive energy sublevels. Each sublevel can accommodate two electrons with opposite direction of spin. For example, $2s$ energy band contains $N$ divisive energy sublevels and can accommodate $2N$ electrons. Three $2p$ energy bands contain $3N$ energy sublevels and can accommodate $6N$ electrons. It is also possible that the energy band is partially filled. For instance, a copper atom contains one electron in $4s$. $N$ copper atoms should contain $N$ $4s$ energy sublevels and accommodate $2N$ electrons. But $N$ electrons are actually available. Only half of the $4s$ energy band are filled.

### 10.1.2 Fermi energy

Different from the classical electron theory, the free electrons which are $10^4$ times higher than gas molecules in the density obey Fermi-Dirac distribution. In thermal equilibrium, the probability of the free electrons in energy state $E$ can be indicated by the following formula:

$$f = \frac{1}{e^{(E-E_F)/(kT)} + 1} \tag{10.1}$$

where $f$ denotes the Fermi–Dirac distribution function, $E_F$ is the Fermi energy at $T$ temperature, which is equal to the free energy increment after adding one electron. $k$ is the Boltzmann constant.

At 0 K, the relationship of $f$ and $E$ is presented with the solid line as shown in Fig. 10.2. In this case, $f = 1$ at $E \le E_F$, while $f = 0$ at $E \ge E_F$. All energy states that are less than Fermi energy are occupied by electrons ($f = 1$). According to the lowest energy principle, the electrons gradually fill up the energy levels below $E_F$ starting from the lowest energy level. Fermi energy is the highest energy level of free electrons at 0 K. All the energy states above $E_F$ have no electrons ($f = 0$), which are empty energy state (also called empty state). If $T > 0$ K, $f = 1/2$ at $E = E_F$, $1/2 < f < 1$

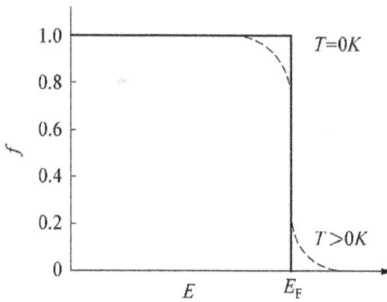

**Fig. 10.2:** Fermi–Dirac distribution.

at $E < E_F$, and $0 < f < 1/2$ at $E > E_F$. These are shown as dashed line in Fig. 10.2. The steep part of $f$ locates in a small energy region (approximately 0.1 eV) near $E_F$. The value of $f$ changes sharply from 1 at $E < E_F$ to 0 at $E > E_F$. In other words, when $T > 0$ K, the energy levels below $E_F$ are filled up by electrons, and those above $E_F$ are almost empty. It can be inferred that the unfilled energy level must be close to $E_F$. The distribution function indicates that a small amount of electrons with energy near $E_F$ can absorb heat to jump into the high energy state above $E_F$. As a result, the empty energy levels above $E_F$ may partially be occupied by the jumped electrons. Because the electrons with the energy far below $E_F$ cannot be excited via absorbing heat, the electrons jumped to the high energy state by this way are very limited.

## 10.2 Electrical property

### 10.2.1 Description of electrical property

The electrical property of materials refers to the response to the external electrical field. This section focuses on the apparent description of the conductive behavior and discussion on the conductance mechanism of materials. The effect of energy band structure on the conductive ability is also highly emphasized. The principles involved are suitable for interpretation of electrical characteristic of various materials including metal, semiconductor, and insulator.

The ability to transfer current is generally an important electrical property of material. The current, $I$ (unit: ampere), depends on the applied voltage, $V$ (unit: volt), according to Ohm's law:

$$V = IR \tag{10.2}$$

where $R$ denotes the electrical resistance of material (unit: ohm). It is influenced by the shape of sample, but independent on the electrical current for most materials.

The electrical resistivity, $\rho$, has nothing to do with the shape of materials, but has a relation to the electrical resistance:

$$\rho = \frac{RA}{l} \qquad (10.3)$$

where $l$ indicates the length of a rod material. $A$ indicates the cross sectional area of this rod. The unit of $\rho$ is ohm-meter $(\Omega \cdot m)$. According to eq. (10.2) it can be expressed as

$$\rho = \frac{VA}{Il} \qquad (10.4)$$

The electrical conductivity, $\sigma$, is another way of describing the electrical property of materials. It is inversely proportional to the electrical resistivity:

$$\sigma = \frac{1}{\rho} \qquad (10.5)$$

The electrical conductivity is a kind of ability of conduction current. The unit of $\sigma$ is the reciprocal of the ohm-meter $(\Omega \cdot m)^{-1}$. So the electrical property described by the electrical conductivity or the electrical resistivity is the same.

Except for eq. (10.2), the Ohm's law can be also expressed as

$$J = \sigma \zeta \qquad (10.6)$$

where $J$ indicates the current per unit area $(= I/A)$, which is also called electrical current density. $\zeta$ indicates the electrical field intensity, equal to the ratio of the voltage between the two ends of a rod material and its length, $l$:

$$\zeta = \frac{V}{l} \qquad (10.7)$$

The equivalency of the two expressions (eqs. (10.2) and (10.6)) of Ohm's law is easy to be proven.

Solid materials exhibit an amazing wide range of electrical conductivity, up to 27 orders of magnitude. According to their conductivity, solid materials are classified into three groups: conductors, semiconductors, and insulators. Metals are good conductors, and have unique electro-conductivity with the order of magnitudes $10^7$ $(\Omega \cdot m)^{-1}$. Insulators have the quite low electro-conductivity with the order of magnitudes $10^{-20}$–$10^{-10}$ $(\Omega \cdot m)^{-1}$. The electroconductivity of semiconductors is moderate, in the range of $10^{-6}$–$10^4$ $(\Omega \cdot m)^{-1}$.

The current is derived from the motion of charges. It is the response to the acting force of external electrical field. The positive charges are accelerated in direction of the electrical field applied, while the negative charges are accelerated in the opposite direction of the electrical field. For most of metallic materials, the current is caused by the motion of electron charges, which is called electron conduction.

Additionally, the motion of ion charges may also generate electrical current in ionic materials, which is called ion conduction. In this section, only electron conduction will be discussed.

### 10.2.2 The conductivity based on the energy band theory

Only the electrons with energy above the Fermi energy can be affected by the electrical field. They are involved in conducting currents, and are called free electrons in metals. In semiconductors and insulators, the vacancies among the charged entities, called holes, can act as a conductive medium. The holes generally have energy below the Fermi energy. The conductivity is a function of the free electrons or/and the number of vacancies. Obviously, the discrepancy among conductors, semiconductors and insulators is the difference in number of free electrons and vacancies.

In metals, the electrons become free only when they are excited into the empty energy states above the Fermi energy. The typical energy band structures of metals are shown in Fig. 10.3(a) and (b). There is an empty state near the highest filled state, $E_F$. The electrons can be activated into the lower empty states by a very little energy. The electrical field can usually activate a large number of electrons to jump into the empty state to conduct.

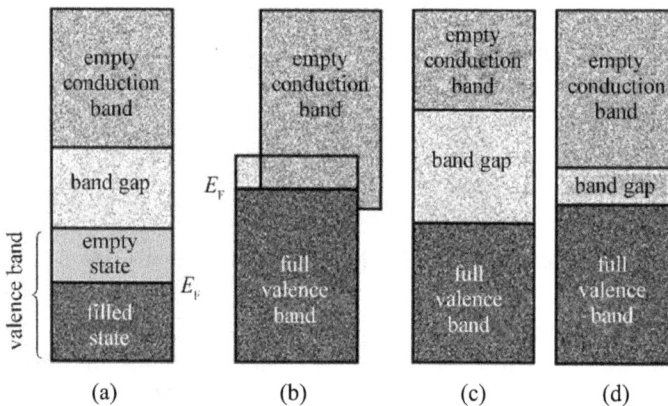

Fig. 10.3: The electron energy band structures of various materials at 0 K: (a) Energy band structures of metals (i.e., Cu), (b) overlap of empty band and full valence for some metals (i.e., Mg), (c) insulator: a wide band gap (>2 eV) exists between full valence band and empty band, and (d) semiconductor: a narrow band gap (<2 eV) exists between full valence band and empty band.

In semiconductors and insulators, there is no empty state on the top of the full valence band. If electrons are to become free, they must be excited and should be able to jump over the band gap to enter into the bottom of the empty conduction band.

It occurs when electrons gain enough energy. This energy should correspond to the energy difference between the two states of the electrons, and be approximately equivalent to the band gap energy $E_g$. Generally, the band gap is about a few electron volts in wide. This indicates that a large electrical field is needed to excite an electron to jump over the band gap. Ordinarily, the excited energy does not always come from the electrical field, does come from light or heat. The amounts of electrons that are heat excited to enter into the empty conduction band are dependent on temperature and the band gap energy. At a certain temperature, the wider energy gap corresponds to the smaller probability of the electrons excited into the conduction band. This means that the material with a wide band gap generally has low electrical conductivity. Therefore, the difference between semiconductors and insulators is that they have different band gap widths as shown in Fig. 10.3(c) and (d). The band gap of semiconductors is relatively narrow, while that of insulators is relatively wide. Obviously, as increasing temperature, both semiconductor and insulator can gain more thermal energy that excites more electrons jumping into the empty conduction band. This causes the enhancement of the electrical conductivity.

### 10.2.3 Electron mobility

When an electric field is applied, the free electrons in material are impacted by a force arisen from the electric field. These electrons with negative charges are accelerated along the opposite direction of the electric field. The accelerated electrons generally do not react with atoms in an ideal crystal in accordance with the principle of quantum mechanics. All the electrons should be accelerated when a electric field exists. This leads to the continuous increase of the electric current with time. However, at the moment of applying an electric field, the electric current reaches a constant value. This indicates that some frictions on the electrons exist to hinder the acceleration of the external electric field. The friction comes from the electron scattering caused by defects in crystal lattice. These defects include vacancies, interstitial and impurity atoms, dislocations, and even vibration of the atoms themselves. All these defects cause kinetic energy loss of the electrons and change of motion direction. After balancing these obstacles, the net electron movement in the opposite direction of the electric field reaches a stable value. The stable electron flow is called electric current.

The electron scattering is considered to be a resistance to the current, which can be measured by the electron drift velocity and electron mobility. The drift velocity, $v_d$, is defined as the average electron velocity in the direction of external electric field force. It is proportional to the electric field intensity, $\zeta$:

$$v_d = \mu_e \zeta \tag{10.8}$$

where the proportional constant $\mu_e$ is called electric mobility. Its unit is m$^2$/V·s.

For most materials, the electrical conductivity can be described as

$$\sigma = n|e|\mu_e \tag{10.9}$$

where $n$ denotes the quantities of free electrons per unit volume. $|e|$ is the absolute value of an electron charge ($1.6 \times 10^{-19}$ C). The electrical conductivity is proportional to the electron mobility and the quantities of free electrons.

## 10.2.4 The electrical resistivity of metal

As mentioned earlier, most metals are good electrical conductors. They exhibit high electrical conductivities at room temperature. The typical values of several common metals are collected in Table 10.1. The high conductivity is attributed to the abundance of free electrons in metals. In other words, the electrons in metals can be easily excited to jump into the empty state above Fermi energy. So the $n$ has a large value in eq. (10.9).

**Table 10.1:** Electrical conductivities of some common metals and alloys at room temperature[1, 2].

| Metals | Electrical conductivity $(\Omega \cdot m)^{-1}$ | Metals | Electrical conductivity $(\Omega \cdot m)^{-1}$ |
|---|---|---|---|
| Silver | $6.9 \times 10^7$ | Brass($Cu_{70}Zn_{30}$) | $1.6 \times 10^7$ |
| Copper | $6.0 \times 10^7$ | Iron | $1.0 \times 10^7$ |
| Gold | $4.3 \times 10^7$ | Mild steel | $0.6 \times 10^7$ |
| Aluminum | $3.8 \times 10^7$ | Stainless steel | $0.2 \times 10^7$ |

Since the electrical resistivity is inversely proportional to the electrical conductivity, it is often used to represent the electrical conductivity of metals. The conduction electrons in metal can be scattered by crystal defects that are main obstacles to electron flow. More defects correspond to large electrical resistivity or small electrical conductivity. The amount of defects in a metal depends on its composition and processing state (e.g., heat treatment and deformation extent). It is also related to temperature. It has been proved experimentally that the total electrical resistivity is attributed from the three contributions: thermal vibration, impurities and plastic deformation. Because each of them affects electron scattering independently, the total electrical resistivity can be expressed as

$$\rho_{total} = \rho_t + \rho_i + \rho_d \tag{10.10}$$

where $\rho_t$, $\rho_i$, and $\rho_d$ represent the contributions of temperature, impurities, and deformation, respectively, to the electrical resistivity. The above equation is also called Matthiessen's rule. As an example, the electrical resistivity of pure copper and its alloys (Cu–Ni) plotted in a resistivity–temperature coordinates clearly shows the respective contributions of the three factors as shown in Fig. 10.4. After adding different amounts of nickel in copper, the curve moves upward for a certain distance accordingly, indicating the impurity contribution superimposed on the basic resistivity of pure copper. The deformed alloy also reveals the contribution of deformation to the resistivity.

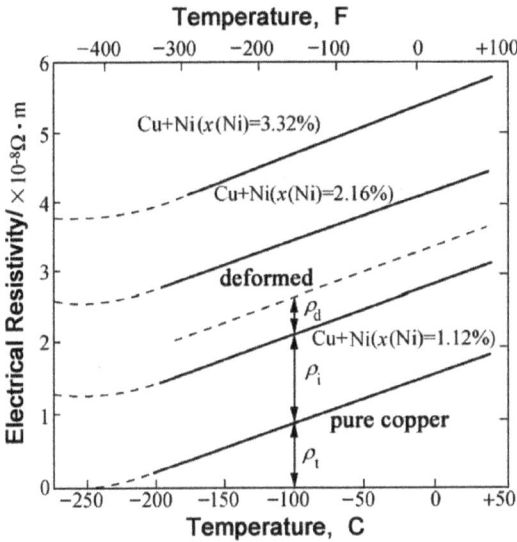

**Fig. 10.4:** Relationships between electrical resistivity and temperature for copper and Cu–Ni alloys.

### 10.2.5 The electrical conductivity of intrinsic and extrinsic semiconductor

Unlike metals, semiconductors exhibit lower electrical conductivity. Their unique electrical characteristics make them of some special applications. These materials are particularly sensitive to a small amount of impurity. High-purity materials exhibit semiconducting behavior via their inherent electron structure, which are called intrinsic semiconductors. The materials with impurity that dominates the semiconducting property are called extrinsic semiconductors.

The electron band structure of the intrinsic semiconductor at 0 K conforms to the characteristics shown in Fig. 10.3(d). Such an electron energy band structure is separated into a fully filled valence band and an empty conduction band by a narrow band gap that is typically smaller than 2 eV. Silicon and germanium are two typical intrinsic semiconductors. Their band gap energy is about 1.1 and 0.7 eV, respectively. They both belong to group IVA in the periodic table, and have covalent bond. Some compound semiconductors also exhibit intrinsic semiconducting behavior.

The elements between groups IIIA and VA can form such compounds, for example, GaAs and InSb. The elements in groups IIB and VA form compounds that also exhibit semiconductor behavior, for example, CdS and ZnTe. Since the two elements are far each other in the periodic table, their atom bonding tends to be ionic. Their band gap energy also becomes larger. Thus, these compounds exhibit more insulative. For easy reference, some electrical properties of the intrinsic semiconductors and compounds discussed above are collected in Table 10.2.

**Table 10.2:** Electrical properties of the intrinsic semiconductors and compounds at room temperature[3].

| Materials | Band gap energy (eV) | Electrical conductivity $(\Omega \cdot m)^{-1}$ | Electron mobility $(m^2 \cdot (V \cdot s)^{-1})$ | Vacancy mobility $(m^2 \cdot (V \cdot s)^{-1})$ |
|---|---|---|---|---|
| Elements | | | | |
| Si | 1.11 | $4 \times 10^{-4}$ | 0.14 | 0.05 |
| Ge | 0.67 | 2.2 | 0.38 | 0.18 |
| III–V compounds | | | | |
| GaP | 2.25 | – | 0.05 | 0.002 |
| GaAs | 1.35 | $10^{-6}$ | 0.85 | 0.45 |
| InSb | 0.17 | $2 \times 10^4$ | 7.7 | 0.07 |
| II–VI compounds | | | | |
| CdS | 2.40 | – | 0.03 | – |
| ZnTe | 2.26 | – | 0.03 | 0.01 |

In the semiconductor, an electron excited to conduction band leaves a vacancy in the covalent band as shown in Fig. 10.5(b). The position of a runaway electron (vacancy) is called hole and confirmed to move under an electric field, which can be realized by other valence electrons filling the incomplete bond (Fig. 10.5(c)). Hole is considered to be a positive charge with about $1.6 \times 10^{-19}$C (same measure as the negative charge in an electron). The holes and the electrons excited in the semiconductor move in opposite directions under the electric field. Both can be scattered by the defects in the semiconductor.

Considering effect of the holes and the excited electrons in the intrinsic semiconductor, the conductivity formula (10. 9) can be modified by adding the contribution of hole current:

$$\sigma = n|e|\mu_e + p|e|\mu_h \qquad (10.11)$$

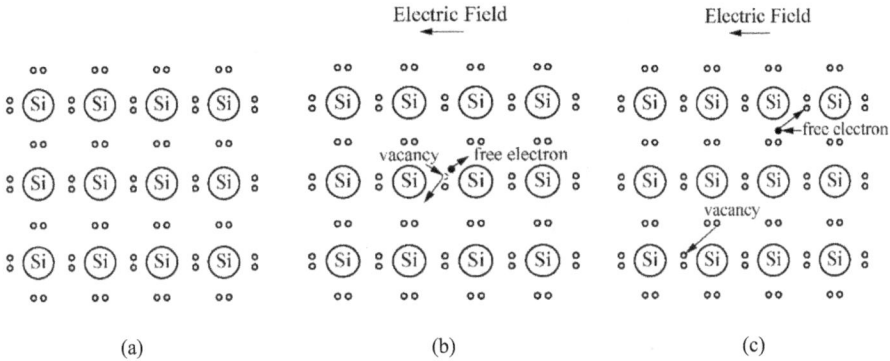

**Fig. 10.5:** Illustration of conductance behavior in intrinsic semiconductor silicon: (a) covalent bond structure of silicon, (b) and (c) motion of electrons and vacancies under an external electric field after excitation.

where $p$ denotes the amount of holes per unit volume. $\mu_h$ indicates the migration rate of the holes. Since each electron excited into conduction band leaves a vacancy in the valence band, one has $n = p$. Thus:

$$\sigma = n|e|(\mu_e + \mu_h) = p|e|(\mu_e + \mu_h) \qquad (10.12)$$

It is found that $\mu_e$ value is always larger than that of $\mu_h$ for those intrinsic semiconductors and compounds discussed above as listed in Table 10.2.

In practice, the commercial products used are always extrinsic semiconductors. Their semiconducting behavior depends on impurities in the material. The small amount of impurities can introduce extra electrons or holes. As an example, even one impurity in $10^{14}$ atoms can make the silicon to be extrinsic.

The silicon semiconductor is considered again in order to explain how the extrinsic semiconductor behaves. There are four valence electrons in a silicon atom. Every valence electron combines with the adjacent silicon atom to form covalent bond. If a pentavalent impurity, from group V A, such as P, As, and Sb, substitutes silicon, only four of the five electrons form covalent bonds with the four adjacent silicon atoms. The remaining one is attracted by weak static electricity around the impurity atoms as shown in Fig. 10.6(a). Because of the relative small bond energy (on the order of 0.01 eV), the electron easily removes from the impurity atom and becomes a free electron or a conducting electron.

The electron with weak bond energy has its own single energy level located in band gap. The level is near the conduction band as shown in Fig. 10.7(a). The bond energy of an electron corresponds to the energy required to excite an electron from an impurity state to an energy state in the conduction band. An electron is donated to the conduction band for every excitation event. The impurities are called donors.

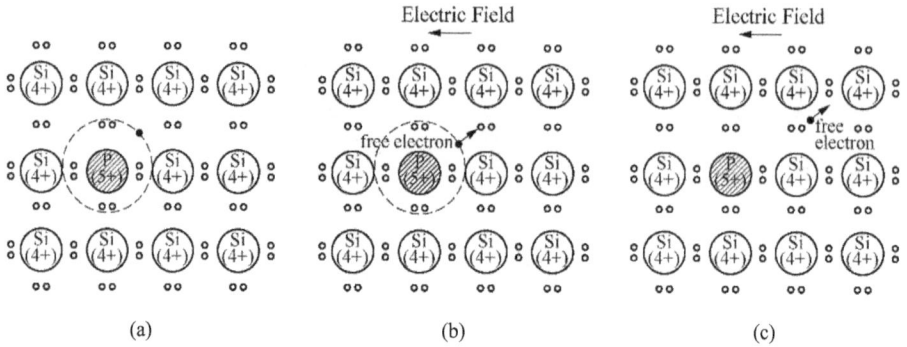

**Fig. 10.6:** Electronic bonding model for nonintrinsic n-type semiconductors. (a) Five-valent phosphorus atom substituted silicon atom, (b) superfluous electrons become free electrons after excitation, and (c) the motion of free electrons in an electric field.

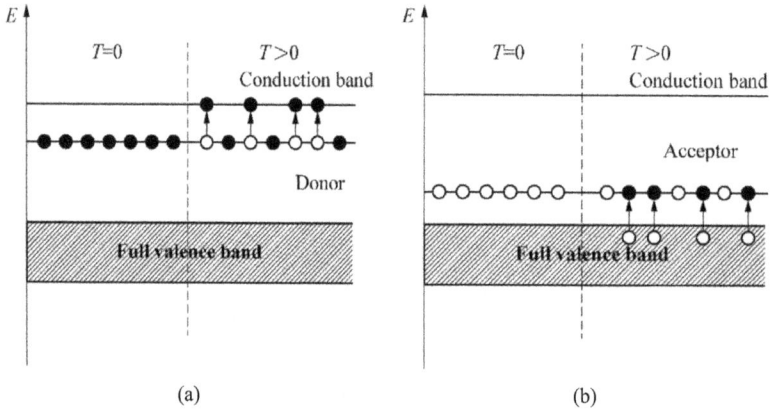

**Fig. 10.7:** Donor and acceptor: (a) n-type and (b) p-type.

Each donor electron comes from the impurity energy state in the band gap. Unlike the intrinsic semiconductors, there is no hole generated in the valence band.

The heat energy acquired at room temperature can excite a large number of electrons from the donor state. The escaping electrons from the valence band in intrinsic semiconductors are very few (Fig. 10.5(b)). Therefore, the amount of free electrons in conduction band is much larger than that of holes in valence band, namely $n \gg p$, thus eq. (10.12) can be approximately rewritten as

$$\sigma \approx n|e|\mu_e \tag{10.13}$$

This kind of material is described as $n$-type extrinsic semiconductor. Their electrical conductivity is mainly determined by the electron concentration.

If trivalence impurities such as Al, B, Ga in the group IIIA in the periodic table substitute the matrix atoms in the intrinsic semiconductors, for example, silicon or germanium, the opposite effect occurs. Lack of an electron in the covalent bonds around each impurity atom can be taken as a hole (vacancy). It is weakly bonded to the impurity atom, and can move by adjacent electron transferring as shown in Fig. 10.8. A hole moves via exchanging positions with an electron. The essence of hole movement is electron movement. So, a hole moving contributes to the conductive process similar to a donor electron mentioned earlier.

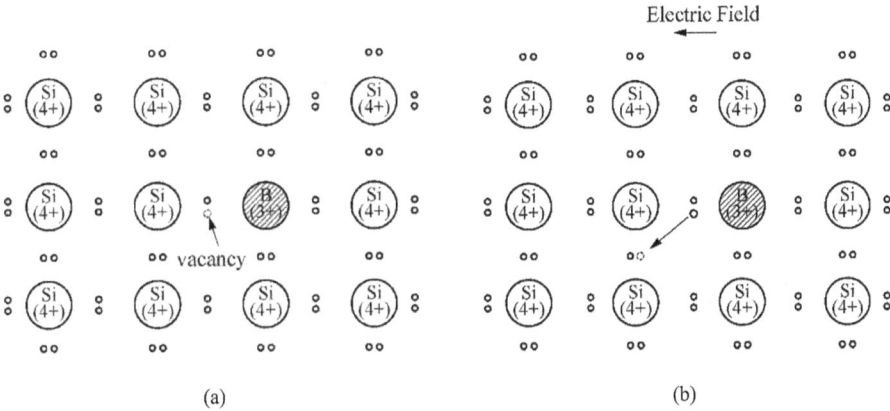

**Fig. 10.8:** Electronic bonding model for intrinsic p-type semiconductors: (a) trivalence boron atom substituted a silicon atom and (b) vacancy movement under an electric field.

The extrinsic activation generating holes can be illustrated by the energy-band model. Every excited impurity atom brings an energy level in band gap. This level is above but near the valence band. Electrons in valence band can be thermally excited into the impurity energy level, and leaves holes in the valence band as shown in Fig. 10.7(b). In this process, only holes are generated in valence band. Free electrons cannot be created in either conduction band or impurity state. This impurity is called acceptor, because it can accept one electron from valence band and leave an electron vacancy. The energy state created by this kind of impurity in the band gap is called acceptor state.

In such extrinsic semiconductor, the number of holes is much larger than that of free electrons (i.e., $p \gg n$), because the holes can be created not only by extrinsic excitation of acceptor, but also by intrinsic excitation of valence band. This material is called p-type semiconductor. Since positive charge carries dominate conductance, holes are main carries contributed to the electrical conductivity. Accordingly, eq. (10.12) can be simplified as

$$\sigma \approx p|e|\mu_h \tag{10.14}$$

Extrinsic semiconductors (including *n*-type and *p*-type) can be firstly made of high purity material (total impurity is under $10^{-7}$ at.%). Then one can mix the impurity into the high-purity material as donors or acceptors. The semiconducting properties may be tailored by adjusting the impurity concentration. This procedure in synthesizing semiconductors is called doping.

In extrinsic semiconductors, the charge carriers can be created by heat activation at room temperature. They exhibit relatively high electrical conductibility at ambient temperature. Most of these semiconductors are designed as electronic components worked at room temperature.

### 10.2.6 Electrical conductivity and dielectricity of insulator

Insulators have a wide band gap between the fully filled valence band and the empty conduction band as shown in Fig. 10.3(c). The band gap energy is generally larger than 2 eV, so that only few electrons can be thermally excited to jump over the band gap at room temperature. This means very small electrical conductivity. In fact, all the ceramic materials and covalent bond polymers are insulators. Some of them are listed in Table 10.3 to show their electrical conductivities at room temperature. In many cases they are used because of their insulativity. In this regard, high electrical resistivity is very much expected. Similar with semiconductors, insulators increase in electrical conductibility as temperature increases.

**Table 10.3:** Electrical conductivities of several nonmetals at room temperature[4].

| Materials | Electrical conductivity $(\Omega \cdot m)^{-1}$ | Materials | Electrical conductivity $(\Omega \cdot m)^{-1}$ |
|---|---|---|---|
| Graphite | $10^5$ | **Polymers** | |
| **Ceramics** | | Phenolic aldehyde | $10^{-9}$–$10^{-10}$ |
| Alumina | $10^{-10}$–$10^{-12}$ | Nylon | $10^{-9}$–$10^{-12}$ |
| Porcelain | $10^{-10}$–$10^{-12}$ | Poly(methyl methacrylate) | $<10^{-12}$ |
| Na,Ca-glass | $<10^{-10}$ | Polyethylene | $10^{-13}$–$10^{-17}$ |
| Mica | $10^{-11}$–$10^{-15}$ | Polystyrene | $<10^{-14}$ |
| | | Polytetrafluoroethylene | $<10^{-16}$ |

Dielectric materials are a kind of nonmetallic materials that have a large electrical resistivity more than $10^8 \ \Omega \cdot m$. Their response to the electrical field is electric induction rather than electrical conductance. Under an electrical field, the dielectric materials exhibit an electric dipole structure. The positive and negative charges tend to be separated at atomic or molecular level. Because of the interaction of dipole and electric field, dielectric materials are usually used as capacitor.

When a voltage is applied on a capacitor, one plate of the capacitor presents positive charge, and the other plate presents negative charge. The positive and negative charges on the two plates are balanced by the external electric field. The capacitance, $C$, depends on the quantity of electricity, $Q$, stored in the plate, which can be expressed as

$$C = \frac{Q}{V} \tag{10.15}$$

where $V$ denotes the voltage applied on the capacitor. The capacitance unit is Coulomb per volt (C/V), that is, farad (F).

Considering a parallel plate capacitor with a vacuum cavity between the plates, its capacitance can be calculated with the below formula:

$$C = \varepsilon_0 \frac{A}{l} \tag{10.16}$$

where $A$ indicates the plate area. $l$ indicates the plate spacing. $\varepsilon_0$ is the vacuum permittivity that is a universal constant, that is, $8.85 \times 10^{-12}$ F/m.

When the dielectric material is placed between the two plates, it becomes

$$C = \varepsilon \frac{A}{l} \tag{10.17}$$

where $\varepsilon$ denotes the dielectric permittivity that is far larger than $\varepsilon_0$. The ratio of the two permittivities is called relative dielectric permittivity $\varepsilon_r$:

$$\varepsilon_r = \frac{\varepsilon}{\varepsilon_0} \tag{10.18}$$

The value of $\varepsilon_r$ is much greater than unity. This means that the storage capacity of charge increases when the dielectric material is inserted into the space between plates. This effect mentioned above was first studied by Faraday in 1873 (Fig. 10.9). The relative dielectric permittivity represents a property of material. It should be considered as a key parameter when designing capacitors.

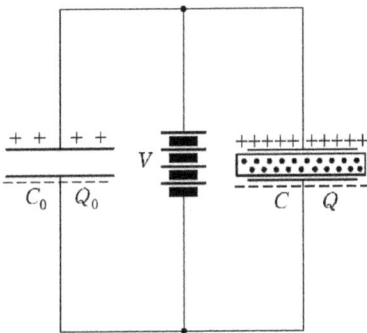

**Fig. 10.9:** Parallel plate capacitance.

The capacitance phenomenon can be well understood with the help of the field vector. The positive and negative charges in a dipole tend to separate. The electric dipole moment, $P$, related to the dipole charge, $q$, and the distance, $d$, between the two poles can be expressed as

$$P = qd \qquad (10.19)$$

There is a potential between the two poles, which starts from the negative pole toward the positive pole. This potential vector is equivalent to the dipole moment. It can rotate toward the direction of the external electric field applied. The directional arrangement of dipoles is called polarization.

For the capacitor, the density of surface charge, $D$, that is, quantity of electric charge per unit area on capacitor plate ($C/m^2$), is determined by the intensity of the electric field, $\zeta$. Under vacuum, it is

$$D_0 = \varepsilon_0 \zeta \qquad (10.20)$$

where $\varepsilon_0$ is the proportional coefficient. There is the similar expression for the dielectric case:

$$D = \varepsilon \zeta \qquad (10.21)$$

$D$ is also named the dielectric displacement.

The polarization model discussed above for the dielectric material can be used to explain the increase of capacitance or dielectric constant. Considering a vacuum capacitor as shown in Fig. 10.9, the dielectric material inserted in the plates is polarized after an electric field is applied. This leads to significant increase in charges on the two plats of the capacitor. In this process, the surface of the dielectric material toward the positive plate of the capacitor accumulates negative charges, while another surface toward the negative plate accumulates positive charges.

In presence of the dielectric material, the surface charge density of the two plates in a capacitor can also be expressed as

$$D = \varepsilon_0 \zeta + P \qquad (10.22)$$

where $P$ represents the contribution of polarization. It can be taken as the increase in charge density after insertion of the dielectric material in the capacitor.

$$P = \frac{Q'}{A}$$

where $Q'$ indicates the increase in charge on the plate of the capacitor after a dielectric material is inserted. $A$ indicates the area of the plate. $P$ and $D$ are in the same unit of $C/m^2$.

The polarization can be considered as a polarization field in the dielectric material, which is caused by the rotation and adjustment of the dipole moments of atoms or molecules under the external electric field. It is simply equivalent to the total dipole moments per unit volume in the material. For many dielectric materials, $P$ is proportional to $\zeta$:

$$P = \varepsilon_0(\varepsilon_r - 1)\,\zeta \tag{10.23}$$

In this case, $\varepsilon_r$ is independent on the electric field intensity. Some dielectric parameters of several dielectric materials are listed in Table 10.4 for reference.

**Table 10.4:** Dielectric constants and dielectric strengths of several dielectric materials.[5, 6].

| Materials | Dielectric constants | | Dielectric strengths $(V \cdot km^{-1})$ |
|---|---|---|---|
| | 60 Hz | 1 Hz | |
| **Ceramics** | | | |
| Titanate ceramics | – | 15–10,000 | 80–483 |
| Mica | – | 5.4–8.7 | 1,609–3,218 |
| Steatite (MgO–SiO$_2$) | – | 5.5–7.5 | 322–563 |
| Na,Ca-glass | 6.9 | 6.9 | 402 |
| Molten silicon oxide | 4.0 | 3.8 | 402 |
| Porcelain | 6.0 | 6.0 | 64–644 |
| **Polymers** | | | |
| Phenolic aldehyde | 5.3 | 4.8 | 483–644 |
| Nylon 66 | 4.0 | 3.6 | 644 |
| Polystyrene | 2.6 | 2.6 | 805–1,126 |
| Polyethylene | 2.3 | 2.3 | 724–805 |
| Polytetrafluoroethylene | 2.1 | 2.1 | 644–805 |

Since ceramics have larger dipole moment, their dielectric constants are mostly larger than that of polymers. The dielectric constants, $\varepsilon_r$, of polymers are generally in the range of 2–5. These materials are usually used in the insulation of electric wire, electric cable, engine, electric generator. In addition, they also can be used in the capacitor.

For example, a parallel capacitor with the plate area of $6.45 \times 10^{-4}$ m$^2$ and the plate spacing of $2 \times 10^{-3}$ m is applied with an electric voltage of 10 V. A dielectric material with relative dielectric constant 6.0 is placed in the zone between the two plates. The followings can be calculated:
a) Capacitance
b) The amount of charge stored in each plate
c) Dielectric displacement
d) Intensity of polarization

Solution:

a) $\varepsilon = \varepsilon_r\varepsilon_0 = (6.0)(8.85\times10^{-12}) = 5.31\times10^{-11}\text{F/m}$

$$C = \varepsilon\frac{A}{l} = 5.31\times10^{-11}\times\frac{6.45\times10^{-4}}{2\times10^{-3}} = 1.71\times10^{-11}\text{F}$$

b) $Q = CV = 1.71\times10^{-11}\times10 = 1.71\times10^{-10}\text{C}$

c) $D = \varepsilon\zeta = \varepsilon\frac{V}{l} = \frac{5.31\times10^{-11}\times10}{2\times10^{-3}} = 2.66\times10^{-7}\text{C/m}^2$

d) $P = D - \varepsilon_0\zeta = D - \varepsilon_0\frac{V}{l} = 2.66\times10^{-7} - \frac{8.85\times10^{-12}\times10}{2\times10^{-3}} = 2.22\times10^{-7}\text{C/m}^2$

## 10.3 Thermal behavior

The thermal behavior of a material refers to the response of the material to heat. When a solid material is heated, the temperature rise and volume expansion will happen. If there is a temperature gradient in a material or medium, the heat will transfer from high temperature to low temperature until the temperature gradient disappears. The thermal property of materials can be characterized by their heat conductivity, heat capacity and thermal expansion coefficient.

### 10.3.1 Heat capacity

The temperature rise of a solid material after heated indicates that some energy is absorbed. The heat capacity indicates the ability of the material to absorb heat energy from the surrounding environment. It is defined as the heat energy needed for the temperature rise of the material by one degree. The heat capacity can be described mathematically as follows:

$$C = \frac{dQ}{dT} \tag{10.24}$$

where $dQ$ indicates the heat energy that is needed to produce a temperature increment, $dT$. The energy needed for 1 mol material to increase 1 K in temperature is called molar heat capacity with the unit of $J\cdot(\text{mol}\cdot K)^{-1}$. The heat capacity per mass is defined as the specific heat capacity, $c$, with the unit of $J\cdot(\text{kg}\cdot K)^{-1}$.

The heat capacity can be measured at either constant volume or constant pressure. The constant volume heat capacity is expressed by $C_v$, while the constant pressure heat capacity is expressed by $C_p$. $C_p$ is always larger than $C_v$. At or below room temperature the difference between $C_p$ and $C_v$ is very small.

In most of solid materials, heat energy is mainly consumed by increasing the vibrational energy of atoms. The vibration occurs with high frequency but low amplitude, which can couple and transfer via the atom bonds. It can be considered as elastic waves or simple harmonic waves with short wavelengths and high frequencies. The vibration waves propagate in sound velocity in crystal solids. Generally, the vibrational heat energy is composed of a series of elastic waves with certain frequency range and distribution. The vibrational wave or heat energy can be quantized. The vibration energy of a single quantum is called a phonon that is similar to a quantum in the electromagnetic radiation. Sometimes the vibration wave is also called phonon.

In the process of electron conduction, the thermal scattering of free electrons occurs via the vibration waves that also contribute to the energy transfer during heat conduction.

For the solid with simple crystal structure, the values of $C_v$ depend on temperature as shown in Fig. 10.10. $C_v$ equals zero at 0 K, but increases rapidly as temperature increases. This is because the vibration wave becomes stronger at relatively higher temperature. At low temperature, the relation of $C_v$ to temperature, $T$, can be written as

$$C_v = AT^3$$

where $A$ is a constant independent of temperature. Above the Debye temperature, $\theta_D$, $C_v$ is independent on temperature, and is approximately equal to $3R$ ($R$: gas constant). Thus, the heat energy needed for increasing 1 K in temperature is constant, although the total energy in material increases with the increase of temperature. Because the value of $\theta_D$ is below room temperature for many materials, their $C_v$ values are about 25 J·(mol · K)$^{-1}$ at room temperature. Table 10.5 collects several thermal parameters including $C_p$ of some common materials for reference.

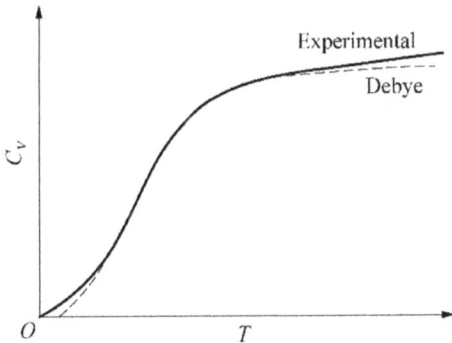

**Fig. 10.10:** Relations of the constant volume heat capacity to temperature.

Heat capacity increase also involves other thermal absorption mechanisms. For example, the electrons can absorb more energy via providing the higher kinetic energy for free electrons. When ferromagnetic materials are heated above Curie point, randomization of electron spin absorbs energy. In most cases, these contributions are relatively smaller than that of the vibration.

**Table 10.5:** Thermal properties of common materials[7, 8].

| Materials | $c_p$ $J \cdot (kg \cdot K)^{-1}$ | $a_l$ $\times 10^{-6}$ $(°C)^{-1}$ | $k$ $W \cdot (m \cdot K)^{-1}$ | $L$ $\times 10^{-8}(\Omega \cdot W) \cdot K^{-2}$ |
|---|---|---|---|---|
| **Metals** | | | | |
| Aluminum | 900 | 23.6 | 247 | 2.24 |
| Copper | 386 | 16.5 | 398 | 2.27 |
| Gold | 130 | 13.8 | 315 | 2.25 |
| Iron | 448 | 11.8 | 80.4 | 2.66 |
| Nickel | 443 | 13.3 | 89.9 | 2.10 |
| Silver | 235 | 19.0 | 428 | 2.32 |
| Tungsten | 142 | 4.5 | 178 | 3.21 |
| 1025 steel | 486 | 12.5 | 51.9 | |
| 316 stainless steel | 502 | 16.0 | 16.3 | |
| Brass (70Cu–30Zn) | 375 | 20.0 | 120 | |
| **Ceramics** | | | | |
| Aluminum oxide (Al$_2$O$_3$) | 775 | 8.8 | 30.1 | |
| Beryllium oxide (BeO) | 1,050 | 9.0 | 220 | |
| Magnesium oxide (MgO) | 940 | 13.5 | 37.7 | |
| Spinel (MgAlO$_4$) | 790 | 7.6 | 15.0 | |
| Ca,Na-glass | 840 | 9.0 | 1.7 | |
| Molten silicon oxide (SiO$_2$) | 740 | 0.5 | 2.0 | |
| **Polymers** | | | | |
| Polyethylene | 2,100 | 60–220 | 0.38 | |
| Polypropylene | 1,880 | 80–100 | 0.12 | |
| Polystyrene | 1,360 | 50–85 | 0.13 | |
| Polytetrafluoroethylene | 1,050 | 100 | 0.25 | |
| Phenolic aldehyde | 1,650 | 68 | 0.15 | |
| Nylon 66 | 1,670 | 80–90 | 0.24 | |
| Polyisoprene | | 220 | 0.14 | |

## 10.3.2 Thermal expansion

Most solid materials are thermal expansion and cold-shrink. Their changes in length are the function of temperature:

$$\frac{l_f - l_0}{l_0} = \alpha_l (T_f - T_0) \qquad (10.25(a))$$

or

$$\frac{\Delta l}{l_0} = \alpha_l \Delta T \qquad (10.25(b))$$

where $l_0$ and $l_f$ are initial length and ultimate length when temperature changes from $T_0$ to $T_f$. The linear expansion coefficient, $\alpha_l$, represents the dilation property when the material is heated. The unit of $\alpha_l$ is $K^{-1}$. In general, heating or cooling affects the three dimensional size of an object, resulting a change in volume. The relationship between volume and temperature can be expressed as the following formula:

$$\frac{\Delta V}{V_0} = \alpha_v \Delta T \tag{10.26}$$

where $V_0$ is the initial volume. $\Delta V$ denotes the volumetric increment. $\alpha_v$ is the coefficient of volume expansion. Its value depending on the crystallographic orientation is often anisotropy. For the materials with isotropic expansion, the value of $\alpha_v$ is about $3\alpha_l$.

On atomic scale, the thermal expansion implies the increase in average distance between atoms. The potential energy of atoms has relations to their distance. At a certain temperature, there is an equilibrium atomic distance, at which the potential energy has a minimum value. At 0 K, the atomic vibration can be ignored. The equilibrium atomic distance is $r_0$ in this case. When heated up to a higher temperature, for example, $T_1$, $T_2$, $T_3$, the vibrational energy becomes $E_1$, $E_2$, $E_3$, respectively. The stronger vibration (i.e., higher potential energy) corresponds to the larger vibration amplitude. The average atomic distance shifts to right side as temperature increases according to the asymmetric curve as shown in Fig. 10.11(a). As the average atomic distance increases, the material expands.

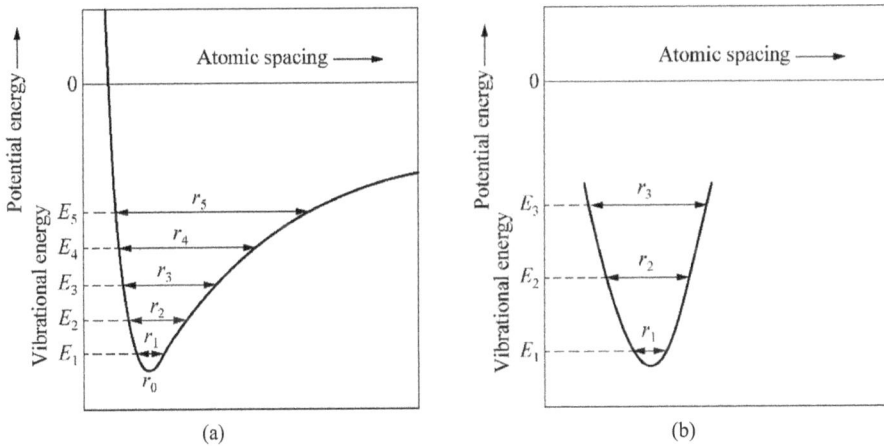

**Fig. 10.11:** The relationship between potential energy and atomic spacing: (a) asymmetric curve and (b) symmetric curve.

As discussed earlier, the larger vibration amplitude does not directly contribute to the increment of the atomic distance. The asymmetric potential energy curve causes larger average atomic distance as temperature increases, and thus results in the thermal expansion. If the curve is symmetric (Fig. 10.11(b)), the average atomic distance keeps unchanged as temperature changes. Thus, the thermal expansion does not occur in this case.

By comparing different types of materials including metals, polymers and ceramics, one can find that the larger atomic bond energy corresponds to the deeper and narrower potential energy curve. This means that the increment of the atomic distance upon heating is relatively small. Thus, the materials with the higher atomic bond energy have the smaller values of $a_l$. The linear expansion coefficients of the three types of materials are listed in Table 10.5.

### 10.3.3 Heat conduction

Heat conduction is a phenomenon that heat transmits from the high temperature zone of a material to the low temperature zone. The heat transfer capability of a material can be described by heat conductivity, $k$ (in W $\cdot$ (m $\cdot$ K)$^{-1}$), which is defined by following formula:

$$q = -k \frac{dT}{dx} \qquad (10.27)$$

where $dT/dx$ indicates the temperature gradient along the heat transfer direction in the material. $q$ indicates the heat flux, which is the heat flowing through a unit area at unit time, that is, W $\cdot$ m$^{-2}$. Eq. (10.27) is valid when the heat flux is independent on time (i.e., steady-state heat transfer). The negative sign in the above formula indicates that the heat flows from high temperature to low temperature. Eq. (10.27) is similar to Fick's first law for the atomic diffusion. Both $k$ and $dT/dx$ are similar to the diffusion coefficient and the concentration gradient in Fick's first law, respectively.

The heat transfer in solid is realized by lattice vibration wave (phonon) and free electrons. Both contribute to the heat conductivity.

$$k = k_1 + k_e \qquad (10.28)$$

where $k_1$ indicates the heat conductivity due to the lattice vibration, while $k_e$ indicates that due to the free electrons. Ordinarily, the heat transfer in a certain material is dominated by one of the above two mechanisms. The contribution of $k_1$ is due to the net movement of phonons from the high temperature region to the low temperature region. In a similar way, that of $k_e$ is due to the movement of free electrons from the high temperature region to the low temperature region. In the process, free electrons can transfer part of kinetic energy to atoms by colliding with phonons and defects in the crystal lattice, becoming vibrational energy. The contribution of $k_e$ increases as the amount of free electrons increases.

If a metal is pure, the contribution of free electrons is far larger than that of phonon. The reason is that the electrons with high velocity, unlike phonons, are not easy to be scattered. Metals are the excellent thermal conductors, because a large number of free electrons in metals contribute to the heat conduction. The heat conductivities of some common metals are listed in Table 10.5. Generally, the heat conductivity of metals is in the range of 20–400 W·$(m \cdot K)^{-1}$.

Because the free electrons dominate the conduction of electricity and heat in metals, the relationship of the two kinds of conductivity is revealed by Wiedemann–Franz law:

$$L = \frac{k}{\sigma T} \qquad (10.29)$$

$L$ is the ratio of the heat conductivity to the product of the electrical conductivity, $\sigma$, and temperature. It can be experimentally determined. One can find that the values of $L$ of metals are very close to a theoretical value: $2.44 \times 10^{-8}$ $(\Omega \cdot W) \cdot K^{-2}$, as listed in Table 10.5. In metals, if the heat transfers are completely attributed to the free electrons, their $L$ values are the same, equal to the above theoretical value.

Impurities in metals lead to the decrease of heat conductivity, and also the decrease of electrical conductivity. The impurity atoms distributed in the metal matrix impede and scatter the moving electrons, which reduces the efficiency of electron motion.

Nonmetallic ceramics with few free electrons are thermal insulators. The thermal conduction in ceramics occurs mainly by phonons. $k_l$ is much larger than $k_e$ in these materials. Additionally, because the phonon is easy to be scattered by the lattice defects, the phonon is less effective than the free electrons during heat transfer. Therefore, the heat conductivities of ceramics are much smaller than that of metals. The data of some ceramic materials are collected in Table 10.5. Their heat conductivities at room temperature are in the range of 2–50 W·$(m \cdot K)^{-1}$. Since the highly disordered atomic structure strongly scatters phonons, the amorphous ceramics and glasses exhibit the smaller heat conductivity than the crystalline ceramics.

Since the scatter of lattice vibration becomes more significant at the higher temperature, most ceramic materials exhibit the lower heat conductivities as temperature increases. This phenomenon can be found at the relatively low temperature range. However, the heat conductivity increases as temperature increases at the relatively high temperature range. This is attributed to the thermal radiation that can transfer heat in ceramics. Its effect is enhanced as temperature increases.

Most polymers exhibit the heat conductivity in a level of 0.3 W·$(m \cdot K)^{-1}$, as listed in Table 10.5. The heat transfer is realized via vibration, migration, and rotation of chain molecules in these materials. The heat conductivity is related to the degree of crystallinity. The high crystallinity and ordered structure always exhibit the larger heat conductivity than the amorphous or disordered one. This is because that the crystalline molecular chain is more effective to coordinate and transfer the heat vibration.

### 10.3.4 Thermal stress

Thermal stress is induced by temperature change in solids. It can cause undesirable plastic deformation and cracks in materials. Understanding the origin and feature of thermal stress is very important for use of materials.

If an isotropic and homogeneous bar is heated or cooled uniformly, it can expand or contract freely without stress. There is no temperature gradient in the bar in this case. However, if the axial motion is constrained by the rigid supporters, a thermal stress, $\sigma$, is generated. If temperature changes from $T_0$ to $T_f$, the thermal stress is

$$\sigma = E\alpha_l(T_o - T_f) = E\alpha_l\Delta T \tag{10.30}$$

where $E$ denotes elastic modulus. $\alpha_l$ denotes the linear expansion coefficient. For the restrained bars, a compressive stress ($\sigma < 0$) arises when heated ($T_f > T_0$). Or, a tensile stress ($\sigma > 0$) is induced when cooled ($T_f < T_0$). The above equation is valid when the bar expands or contracts elastically in response to the temperature change.

In engineering, the temperature inhomogeneity in a material depends on its size, shape, heat conductivity and heating/cooling rate. Thermal stress can be resulted from the temperature gradient inside the material. The existence of temperature gradients is generally due to rapid heating or cooling. In this case, the surface temperature of the material changes faster than inside, which causes inhomogeneous expansion or contraction. In the process, each regions in the material response to the temperature change differently, which constrain the free expansion or contraction of its adjacent regions. For example, the surface layer of a material has the higher temperature and tends to expand more significant than inside when heated. In this process, the surface produces compressive stress, which is balanced with the internal tensile stress. Conversely, the surface produces tensile stresses when cooled rapidly.

For ductile metals and polymers, the stress induced by heat can be relaxed by the plastic deformation. But, most ceramics without ductility are easily broken under thermal stress, and exhibit brittle fracture mode. A brittle object is much easier to be broken at rapidly cooling than at quickly heating, because its surface suffers tensile stress when cooled. It is easier for cracks to originate from and propagate on the surface.

The ability to withstand this heat shock is called the thermal shock resistance (TSR). When ceramics undergo rapid cooling, their TSR depends not only on the temperature fall, but also on their thermal and mechanical properties. Generally, the large heat conductivity and small thermal expansion coefficient are beneficial to TSR. The higher fracture strength, $\sigma_f$, and lower elastic modulus are also favorable to the TSR. The TSR of materials can be approximately evaluated by the following:

$$\text{TSR} \approx \frac{\sigma_f k}{E\alpha_l} \tag{10.31}$$

The thermal shock can be prevented by changing external conditions, for example, decreasing cooling or heating rate and reducing the temperature gradient in materials. Although both the thermal and mechanical properties together determine the TSR, the thermal expansion coefficient is relatively most easily adjusted. Reducing the thermal expansion coefficient is an effective way to improve TSR.

## 10.4 Magnetic performance

The magnetism was first discovered in ferromagnetic minerals with strong magnetism. In 2500 BC, magnetite was used in the guide in China. Magnetism has been known thousands of years ago, but its principle and mechanism is quite complex and subtle. Magnetism puzzled scientists for a long time. The establishment of modern magnetism and the development of magnetic materials have occurred in nearly a hundred years, which is related to the modern industry. Many modern electrical, electronic, and computer technologies rely on magnetism and magnetic materials. This section will introduce the origin of magnetic field, and discuss various magnetic properties and their characterization. Some magnetic materials will also be involved.

### 10.4.1 Representation of magnetic property

The moving charged particles produce magnetic force. It is convenient to consider the magnetic force as a field. The position and orientation of a magnetic field can be imagined as be composed of a series of magnetic lines of force. For example, the magnetic field distribution created by an electrical current loop and a magnetic bar can be clearly revealed by using the magnetic lines of force.

The magnetic dipole discovered in the magnetic materials is similar to the electric dipole discussed earlier. The magnetic dipole can be regarded as a small magnetic bar with south and north poles instead of the positive and negative charges in electric dipole. The magnetic dipole moments can be indicated by arrows (from S pole to N pole). The magnetic dipole is affected by the magnetic field, similar to the electric dipole behavior under an electric field. The torque produced by the magnetic force makes the dipoles to deflect toward the magnetic field direction.

The magnetic behavior can be described by several magnetic field vectors. The external magnetic field, also called magnetic field intensity, is denoted by $H$. If the magnetic field is produced by a spiral coil made up of $N$ turns, its intensity is

$$H = \frac{NI}{l} \qquad (10.32)$$

where $l$ is the length of the coil. $I$ indicates the electric current flowing through the coil. $H$ has the unit of A/m.

Magnetic induction intensity denoted by $B$, also called magnetic flux density, indicates the internal magnetic field intensity produced by an external magnetic field. $B$ has the unit of tesla (T). Like $H$, $B$ is also a magnetic field vector. The relationship between the magnetic induction intensity and the magnetic field intensity is:

$$B = \mu H \tag{10.33}$$

where $\mu$ reveals the characteristics of a specific medium in a magnetic field, is called magnetic permeability. The unit of $\mu$ is Wb/(A · m).

In vacuum:

$$B_0 = \mu_0 H \tag{10.34}$$

where $\mu_0$ denotes the vacuum magnetic permeability. It is a universal constant; equal to $1.257 \times 10^{-6}$ H/m. $B_0$ denotes the magnetic flux density in vacuum.

The ratio of the magnetic permeability and the vacuum magnetic permeability, called relative magnetic permeability and denoted by $\mu_r$, is also often used in magnetism:

$$\mu_r = \frac{\mu}{\mu_0} \tag{10.35}$$

The relative magnetic permeability is a measure to reveal the extent to which the material can be magnetized under an external magnetic field $H$.

Magnetization, another magnetic parameter of materials, denoted by $M$, is defined by the following relation:

$$B = \mu_0 H + \mu_0 M \tag{10.36}$$

When an external magnetic field is applied, the magnetic moments in the material intend to rotate toward the external field direction. The induced magnetism overlapped on the external magnetic field strengthens the total magnetic flux density. Its contribution is indicated by $\mu_0 M$ in eq. (10.36).

The value of $M$ depends on the external magnetic field intensity:

$$M = \chi_m H \tag{10.37}$$

where the scale coefficient, $\chi_m$, is dimensionless, named as the magnetic susceptibility. Its relation to the relative permeability can be easily derived:

$$\chi_m = \mu_r - 1 \tag{10.38}$$

These magnetic parameters are similar to the dielectric parameters. $B$ is like $D$ (dielectric displacement); $H$ is like $\zeta$ (electric field intensity); $\mu$ is like $\varepsilon$ (electronic conductivity); and $M$ is like $P$ (polarization intensity).

The magnetic units are easily confused, because they have two systems of units in actual use. One is "International System of Units" (SI) (i.e., $M \cdot K \cdot S$ (meter kilogram second)). The other is the centimeter gram second (c g s) unit system. The conversion of the two unit systems is summarized in Table 10.6.

**Table 10.6:** Magnetic units and conversion factors in the international system of units and the centimeter–gram–second system of units.

| Physical quantity | Symbol | International System of Units | | Gaussian-CGS units | SI–CGS unit conversion |
|---|---|---|---|---|---|
| | | Derived unit | Elementary unit | Electromagnetic unit | |
| Magnetic induction intensity | $B$ | $W_b/m^2$ | $kg/s \cdot C$ | Gs | $1\,T = 10^4 Gs$ |
| Magnetic field intensity | $H$ | $A \cdot t/m$ | $C/m \cdot s$ | Oe | $1\,A/m$ $= 4\pi \times 10^{-3}$ Oe |
| Magnetization | $M$ (Int. unit) $I$ (CGS unit) | $A \cdot t/m$ | $C/m \cdot s$ | $Mx/cm^2$ | $1\,A/m = 10^{-3}$ $Mx/cm^2$ |
| Permeability in vacuum | $\mu_0$ | $H/m^6$ | $kg \cdot m/C^2$ | (1 emu) | $4\pi \times 10^{-7}\,H/m$ $= 1$ emu |
| Relative permeability | $\mu_r$ $\mu$ | – | – | – | $\mu_r = \mu'$ |
| Magnetic susceptibility | $X_m$ (Int. unit) $X'_m$ (CGS unit) | – | – | – | $X_m = 4\pi X'_m$ |

## 10.4.2 Origin of magnetic moment

The macroscopic magnetic properties of materials originate from the atomic magnetic moment. It is produced by the movement of electrons, protons, and neutrons in atoms. Electrons contribute to the magnetic moment by the orbital motion around the nucleus and the rotation around the spin axis. The former generates the magnetic moment of the electronic orbit with the direction toward the axis of the orbit rotation. The latter produces the electron spin magnetic moment with the direction toward the spin axis. Protons and neutrons also contribute to the magnetic moment by moving inside the nucleus. But, their contribution is about 2,000 times smaller than that of electrons, and can be ignored in the macroscopic magnetism.

Each electron in atoms can be regarded as a small magnet permanently with the orbital and spin magnetic moments. The magnetic moment produced by a single free electron with spin is $9.27 \times 10^{-24}$ A $\cdot$ m$^2$, which is called Bohr magneton, denoted by $\mu_B$. Each electron in atoms generates the spin magnetic moment and the orbital magnetic moment. The former is $\pm\mu_B$ (the up is positive, the down is negative), while the latter is equal to $m_1\mu_B$. $m_1$ is the magnetic quantum number of electrons.

Orbital magnetic moments of some electron pairs in an atom counteract each other if they have opposite directions. This counteraction also occurs for spin magnetic moments. For instance, the spin-up magnetic moment can be offset by the spin-down magnetic. After deducting the cancelled magnetic moment, the accumulation of the residual orbital and spin magnetic moments is the final net magnetic moment in an atom. If all the atom shells or sub-shells are completely filled with electrons, it is possible that all the orbital magnetic moment and spin magnetic moment are completely eliminated by offsetting each other. Therefore, the materials with such atomic structure (e.g., helium, neon, argon, and some ionic materials) cannot be permanently magnetized. In general, the magnetic behavior of materials depends on the magnetic dipoles in atoms and their response to the external magnetic field.

### 10.4.3 Classification of magnetism

Diamagnetic (or antimagnetic) materials exhibit very weak magnetic response under an external magnetic field. Their weak magnetism can only exist with an external magnetic field, thus is nonpermanent. In the atomic structure, the electronic orbital motion can be affected by the external magnetic field, which induces a small magnetic moment. The induced dipoles turn to opposite direction of the external field in this material, as shown in Fig. 10.12(a). In this case, the value of $\mu_r$ is less than 1. $\chi_m$ is negative according to eq. (10.38), which is in the order of $-10^{-5}$ for the antimagnetic material. Magnetic induction intensity in the antimagnetic material is less than that in vacuum. When antimagnetic materials are placed in a strong magnetic field, they will be pushed to the weak magnetic field area.

For the antimagnetic materials, the magnetic induction intensity in the materials is negative, which increases in absolute value as the external magnetic field intensity increases as shown in Fig. 10.13. Some materials with typical antimagnetic behavior are listed in Table 10.7 for reference. All their magnetic susceptibilities are negative. Antimagnetism is very weak, but exists in all materials. In ferromagnetic and even paramagnetic materials, the antimagnetism is completely offset by the magnetism, so that it cannot be observed.

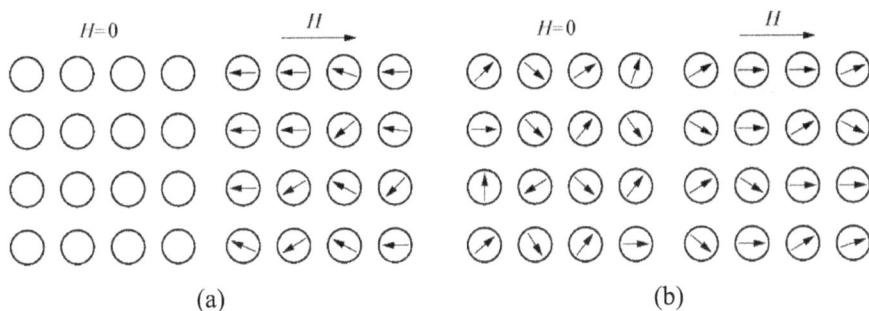

Fig. 10.12: Atomic dipole configurations of diamagnetic and paramagnetic materials with/without external magnetic field. (a) No dipoles exist without external magnetic field, and the induced dipoles with an external magnetic field, in which the dipoles turn to opposite direction of the external field. (b) Atomic dipole configurations of paramagnetic materials with/without external magnetic field.

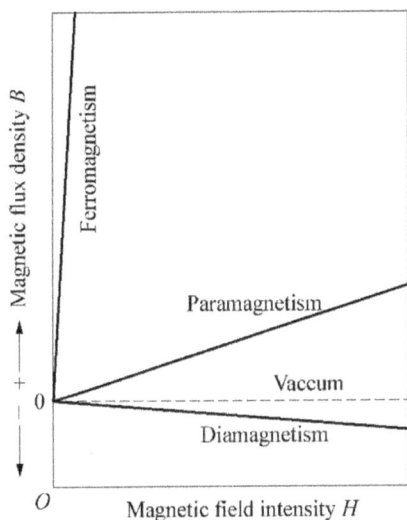

Fig. 10.13: Magnetic induction intensity vs. magnetic field intensity for antimagnetism, paramagnetism and ferromagnetism.

Unlike antimagnetic materials, some solids have permanent dipole moment due to unbalanced spin and orbital magnetic moments. These atomic magnetic moments are in arbitrary orientation, so that the materials do not exhibit macroscopic magnetic behavior. Under an external magnetic field, these atomic magnetic moments are inclined toward the external field, and then produce paramagnetism as shown in Fig. 10.12(b). There is no interaction between these magnetic dipoles in the process. The induced magnetism superposed with the external magnetic field leads to the relative magnetic conductivity slightly greater than one. For the paramagnetic materials, the magnetic susceptibility is positive, and is in the order of $10^{-5}$–$10^{-2}$ (Table 10.7).

**Table 10.7:** Magnetic susceptibility of antimagnetic and paramagnetic materials at room temperature.[5].

| Antimagnetic | | Paramagnetic | |
|---|---|---|---|
| Materials | Susceptibility $\chi_m$ | Materials | Susceptibility $\chi_m$ |
| Aluminum oxide | $-1.81 \times 10^{-5}$ | Aluminum | $2.07 \times 10^{-5}$ |
| Copper | $-0.96 \times 10^{-5}$ | Chromium | $3.13 \times 10^{-4}$ |
| Gold | $-3.44 \times 10^{-5}$ | Chromic chloride | $1.51 \times 10^{-3}$ |
| Mercury | $-2.85 \times 10^{-5}$ | Manganese sulfide | $3.70 \times 10^{-3}$ |
| Silicon | $-0.41 \times 10^{-5}$ | Molybdenum | $1.19 \times 10^{-4}$ |
| Silver | $-2.38 \times 10^{-5}$ | Sodium | $8.48 \times 10^{-6}$ |
| Sodium chloride | $-1.41 \times 10^{-5}$ | Titanium | $1.81 \times 10^{-4}$ |
| Zinc | $-1.56 \times 10^{-5}$ | Zirconium | $1.09 \times 10^{-4}$ |

The magnetic induction intensity in the paramagnetic material under an external magnetic field behaves following the paramagnetic $B$–$H$ relation as shown in Fig. 10.13.

Both the antimagnetic and paramagnetic solids are regarded as nonmagnetic materials. They only exhibit magnetism under the action of external magnetic field. The magnetic flux density in the two types of materials is almost the same as in vacuum.

Some metallic materials have the permanent magnetic moment without external magnetic field, and exhibit very strong permanent magnetization. This is the characteristics of ferromagnetism. The transition metals, iron, cobalt, nickel, and rare earth metal, gadolinium, are ferromagnetic materials. The magnetic susceptibility of ferromagnetic materials can reach $10^6$. Thus $H \ll M$. The magnetic induction intensity can be simplified to the following formula according to eq. (10.36):

$$B \approx \mu_0 M \tag{10.39}$$

The permanent magnetic moment of ferromagnetic materials originates from the uncanceled spin magnetic moments. It depends on their electron structures. The orbital magnetic moment in ferromagnetic materials also contributes to magnetism. But, this contribution is very small compared with that of spin magnetic moment. The latter involves coupling interaction between adjacent atoms, which leads to an alignment of the spin magnetic moments in the same direction. This process can occur without external magnetic field as shown in Fig. 10.14. The coupling interaction is induced by the electrostatic interaction between the electrons in the adjacent atoms. The electrostatic energy makes the magnetic moments in parallel or antiparallel. This phenomenon has been proved by Heisenberg and Frenkel based on quantum theory. The parallel spin magnetic moments can gather in a certain region in material. The region is called domain.

$H=0$

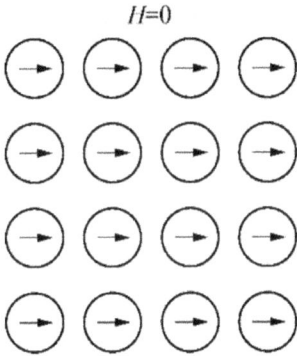

**Fig. 10.14:** Spontaneous arrangement of atomic dipoles in ferromagnetic materials.

If the magnetic dipoles in ferromagnetic materials are all rotated toward the external magnetic field, the magnetization approaches the maximum value that is called saturation magnetization $M_s$. It corresponds to the saturation magnetic induction intensity $B_s$. If the net magnetic moment in each atom is known, the saturation magnetization can be simply calculated by multiplying the magnetic moment and the number of atoms. For the well known ferromagnetic metals, Fe, Co, and Ni, their Bohr magnetons are 2.22, 1.72, and 0.60, respectively.

Unlike ferromagnetism, some materials with magnetic moments aligning in antiparallel do not exhibit magnetism. It occurs due to the magnetic moment coupling between adjacent atoms. These materials are antiferromagnetic. For example, the ceramic MnO composed of $Mn^{2+}$ and $O^{2-}$ exhibits antiferromagnetic behavior. Ion $O^{2-}$ has no net magnetic moment, because its spin and orbital magnetic moments are completely canceled. But ion $Mn^{2+}$ has the net magnetic moment that aligns in antiparallel as shown in Fig. 10.15. The opposite magnetic moments counteract each other. The material does not have net magnetic moment.

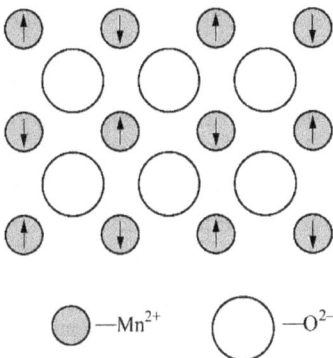

$—Mn^{2+}$  $—O^{2-}$  **Fig. 10.15:** Antiparallel adjustment of spin moment in antiferromagnetic MnO.

Similar to ferromagnetism, some ceramics also exhibit permanent magnetization, are called ferrimagnetism. But they originate from different net magnetic moment.

To understand ferrimagnetism, one can start from $Fe_3O_4$, a mineral magnetite. Its ionic crystal lattice is composed of four $O^{2-}$ ions, two $Fe^{3+}$ ions and one $Fe^{2+}$ ion. The $O^{2-}$ ions without net magnetic moment is neutral in magnetism. The $Fe^{3+}$ ions occupy both the octahedral interstices and tetrahedron interstices. They spin in opposite direction, and thus cancel the magnetic moment between them. The $Fe^{2+}$ ion in octahedral interstices generates the net spin magnetic moment. In the ion structure, the coupling interaction between Fe ions also causes partial antiparallel alignment, similar to antiferromagnetic property, as shown in Fig. 10.16. The incomplete cancellation of the spin magnetic moments results in the net magnetic moment.

$\bigcirc$ —$O^{2-}$    $\bullet$ —$Fe^{2+}$
                Octahedral

$\oslash$ —$Fe^{3+}$    $\oslash$ —$Fe^{3+}$    **Fig. 10.16:** Spin magnetic moment configuration of $Fe_3O_4$ ionic
  Octahedral    Tetrahedral    crystal.

In addition, the hexagonal ferrimagnetic ceramics and the garnet with complex crystal structure exhibit ferrimagnetism. Their saturation magnetization is generally less than that of ferromagnetic materials. However, the ferrimagnetic ceramics, unlike the conductive magnetic material, are electrically nonconductive materials that can be used in some special situations.

The magnetic characteristics of materials can be affected by temperature. Increasing temperature leads to aggravation of the atomic thermal vibration that may eliminate magnetism of materials. This is because the aligned magnetic moments can be disrupted due to the strong atomic vibration. For ferromagnetic, antiferromagnetic and ferrimagnetic materials, the atomic thermal motion can interfere with the coupling force of adjacent atomic dipole moments, leading to misalignment of dipoles regardless of presence of an external magnetic field or not. Obviously, increasing temperature reduces saturation magnetization of ferromagnetic and ferrimagnetic materials. At 0 K, the minimum thermal vibration corresponds to the maximal saturation magnetization. Generally, the saturation magnetization decreases gradually as temperature increases, but quickly drops to zero at a certain temperature. At and above this temperature, the ferromagnetic and ferrimagnetic materials become paramagnetic because the spin coupling in the atomic or ionic structures is completely destroyed. This temperature is called Curie temperature, $T_c$. Each magnetic material has its own Curie temperature. For example, Fe, Co, Ni, and $Fe_2O_3$

exhibit the Curie temperature at 1,041, 1,393, 608, and 858 K, respectively. The antiferromagnetic properties are also affected by temperature. The temperature at which the antiferromagnetism vanishes is called Neel temperature, $T_n$. The antiferromagnetic material becomes paramagnetic material above $T_n$.

### 10.4.4 Domain and hysteresis

Below Curie temperature, the ferromagnetic or ferrimagnetic material can be divided into small regions in microstructure, in which all magnetic dipoles incline toward the same direction as shown in Fig. 10.17. These regions are called domain. Each domain has its own magnetization direction. The adjacent domains may be in different magnetization direction. The domain direction is gradually transitioned to the adjacent domain direction in the domain boundaries as shown in Fig. 10.18. The scale of domain is usually microscopic. One grain in a polycrystalline material generally has more than one domain. Macroscopically, a magnetic material is composed of a large number of domains with various magnetization directions. The total magnetization is the vector sum of all magnetic domains. The contribution of each domain depends on its volume fraction. If the material is not magnetized, the sum of magnetic domain vectors is zero. It is at the lowest energy state.

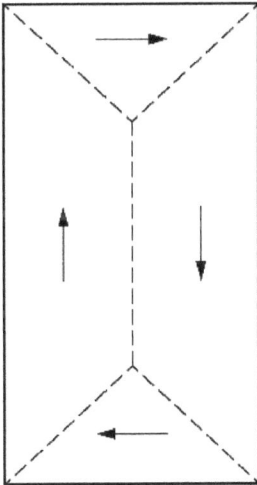

**Fig. 10.17:** Magnetic domains in ferromagnetic or ferrimagnetic materials.

For ferromagnetic and ferrimagnetic materials, the magnetic induction intensity is not proportional to the magnetic field intensity. If material is not magnetized, the magnetic induction intensity changes nonlinearly with the magnetic field intensity as shown in Fig. 10.19. The magnetic induction intensity starts from zero and rises slowly as the magnetic field intensity increases. Then it rises quickly until

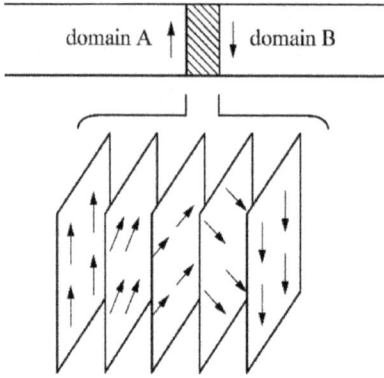

**Fig. 10.18:** The gradual change of magnetic dipole orientation over domain boundary.

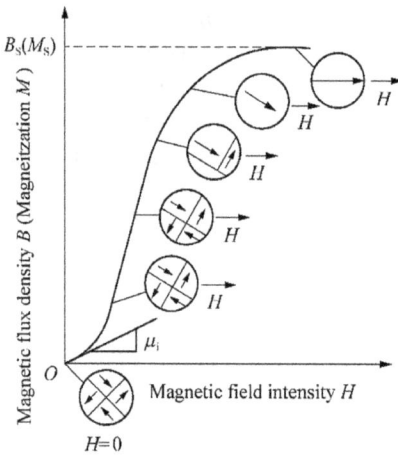

**Fig. 10.19:** $B(M)$–$H$ curve of ferromagnetic or ferrimagnetic material without initial magnetization.

approaching saturation magnetization at which the magnetic induction intensity does not increase as the magnetic field intensity increases.

The magnetic induction intensity at saturation state, $B_s$, corresponds to the saturation magnetization intensity, $M_s$. According to eq. (10.33), the magnetic permeability, $\mu$, is the slope of the B–H curve. It changes with the magnetic field intensity. The slope of B–H curve at $H = 0$ is called initial permeability, $\mu_i$, that sometimes is used to indicate the magnetic property of materials.

Under the action of external magnetic field $H$, all the magnetic moments in domains incline to the $H$. This process drives motion of the domain boundaries until they all disappear and all the magnetic moments are directed toward the external magnetic field as illustrated in Fig. 10.19. Obviously, the domain evolution described above depends on the magnetic field intensity. It occurs continuously as the magnetic field intensity increases until the material becomes a single domain.

When the magnetic moment direction of the single domain is exactly the same as that of the external field, the saturation magnetization is realized.

Once the material is magnetized to saturation, further magnetization is meaningless. If the external magnetic field intensity decreases from the saturation magnetization, that is, point S in the $B$–$H$ curve, as shown in Fig. 10.20, the demagnetization curve changes along SR traces instead of the original magnetization curve (dotted line OS in Fig. 10.20). It is obvious that the degradation of the magnetic induction intensity lags behind the reduction of the external magnetic field. The magnetic hysteresis effect occurs in this case. When $H = 0$, the residual magnetic induction intensity, $B_R$, is called remanence. This means that the material maintains magnetization without external magnetic field.

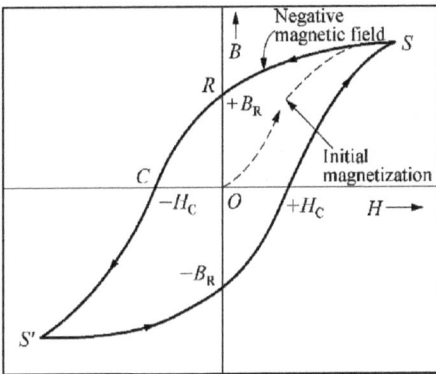

**Fig. 10.20:** Change of magnetic flux density of ferromagnetic materials under positive and negative magnetic field.

The permanent magnetization and the hysteresis behavior can be interpreted by assessing the motion of domain boundaries. The domain evolution during the demagnetization and the reverse magnetic field loading is opposite to that during the magnetization. The magnetic moments in the single domain incline toward the reverse magnetic field. This process occurs via motion of the domain boundaries. The resistance for the motion leaves some domains maintained their original magnetic field direction, which contribute to the remanence, $B_R$. This also induces the hysteresis effect.

If the reverse magnetic field increases, the residual magnetic induction intensity decreases, and reaches zero at $H = -H_c$ (point C in Fig. 10.20). $H_c$ is called coercive force. When the reverse magnetic field continues to be applied, the material can be magnetized in the reverse direction to saturation (corresponding to point $S'$ in Fig. 10.20). After that, if a positive external magnetic field is applied again, the magnetization curve extends across the $B$ axis at $B = -B_R$ and $H$ axis at $H = H_c$, in turn, and then reaches the saturation magnetization (point $S$). A magnetic hysteresis loop is completed.

### 10.4.5 Soft magnetic and hard magnetic materials

For ferromagnetic and ferrimagnetic materials, their hysteresis loop provides very important information for the magnetic properties. The area in the hysteresis loop represents the energy loss in unit volume of the material in each of the magnetization and demagnetization cycles. The lost energy is converted into heat, which increases the temperature of the material.

According to the hysteresis characteristic, the ferromagnetic and ferrimagnetic materials can be classified as soft magnetic and hard magnetic materials. If the magnetic material has a relatively high initial permeability and small coercive force, it can be easily magnetized and demagnetized. Generally, this material has a narrow hysteresis loop, that is, relatively small energy loss. Therefore, it is particularly suitable for operating in an alternating magnetic field environment, such as a transformer core. This kind of magnetic materials is called soft magnetic material.

It has been found that the chemical composition mainly affects the saturation magnetization of materials, while the crystal structure mainly affects the permeability and coercive force. For instance, the magnetization of the cubic ceramic $Fe_3O_4$ exhibits obvious composition dependence when its $Fe^{2+}$ ion is replaced by $Ni^{2+}$ ion. The coercive force may increase if there are crystal defects, such as nonmagnetic phase or voids, because they all hinder the movement of domain boundaries. In soft magnetic materials, such crystal defects are not allowed.

In addition to the hysteresis loss, the energy loss of the soft magnetic material can also be derived from the electric current induced by the alternating magnetic field. This current is called eddy currents or vortex. The eddy current loss can be reduced by increasing electrical resistivity of the material. The addition of alloying elements in the ferromagnetic material may increase the resistivity. For example, adding silicon or nickel in iron, one can obtain Fe-Si and Fe-Ni soft magnetic alloys. Ceramics with high electrical resistivity can be used as the soft magnetic materials. But their application is limited because they have relative small magnetic susceptibility.

According to their hysteresis behavior, hard magnetic materials have the high remanence and high hysteresis loss. The product of $B_R$ and $H_c$ can be roughly taken as an indicator of the hard magnetic feature. Larger the product is, harder the magnetism. Hysteresis effect depends on the motion obstacle of the domain boundaries, because the coercive force and magnetic susceptibility can be enhanced by hindering the movement of domain boundaries. In hard magnetic materials, the fine precipitates in microstructure can effectively hinder the motion of the domain boundaries. Several typical hysteresis loops of some ferromagnetic and ferrimagnetic materials are shown in Fig. 10.21.

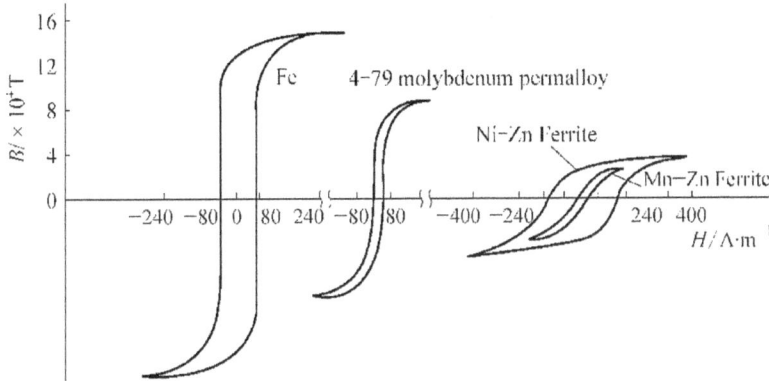

**Fig. 10.21:** Comparison of hysteresis loops of several ferromagnetic and ferrimagnetic materials.

## 10.5 Optical performance

The optical performance refers to the respondence of materials to the electromagnetic radiation, especially to visible light. This section will introduce and discuss the related concepts and the involved fundamentals that have been established in the field of the electromagnetic radiation and the possible interaction between radiation and materials. The optical behaviors of metallic materials and nonmetallic materials are then described according to the absorption, the reflection and the transmission properties of materials.

### 10.5.1 Electromagnetic radiation

Based on classical meaning, electromagnetic radiation is considered as a kind of wave, and composed of mutually overlapping electric and magnetic fields. It propagates over and over again between the electric field and the magnetic field. Both the electric field and the magnetic field are perpendicular to the propagating direction. Commonly, light, rays, radio waves, and radiation heat belong to the form of electromagnetic radiation. Every kind of radiation is characterized by specific range of wavelength. The electromagnetic wave spectrum covers the full band from ultrarays with the shortest wavelength to radio wave with the longest wavelength. If arranged in wavelength order, it is: ultrarays (from about $10^{-12}$ μm) → X-ray → ultraviolet → visible light → infrared ray → long waves (up to about $10^{10}$ μm).

Visible light locates in a very narrow zone of radiation spectrum, and the wavelength is in the range of 0.4–0.7 μm. The colors distinguished by eyes depend on the wavelength. The wavelength of about 0.4 μm corresponds to purple. About 0.5 and 0.65 μm wavelengths correspond to green and red, respectively.

The wavelength intervals corresponding to the seven basic colors are marked in Fig. 10.22. Mixture of these basic colors shows white color. All these colors belong to visible light.

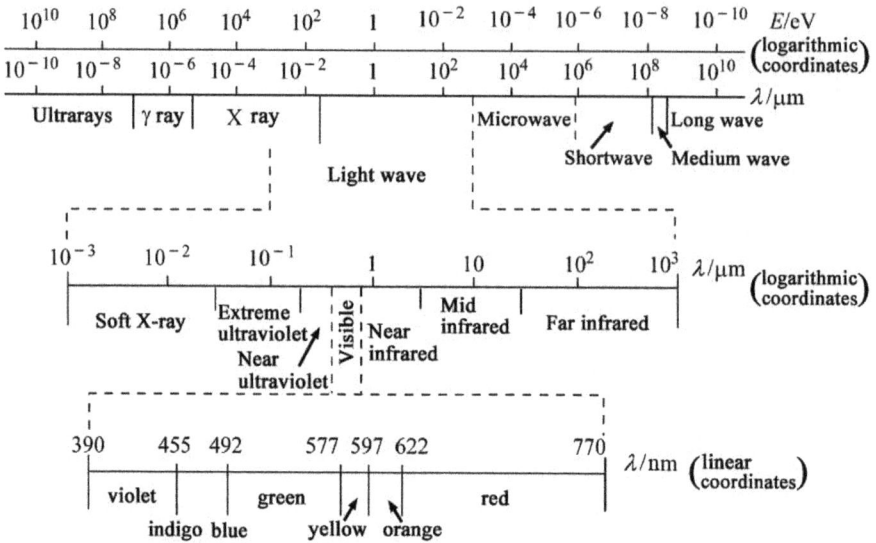

**Fig. 10.22:** Electromagnetic wave spectrum.

All electromagnetic radiations propagating through a vacuum exhibit a constant propagation speed, that is, about $3.0 \times 10^8$ m/s. The relationship between the velocity of light, $c$, and the vacuum electric permittivity, $\varepsilon_0$, and vacuum permeability, $\mu_0$, is as follows:

$$c = \frac{1}{\sqrt{\varepsilon_0 \mu_0}}$$

(10.40)

Thus, there is a correlation among electricity, magnetic constant and the velocity of light. In the other hand, the wavelength, $\lambda$, and frequency, $\nu$, of the electromagnetic radiation have a fixed functional relationship with the velocity of light:

$$c = \lambda \nu$$

(10.41)

The unit of frequency is hertz (Hz).

Electromagnetic radiation can also be considered to consist of photons according to wave-particle duality of microparticles. The photon energy can be quantized, and defined as:

$$E = h\nu = \frac{hc}{\lambda}$$

(10.42)

where $h$ indicates the Planck constant that is $6.63 \times 10^{-34}$ J $\cdot$ s. This expression reveals that the photon energy is proportional to the radiation frequency and the light velocity, but inversely proportional to the wavelength.

When referring to the optical phenomena of interaction between radiation and matter, it is convenient to interpret from the perspective of the photon. In this case, the optical wave (i.e., light) can be treated as photons. In other cases, it is more appropriate to interpret from the perspective of the wave. These two approaches will be used in this section.

### 10.5.2 Interaction between light and solid

When light travels from one medium to another (e.g., from air to solid), several optical phenomena can be observed. Some of the lights are absorbed, while some are reflected by the interface between the two media. Only partial lights can pass through the second medium. Radiation beam intensity, $I_0$, incident to solid interface must be equal to the sum of the transmission intensity, $I_T$, the absorption intensity, $I_A$, and reflection intensity, $I_R$:

$$I_0 = I_T + I_A + I_R \qquad (10.43)$$

where the unit of the above intensity is W/m$^2$, corresponding to the energy per unit area per unit time perpendicular to the propagation direction.

Equation (10. 43) is also expressed in another form:

$$T + A + R = 1 \qquad (10.44)$$

where $T = I_T/I_0$, indicating the transmissivity. $A = I_A/I_0$, indicating the absorptivity. $R = I_R/I_0$, indicating the reflectivity.

If the absorptivity and reflectivity are very small, the material has large transmissivity, and is therefore transparent. For a transparent matter, light is scattered in the interior of the material and diffuses through the material. The material that cannot be transmitted by light is called opaque object.

A bulk metal is completely opaque to visible light. In other words, all of the light that is irradiated on the metal either is absorbed or reflected. Insulation materials generally have small absorptivity and reflectivity, so that they can be made into transparent materials. Semiconductor materials are different: some of them are transparent, but some are opaque.

### 10.5.3 Interaction between atoms and electrons

Optical responses in solid involve interaction between the light and the atoms (or ions, or electrons) in the material, which results in electron polarization and electron energy transition.

Electron polarization occurs because the electric field of the electromagnetic wave interacts with electronic cloud around the atom. This interaction can shift and deflect the electron cloud around the nucleus. Electron polarization consumes a certain amount of energy of the light, leading to absorption of some of the electromagnetic radiation. The electron polarization may also decrease the velocity of light when passing through the medium.

Absorption and emission of electromagnetic radiation refer to the transition of electrons from one state of energy to another. For simplicity, an isolated atom is first considered, and atomic electron energy level is shown in Fig. 10.23. Electrons are excited from the occupied energy state $E_2$ to higher empty energy state $E_4$ by absorbing photon's energy. The energy change described by electrons $\Delta E$ depends on the frequency of radiation:

$$\Delta E = h v \tag{10.45}$$

where $h$ denotes the Planck's constant. Several concepts involved should be emphasized. The first concept is that the energy state of an atom is split. Only specific energy $\Delta E$ between the energy levels is valid. The electrons around the atoms with frequency of $\Delta E$ can be absorbed through electronic transition. The second concept is that the excited electrons cannot remain excited state for a long time. The fact is that the electrons will return to the ground state after a short period of time. The energies of absorption and emission electron transitions are conserved in any case.

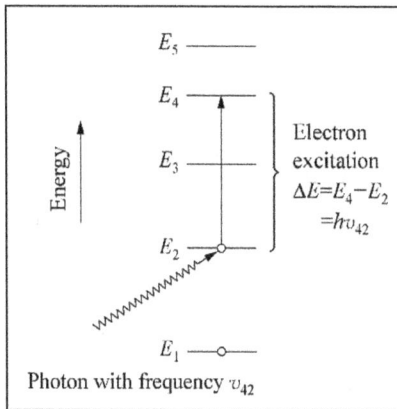

**Fig. 10.23:** The electron absorbs the incident photon energy in an isolated atom and is excited from one energy state to another high energy state.

According to two electron energy band structures of metals (Fig. 10.3(a) and (b)), their valence bands are only partially occupied with electrons. The incident radiation with visible light frequency can excite electrons to the empty energy state above the Fermi energy as shown in Fig. 10.3(a). This means that the incident radiation is absorbed, and thus the metals are opaque. The total absorption happens on the thin layer (its thickness is usually less than 0.1 μm) on the surface of the solid. This indicates that only the metal film less than 0.1 μm can be transmitted by the visible light.

Metals can absorb visible light at all wavelengths because they have un-full valence bands or continuous empty bands that provide sites for electronic transitions, as shown in Fig. 10.24(a). In addition, metals are opaque to all electromagnetic radiations with wavelength from the middle of the ultraviolet light to the long waves. But it is transparent to X-ray and γ-ray radiation with very small wavelength.

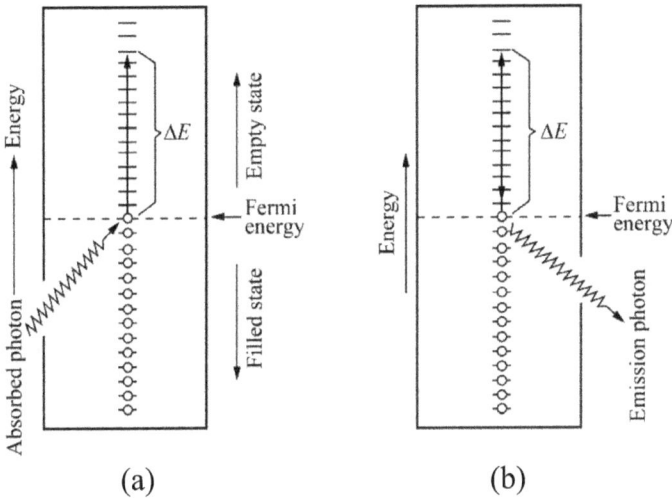

**Fig. 10.24:** Mechanism of photon absorption and reemission in metal materials: (a) absorption mechanism and (b) reemission mechanism.

The absorbed radiations can be reemitted from the surface with the same wavelength in the form of visible light. The electronic transition accompanied by reemision is shown in Fig. 10.24(b). The reflectivity of most metal is in the range of 0.90–0.95. In the process of emitting photons, a small part of the energy is lost in the form of heat.

The larger reflectivity of metals implies that the reflected light is dominant to show the visible color of the metals, which is dependent on the frequency of the reflection rather than absorption. A bright silvery appearance indicates the metal is highly reflective in the whole visible spectrum when the metal is exposed to white light. In other words, the frequency and number of photons reemitted for reflected beam are very close to the incident light. Both aluminum and silver are examples that display such reflection behavior to the visible spectrum. Because the energy of some photons with shorter wavelength cannot be reemitted as visible light, copper, and gold present orange and yellow, respectively.

Insulators (e.g., ceramics) and some semiconductors can be transparent to visible radiation due to their special electron energy band structures (Fig. 10.3(c) and (d)). In these materials, the refraction and transmission phenomena must be considered in addition to reflection and absorption.

### 10.5.4 Refraction

When a light beam enters into a transparent material, the velocity of the light reduces. This leads to bending of the beam at the incident interface. This behavior of light is called refraction. The refractive behavior can be measured by using the refractive index, $n$, that is expressed as the ratio of two speeds of light:

$$n = \frac{c}{v} \tag{10.46}$$

where $c$ indicates the velocity of light in vacuum. $v$ indicates the velocity of light in a medium. The value of $n$ can be directly measured by dispersing a white light beam entering a glass prism and leaving it. Obviously, the index, $n$, and the velocity of light, $v$, in the medium depend on frequency of the light. The refractive index not only reflects the optical path of the light, but also reflects incident light fraction on the surface.

Similar to the vacuum velocity of light defined in eq. (10.40), the velocity of light in a medium is defined as:

$$v = \frac{1}{\sqrt{\varepsilon\mu}} \tag{10.47}$$

where $\varepsilon$, $\mu$ is the permittivity and permeability of medium, respectively. It can be obtained from eq. (10.46):

$$n = \frac{c}{v} = \frac{\sqrt{\varepsilon\mu}}{\sqrt{\varepsilon_0\mu_0}} = \sqrt{\varepsilon_r\mu_r} \tag{10.48}$$

where $\varepsilon_r$, $\mu_r$ is relative permittivity and relative permeability, respectively. Most materials have only very weak magnetic, so $\mu_r \cong 1$. Then:

$$n \approx \sqrt{\varepsilon_r} \tag{10.49}$$

Thus, there exists a relationship between refractive index and dielectric constant for transparent materials. As mentioned above, the refraction phenomenon for the visible light with high frequency can be attributed to the electronic polarization. The relative permittivity (dielectric constant) can be calculated by the refractivity measured by eq. (10.49). Electronic polarization prevents the electron radiation in the medium. Therefore, the size of atoms and ions must influence the refractive effect. The larger atoms and ions generate the stronger electron polarization that decreases the velocity of the electromagnetic radiation. This corresponds to the larger refractive index. For the ceramics and glass with cubic crystal structure, the refractive index is independent of the crystal orientation, that is, isotropy. But it is anisotropic for those with noncubic crystal structures. The largest refractive index occurs in the densest crystal orientation of the materials.

### 10.5.5 Reflection

When a light beam enters from one medium into another with a different refractive index, the interface between the two media can scatter the light. This always occurs even if the two media are transparent. The interface acts as a reflector to partially reflect the incident light. Reflectivity represents the fraction of reflected light occupied in the incident light:

$$R = \frac{I_R}{I_0} \tag{10.50}$$

where $I_R$ and $I_0$ express the intensities of the reflected light and the incident light, respectively. Since light is a transverse wave, electric vector can be arbitrary orientation on the plane perpendicular to the propagating direction of the wave. It can be divided into two linear polarization components. One is called $S$ component or $S$ wave perpendicular to the incident surface of the light. The other is called $P$ component or $P$ wave parallel to the incident surface of the light. For the polarized light with vibration perpendicular to the incident plane, reflectivity can be derived:

$$R_s = \frac{\sin^2(\alpha - \gamma)}{\sin^2(\alpha + \gamma)} \tag{10.51}$$

where $\alpha$ and $\gamma$ is the incident angle and refracted angle, respectively. For the polarized light parallel to the incident plane, the reflectivity:

$$R_p = \frac{\tan^2(\alpha - \gamma)}{\tan^2(\alpha + \gamma)} \tag{10.52}$$

If the incident beam is perpendicular to the interface, $\alpha = \gamma = 0$, the above two formulas are equal, and then:

$$R = \left(\frac{n_b - n_a}{n_b + n_a}\right)^2 \tag{10.53}$$

where $n_a$ and $n_b$ are the refractive indexes of two media, respectively. The above formula indicates that the greater the difference of refractive index of the two kinds of medium, the greater the reflectivity.

When light enters into the solid from the air or vacuum (the refractive index of air is close to 1, namely $n_{air} \approx 1$), the above formula becomes:

$$R = \left(\frac{n_{solid} - 1}{n_{solid} + 1}\right)^2 \tag{10.54}$$

Since the refractive indexes of most common solids are larger than one, it can be evaluated that the larger reflectivity of the solid corresponds to the larger refractive index according to the above formula.

## 10.5.6 Absorption

Nonmetallic materials may be transparent to visible light and may be opaque. For transparent materials, they tend to be colored. In principle, these materials absorb the light radiation by three mechanisms that also influence the transmission characteristics of the nonmetallic materials. One is the electronic polarization. The electronic polarization absorption mechanism only plays an important role in the optical frequency near the relaxation frequency of the component atoms. Electron transition is also a key factor affecting the light absorption. It is involved by two ways. One is that the electrons are excited across the band gap accompanying absorption of photon. The other is that the electrons jump to the defect energy level in the band gap.

The electrons in near full valence band excited by light radiation can jump over band gap into the conduction band, as shown in Fig. 10.25(a). This process produces holes in the valence band and free electrons in the conduction band. The absorbed photon frequency depends to the excitation energy according to eq. (10.45). The excitation causing absorption occurs only when the photonic energy is greater than the band gap energy, $E_{g}$. that is:

$$hv > E_g \tag{10.55}$$

or

$$\frac{hc}{\lambda} > E_g$$

Since $h$ and $c$ are constants ($h = 4.13 \times 10^{-15}$ eV $\cdot$ s, $c = 3.0 \times 10^8$ m/s), The minimum wavelength of visible light (about 0.4 μm) corresponds to the maximum band gap energy:

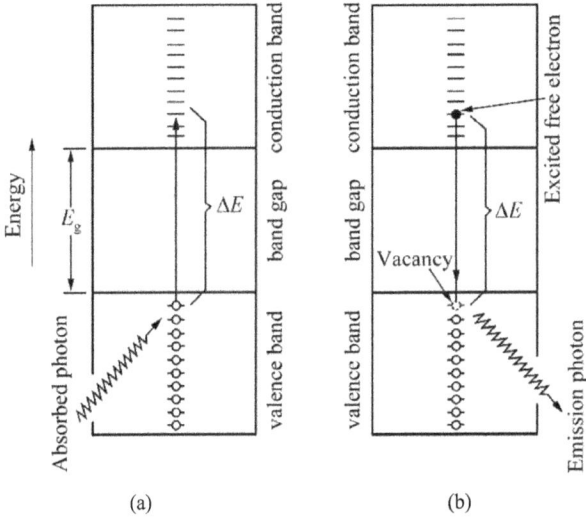

**Fig. 10.25:** Photonic absorption and emission mechanism of nonmetallic materials: (a) photonic absorption mechanism and (b) photonic emission mechanism.

$$E_{gmax} = \frac{hc}{\lambda_{min}} = \frac{4.13 \times 10^{-15} \times 3 \times 10^8}{4 \times 10^{-7}} = 3.1 \text{ eV}$$

In other words, the nonmetallic materials with band gap energy above 3.1 eV do not absorb visible light. If they are of high purity, they will be colorless and transparent.

The minimum band gap energy can also be calculated by substituting the maximum wavelength value of visible light, 0.7 µm, into eq. (10.56):

$$E_{gmin} = \frac{hc}{\lambda_{max}} = \frac{4.13 \times 10^{-15} \times 3 \times 10^8}{7 \times 10^{-7}} = 1.8 \text{ eV}$$

It means that the materials with band gap energy less than 1.8 eV can absorb all the visible light via the electron transition from valence band to the conduction band. Semiconductors are in accordance with such a feature and are therefore opaque. The materials with band gap energy between 1.8 and 3.1 eV can absorb only partial visible spectra, so they present colors.

There is a critical wavelength for transparency for the nonmetallic material, which can be determined by the band gap energy of the material. For example, diamond with the band gap energy of 5.6 eV is opaque to the radiation with wavelengths less than 0.22 µm.

Dielectric materials with wide band gap also can absorb visible light. Its mechanism is different from the electron transition from valence band to conduction band. It is possible that some electronic energy levels (e.g., donor and acceptor levels) are located in the band gap when impurities or defects with electroactivity are present. The electron transitions between levels in the band gap lead to absorption of light radiation with special wavelengths, as shown in Fig. 10.26(a).

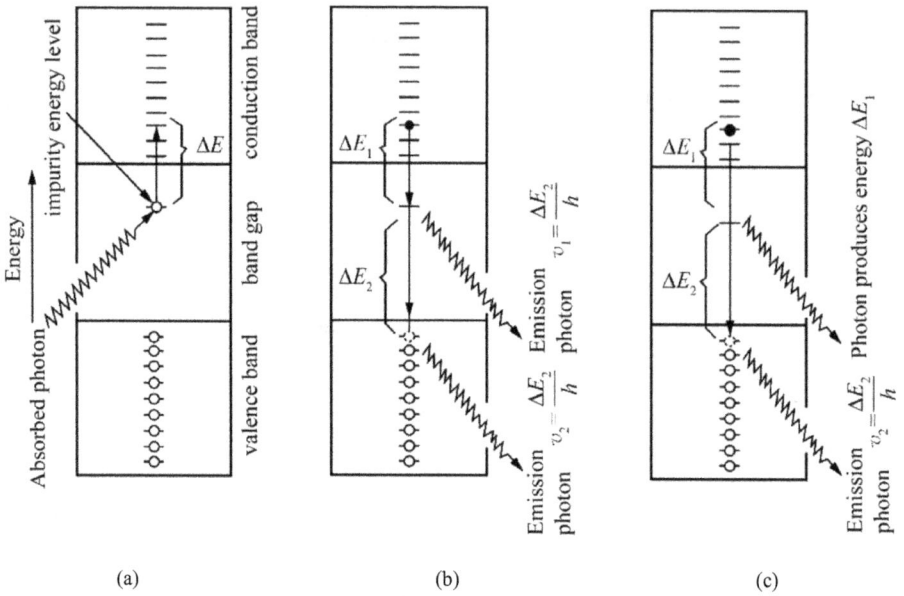

**Fig. 10.26:** Photon absorption and emission from impurity levels: (a) photon absorption, (b) emission of two photons, and (c) emission of two photons, one of which is accompanied by the release of heat energy.

Generally, electromagnetic energy absorbed via exciting electrons must be released in some way. There are several possible mechanisms for the release. If the excitation occurs via electronic transition from valence band to conduction band, the energy release can be completed directly by the recombination of electron and vacancy. The reaction process is that electrons + vacancies = energy ($\Delta E$). This process is illustrated in Fig. 10.25(b). Besides, multiple-step transitions can also occur, in which the impurity levels in the band gap are involved (see Fig. 10.26(b)). A possible way is that two photons emit through different routes. One photon emits via electron in the conduction band falling down to impurity energy level in the band gap. Another one emits via electron in the impurity energy levels falling down to the valence band. Or, one of the transitions can emit one photon, while the other generates heat, as shown in Fig. 10.26(c).

The radiation intensity of net absorption is dependent on the energy band structure and the path length in the medium. Transmission intensity, $I_T'$, decreases as the distance, $x$, in the medium increases:

$$I_T' = I_0' e^{-\beta x}$$

(10.57)

where $I_0'$ indicates the intensity of the incident light without reflection. $\beta$ indicates the absorption coefficient $(mm^{-1})$, which is related to the properties of material. $\beta$ varies with the wavelength of the incident radiation. Materials with large $\beta$ are considered to be high absorption materials.

### 10.5.7 Transmission

Transmission, absorption, and reflection phenomena and the relationship among them can be clearly illustrated by considering a light beam passes through a transparent solid, as shown in Fig. 10.27. When an incident light with intensity of $I_0$ impacts on the front surface of a transparent sample, the transmission intensity in the back face of the sample is

$$I_T = I_0(1-R)^2 e^{-\beta l} \tag{10.58}$$

where $R$ indicates the reflectivity. $l$ is the thickness of the sample. The above formula assumes the same medium outside both front and back surfaces. The proportion of transmission intensity in incident light intensity depends on the loss of reflection and absorption. The sum of transmittance, $T$, reflectivity, $R$, and absorptivity, $A$, is equal to one according to eq. (10.44). Since all the three optical parameters, $T$, $R$, and $A$, are functions of wavelengths of light radiation, the fractions of the three items varies at different wavelength. As an example, the fractions of $T$, $R$, and $A$ of green glass in the visible wavelength range are shown in Fig. 10.28. It can be found that the transmittance, reflectivity and absorptivity are 0.90, 0.05, and 0.05, respectively, at the wavelength of 0.4 µm, while they become 0.50, 0.02, and 0.48, respectively, at the wavelength of 0.55 µm.

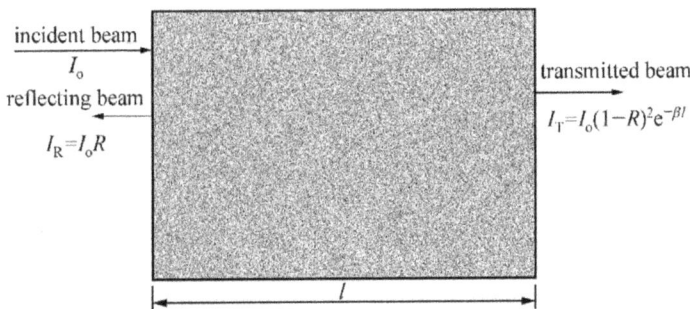

**Fig. 10.27:** Transmission intensity of light passing through a transparent solid.

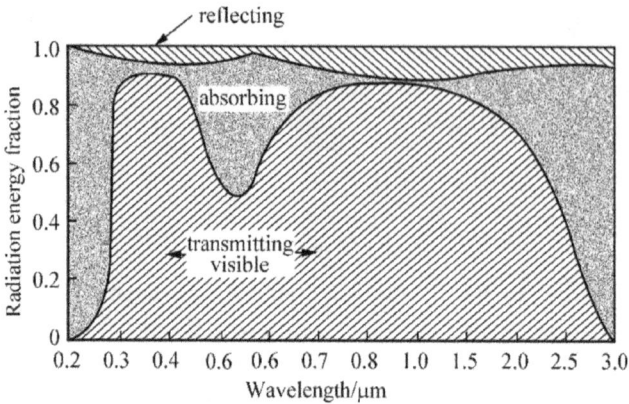

**Fig. 10.28:** Fractions of *T*, *R*, and *A* of a transparent green glass in the visible wavelength range.

## 10.5.8 Color

If some wavelengths of light can be selectively absorbed by a transparent material, it will present color. The distinguishable color depends on the combination of the transmitted wavelengths. If visible light are absorbed evenly, the material is colorless. Single crystal diamond and highly pure inorganic glass exhibit such optical properties.

Selective absorption generally takes place due to electron excitation. It occurs in semiconductive materials with band gap energy in the range of visible photon energy (1.8–3.1 eV). The visible light energy above $E_g$ can be selectively absorbed via electronic transition from valence band to conduction band. If the excited electron returns to its initial low energy state, some absorbed radiation is reemitted. In this case, both the reemitted radiation and the transmitted light jointly determine the color of the material.

For instance, CdS with the band gap energy of about 2.4 eV can absorb photon energy between 2.4–3.1 eV (corresponding to the range between blue and purple in visible light). Photon energy in the range of 1.8–2.4 eV cannot be absorbed. Transmission beam is about 0.55–0.65 μm in wavelength, thus CdS presents saffron yellow.

In addition, the special impurity can bring electron energy levels in the band gap in the insulator ceramics. Because impurity atoms or ions involve electron excitation, the photons with energy smaller than the band gap energy can also be absorbed. In such situation, some reemission is also possible. The color of material is dependent on the wavelength distribution of transmission beam. For example, the single crystal $\alpha$-$Al_2O_3$ is colorless. After adding 0.5 to 2 w.t.% of chromium oxide ($Cr_2O_3$), it becomes bright red (so called Ruby). This is because the ion $Al^{3+}$ in the $\alpha$-$Al_2O_3$ crystal structure is partially replaced by ion $Cr^{3+}$ that brings impurity energy level in band gap. The impurity energy levels are involved in the electron transitions, and thus lead to preferential absorption at some specific wavelengths. Sapphire and ruby containing a small quantity of $Ti^{4+}$ and $Fe^{2+}$ exhibit some preferential absorption as shown in Fig. 10.29. Sapphire shows a relative small fluctuation in transmittance in whole range of visible light wavelength, indicating that this material is almost colorless. Ruby, however, exhibits two strong absorption peaks. One happens in blue purple zone (wavelength of about 0.42 μm), which corresponds to the minimum value of the transmittance. The other happens in yellow green zone (wavelength of about 0.60 μm). The mixture of transmitted light and reemitted light makes ruby crimson.

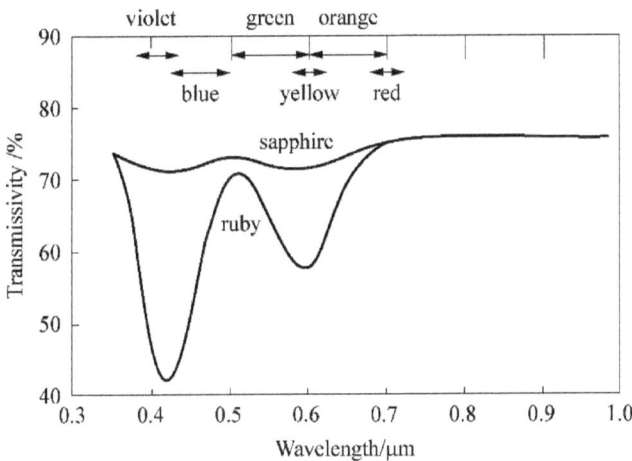

**Fig. 10.29:** The change of transmittance with wavelength of light radiation passing through sapphire and ruby.

### 10.5.9 Stimulated emission and light amplification

At a certain temperature, most of the atoms in matter are in the ground state, and only a few atoms are excited. The excited atom is unstable, and tends to transit to a low energy state and emit photon simultaneously. This phenomenon is called spontaneous emission (spontaneous radiation).

When an atom is irradiated by the incident light with a frequency of $v$, the atom absorbs a photon and jumps from the low energy state $E_m$ to the high energy state $E_n$. This process takes place at the condition of $hv = E_n - E_m$, and is called stimulated absorption. If the incident light has high dense homomorphic photons, the atom can absorb $n$ photons to jump from the low energy state $E_m$ to the high energy state $E_n$ at the condition of $nhv = E_n - E_m$. This process is called stimulated absorption of multiphotons.

Besides, an atom in the high energy state can also be stimulated to the low energy state under the incident light, and emit photon with the same frequency as the incident light. This process is called stimulated emission (stimulated radiation). In general, the three processes discussed earlier (as shown in Fig. 10.30) can occur simultaneously. It is possible that one process of these three is dominant under a certain condition.

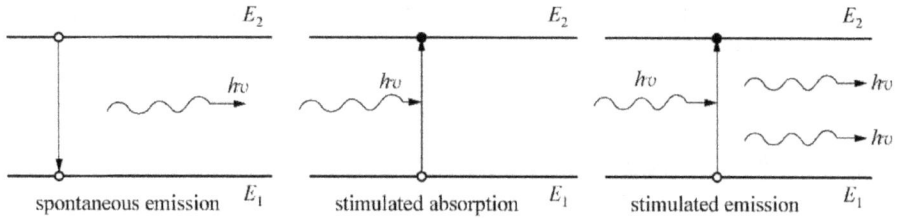

Fig. 10.30: Spontaneous emission, stimulated absorption and stimulated emission.

It should be noted that the spontaneous emission is a random process. Emissions are usually different in photon frequency, polarization state, phase position and direction of photonic motion. Thus, spontaneous emission is a kind of disorder emission. Stimulated emission is also a kind of random emission. But, the photons emitted are the same as that of incident light in their optical characteristics including frequency, polarization state, phase position, and motion direction. Thus, this is a more orderly emission. The light generated by stimulated emission has good coherence.

When the incident light with a certain frequency enters into matter, stimulated absorption and stimulated emission may occur simultaneously. The occurrence of the two opposite processes has the same probability. In general, the atoms in the low energy state are much more than those in the high energy state, so the matter always exhibits stimulated absorption. If one can make the atoms in the high energy state greater than that in the low energy state, the stimulated emission will dominate. In this situation, matter emits the light with the same frequency as the incident light. The emitted light is superimposed on the incident light, and enhances the incident light. It means that the light is amplified. If the atoms in the high energy state are greater than that in the low energy state in a matter, it is called inverse distribution or inversion of atomic quantity. This matter is called activation medium. Laser was developed according to the principle described above.

For example, the first ruby laser was made by Maiman et al. in 1960. The ruby used was doped with 0.05% chromium oxide. Partial ion $Al^{3+}$ was replaced by ion $Cr^{3+}$ in the $\alpha$-$Al_2O_3$ crystal structure, inducing impurity energy level in band gap. In this situation, there are three energy states in the crystal, that is, ground state ($E_1$), excited state ($E_3$), and metastable intermediate (impurity) state ($E_2$). Under the radiation of incident light, ions in the ground state $E_1$ that absorbed green and blue light (crystal presents pale red) are excited to the energy level $E_3$. But, ions $Cr^{3+}$ maintain in the energy level $E_3$ for very short time ($10^{-8}$ s). They transfer partial energy to ion $Al^{3+}$ by inelastic collision, and fall down to the energy level $E_2$. It is possible that partial ions $Cr^{3+}$ fall back to the ground state, and reemit green and blue light.

In overall, $Cr^{3+}$ ions in the ground state $E_1$ decrease continuously under the irradiation of the external incident light, while those in the state $E_2$ continue to increase, because the $Cr^{3+}$ ions remain in $E_2$ for longer (about $10^{-3}$ s). As a result, an inversion of atomic quantity between the impurity energy level and the ground energy level is established. In this case, ion transition from level $E_2$ to level $E_1$ can be induced by radiation of a photon with the frequency of $(E_2 - E_1)/h$. The transition emits another photon, and leading to a chain reaction. Such avalanche effect can trigger strong stimulated emission. By this way, the incident light signal is amplified, forming laser.

## 10.6 Other functional properties

### 10.6.1 Internal friction and damping behavior

Internal friction is a phenomenon that reflects the dissipation of mechanical energy inside a gaseous, liquid or solid medium. Damping is the ability of a system or a material to reduce, limit, or prevent its oscillation. Damping behavior is very useful for dragging or slowing down the oscillation. This has been used in mechanical systems as dampers, and in electronic oscillators as resistances. When vibration energy transmits through a solid, it will be decayed. The attenuation occurs because of the internal friction inside the solid. The amplitude of attenuation indicates damping ability of the solid, so that the "internal friction" or "damping" of a solid material usually stands for the same meaning.

From the view-point of materials science and mechanical engineering, internal friction provides the resistance to lattice slip and deformation in solid materials under loading. Internal friction is responsible for the damping properties of materials. It usually refers to energy dissipation inside materials, which leads to obvious stress-strain hysteresis behavior (Fig. 10.31). This means that the materials deviate from Hooke's law. The energy dissipated due to internal friction can be evaluated by the area enclosed by the hysteresis loop during one cycle. Thus, a common measure of the internal friction or damping can be expressed as the ratio of the energy dissipated during one cycle to the total energy stored in the material via loading:

$$\psi = \frac{\Delta W}{W} \text{ or } Q^{-1} = \frac{\Delta W}{2\pi W} \tag{10.59}$$

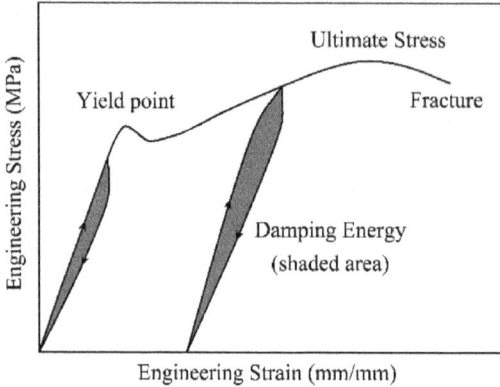

**Fig. 10.31:** Stress–strain curve with hysteresis loops under cyclic loading.

$$\Delta W = \oint \sigma d\varepsilon \tag{10.60}$$

$$W = \int_{\omega t = 0}^{\omega t = \pi/2} \sigma d\varepsilon \tag{10.61}$$

where $\psi$ is defined as specific damping capacity. $Q^{-1}$ denotes damping or internal friction. $\Delta W$ denotes the energy absorbed during one stress-train cycle. $W$ denotes the maximum elastic energy stored during the cycle. $\sigma$ denotes stress, $\varepsilon$ denotes strain, $t$ is the time, $\omega$ denotes the circular frequency.

The internal friction is generally small for most metallic materials, that is, $Q^{-1} \ll 1$ (for most of metals or alloys, $Q^{-1}$ is in range of $10^{-5}$–$10^{-1}$; for some metallic glasses, for example, Pd-based or Zr-based metallic glasses, $Q^{-1}$ reaches ~1, or even >1). For polymers, $Q^{-1}$ is in the range of $10^{-2}$–$10^{0}$; some are larger than 1. The materials with larger internal friction (significant hysteresis), such as rubbery polymers and some viscoelastic metallic materials, are also called as "anelastic solid."

For an ideal elastic solid, the strain response to a rectangular stress (with abrupt loading and unloading) exhibits strict consistency as shown in Fig. 10.32(a) and (b). The internal friction is zero in this case. For anelastic solid, after an initial abrupt elastic deformation (unrelaxed), strain becomes time dependent, exhibiting Newton viscous behavior, comparing Fig. 10.32(a) with (c). Under the sustained stress, strain increases obeying Newtonian viscoelastic relation: $\sigma = \mu \dot{\varepsilon}$. Such "creep" strain $\varepsilon(t)$ after loading becomes saturated with time (relaxed). Considering a single relaxation time $\tau$, the time-dependent strain is of the form $e^{-t/\tau}$. The relaxed and unrelaxed strain values, denoted as $\varepsilon_R$ and $\varepsilon_U$, indicate the relaxation strength: $\Delta = (\varepsilon_R - \varepsilon_U)/\varepsilon_U$, which depends on the internal friction. After unloading, the "relaxed" strain can be completely recovered with time (Fig. 10.32(c)). For some anelastic–viscoplastic

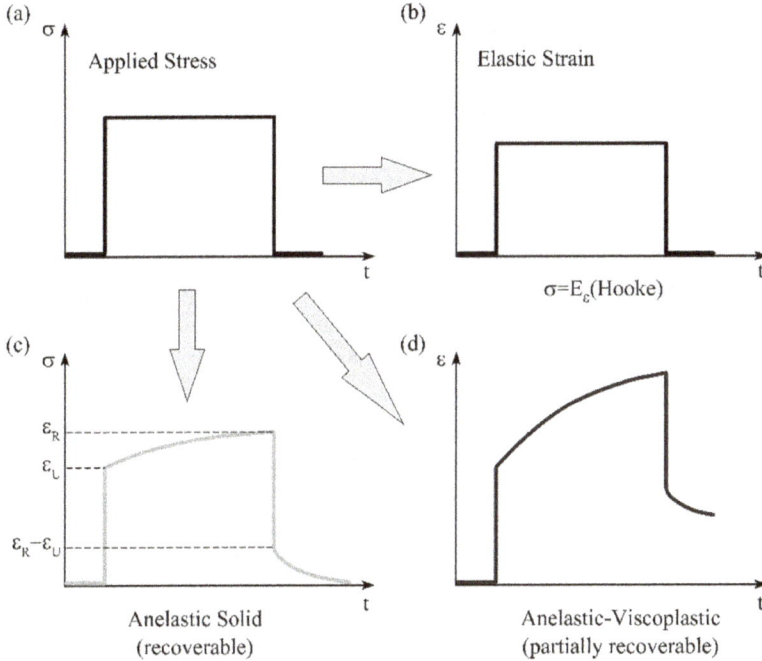

**Fig. 10.32:** The strain response to a rectangular stress: (a) applied stress, (b) ideal elastic solid, (c) anelastic solid, and (d) anelastic–viscoplastic materials.

materials, the anelastic strain occurs accompanied with viscoplastic strain. After unloading, the strain can only partially recovered (Fig. 10.32(d)).

In engineering practice, cyclic loading is of great importance. According to the linear theory of anelasticity, the sinusoidally varying stress and strain can be expressed mathematically by using complex notation:

$$\sigma^* = \sigma_0 e^{i\omega t} \tag{10.62}$$

$$\varepsilon^* = \varepsilon_0 e^{(i\omega t - \delta)} \tag{10.63}$$

where $\sigma_0$ denotes initial stress amplitude, and $\varepsilon_0$ denotes strain amplitude. $\omega = 2\pi f$, $f$ is the vibrational frequency. The phase lag, $\delta$, denotes the loss angle. Then, a complex modulus is

$$E^* = \frac{\sigma^*}{\varepsilon^*} = \frac{\sigma_0}{\varepsilon_0} e^{i\delta(\omega)} = E(\omega)(\cos\delta + i\sin\delta) = E'(\omega) + iE''(\omega) \tag{10.64}$$

$E'(\omega)$ and $E''(\omega)$ represent storage modulus and loss modulus, respectively. The real quantity, $E(\omega)$, is called absolute dynamic modulus. Using the above equation, the loss tangent tan $\delta$ (also called loss factor) can be written as

$$\tan \delta = \frac{E''}{E'} \tag{10.65}$$

The loss tangent is another expression of "internal friction" or "damping" (10.59) introduced earlier. Thus,

$$\tan \delta = \frac{\Delta W}{2\pi W} = Q^{-1} \tag{10.66}$$

Damping behavior and mechanism in metallic materials have been comprehensively investigated. Dynamic hysteretic damping is usually controlled by diffusion, which is characterized by the complete recovery of small residual strain when low stress is applied and eliminated. Static hysteretic damping occurs in systems with transient stress relaxation and small permanent residual strain that can be eliminated only with reverse stress. Mechanisms of damping may be interpreted via the physical characteristics of the materials, such as thermoelastic, viscous, as well as magnetic properties, or the sources in the materials, such as metallurgical or lattice defects, which all are responsible for the damping behavior. Defects in crystal lattice, including vacancies, interstitial/substitutional atoms, dislocations, grain-boundaries and interfaces, generate internal friction when responding to any external loading, and then serve as intrinsic damping source in crystalline materials. The damping behavior, occurred due to the macroscopic physical property of a material, can be considered that it originates from extrinsic damping sources.

In addition to stress/strain amplitude and cyclic loading frequency, temperature is the most significant factor affecting damping performance of the materials. For most metallic materials, the damping capacity can be greatly influenced by changing temperature. This is because the intricate dislocations can be free again by unpinning upon increasing temperature, leading to smaller internal friction resistance. In other hand, as temperature increases, some microstructural recovery and/or recrystallization may occur for deformed metallic materials, resulting in lower defect density which decreases damping capacity. Besides, the strain-induced phase transformation (e.g., martensitic transformation) can also significantly influence the damping property of the materials.

For metallic glasses, the damping property becomes significant as temperature increases. In the temperature window of $T_x - T_g$ ($T_x$ and $T_g$ indicate crystallization temperature and glass transition temperature, respectively), the metallic glassy alloys become viscous, and their behaviors like a fused silica glass. Their damping capacity, $Q^{-1}$, can reach up to 2–3.

In order to have a quantitative measurement of the damping property, Table 10.8 collects the damping capacities of some representative metals, alloys, ceramics, and polymers. For most metals and alloys, their damping capacities are generally far less than one, that is, $Q^{-1} \ll 1$. But some metals and alloys exhibit larger damping capacities, such as Mg and Mn–7Cu alloy. Rubber and polymers generally demonstrate

Table 10.8: Damping capacity of some representative metallic materials and ceramics.[9–11]

| Materials | Test | $\varepsilon_0$ (×10⁻⁶) | T (°C) | f (Hz) | $Q^{-1}$ (×10⁻³) | Remarks |
|---|---|---|---|---|---|---|
| **Pure metals** | | | | | | |
| Al | Axial | 0.2–50 | 25–440 | – | 0.03–6 | C.P. |
| Cu | Torsional | 20–100 | 20–250 | – | 5–100 | C.P., 44% reduction |
| Fe | Torsional | ~80 | –150–70 | 0.83 | 2–20 | Heat treated |
| Mg | Bending | – | – | 60–400 | 14–60 | Purity 99%, casted |
| Re | Torsional | | 900–1,600 | | 20–100 | Heat treated |
| Ti | Torsional | | 400–650 | 0.5 | 1–70 | Grain size = 19 μm |
| **Alloys** | | | | | | |
| 2014 Al alloy | Torsional | – | 10–130 | 2.2–124 | 0.1–0.6 | Heat treated |
| Brass | Bending | | | 50–600 | 3–6 | Drawn |
| Co–20Fe | Torsional | 100–600 | | | 15–30 | Heat treated |
| Fe–13Cr | Torsional | 150–900 | | | 6–60 | Forged and swaged |
| Gray cast iron | Torsional | 60–540 | | | 30–90 | |
| Mn–7Cu | Axial | Low strain | | | 100–700 | Heat treated |
| Ti alloy | Bending | 60–1,800 | –200 to 250 | | 0.08–1 | |
| Nitinol (55Ni–45Ti) | | 350 | RT | Low freq. | 41.3 | |
| **Ceramics** | | | | | | |
| $Al_2O_3$ | Axial | | 0–1,200 | | 0.01–1 | Single crystal and predeformed |
| $Al_2O_3$–0.25% $La_2O_3$ | | | 900 1,250 | | 24–60 | |
| BN | Bending | 260 | 25–250 | | 28–40 | |
| Carbon/Carbon | | | | 14,900 | 92 | composite |
| Graphite | Bending | 1–100 | ~25 | ~1,000 | 5–15 | |
| SiC | Bending | | 26 | 1–5,000 | 1.1–2.5 | Whisker |
| $ZrO_2$ | | | 0–230 | 4 | 2.0–10.5 | Refractory |

(continued)

Table 10.8 (continued)

| Materials | Test | $\varepsilon_0$ ($\times 10^{-6}$) | $T$ (°C) | $f$ (Hz) | $Q^{-1}$ ($\times 10^{-3}$) | Remarks |
|---|---|---|---|---|---|---|
| **Metallic glasses** | | | | | | |
| $Co_{70}Fe_5Si_{15}B_{10}$ | | | 27–427 | 2–6 | 1–70 | As quenched |
| $Fe_{40}Ni_{40}P_{14}B_6$ | | | 20–500 | ~1 | 0.5–20 | Melt-spun amorphous + annl. |
| $Pd_{77.5}Ag_6Si_{16.5}$ | | | 370 | ~0.1 | 900 | Melt-spun amorphous |
| $Pd_{40}Ni_{40}P_{20}$ | | | 326–331 | 0.08 | ~2,000 | Melt-spun amorphous |
| $Zr_{52}Cu_{18}Ni_{15}Al_{10}Ti_5$ | | | 27–527 | 90–2,500 | 0.1–8 | Melt-spun amorphous |
| $Zr_{41}Be_{22}Ti_{14}Cu_{13}Ni_{10}$ | | | 427–452 | 1 | 500–3,300 | Bulk amorphous |
| **Polymers and rubber** | | | | | | |
| Neoprene rubber | | | | | 670 | |
| PTFE | | | | | 189 | |
| PMMA | | | | | 90 | |
| PA-66 | | | | | 40 | |
| Acetal | | | | | 30 | |
| Epoxy | | | | | 30 | |

larger damping capacities than most of metallic materials. It should be noticed that the very large damping capacities ($Q^{-1} > 1$) obtained in metallic glasses are attributed to their viscoelastic nature above their glass transition temperature.

Typical applications of high damping materials are to control or reduce vibration and noise in mechanical systems, moving vehicles, naval ship and submarine, aircrafts and space flights, and so on. The high damping materials can absorb and obstruct vibration and noise to the greatest extent. In other hand, some high-precision instruments or vibration sensors should be made up by using low damping materials. For engineering applications, strength and stiffness of damping materials are also very important. In general, metallic materials with high strength and large stiffness exhibit relatively low damping capacities. Rubbers and some polymers having larger damping capacities are always flexible. The relationships between the damping capacity and the stiffness of several kinds of common materials are illustrated in Fig. 10.33. The material with both large damping capacity and high stiffness (top right corner area of Fig. 10.33) is rare and the goal of new material research.

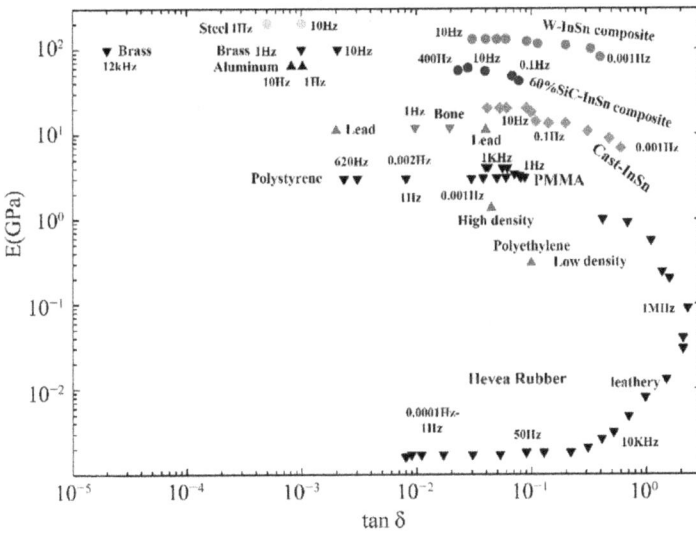

**Fig. 10.33:** The relationships between the damping capacity and the stiffness of several kinds of common materials.

## 10.6.2 Piezoelectricity

Some solid materials can be charged instantaneously when a mechanical stress is applied. This phenomenon is called piezoelectricity. The piezoelectric effect is a reversible process in which a material is charged when a stress/strain is applied, or a stress/strain in the material is generated when a voltage is imposed. The former is called "direct piezoelectric effect"; and the latter is called "reverse piezoelectric effect." The significant piezoelectric response occurs in certain ceramics that have been used as sensors and actuators for mechanical–electrical energy transformation. Besides the piezoelectric ceramics, piezoelectric effect is also found in other crystals and biological materials including proteins, DNA and osseous tissue.

The direct piezoelectric effect was first found in crystals of some minerals (including quartz, topaz, and tourmaline), cane sugar and tartrate (i.e., potassium sodium tartrate) in 1880 by Curie brothers. Among the materials tested, quartz and sodium potassium tartrate exhibited the strongest piezoelectricity. Accordingly, they first established some of the relationships between piezoelectric and crystal structures. Voigt strictly fixed those relations in 1894. As early as in 1881, Lippmann theoretically predicted the reverse piezoelectric effect according to the principle of thermodynamics.

The origination of the piezoelectricity is attributed to the occurrence of electric dipole moment in materials, which is equal to the product of the charge and the distance between the positive and negative charges. If ions in crystal lattice are surrounded by asymmetric charges, electric dipole moments can be caused. Typical examples are $BaTiO_3$ and $Pb[Zr_xTi_{1-x}]O_3$. In low symmetry crystals, the relative displacement of positive and negative ions in the crystallographic unit cell causes an unequal movement of the positive and negative charges when a deformation occurs under external force, resulting in macroscopic polarization of the crystals. The surface charge density of a crystal is equal to the projection of the polarization intensity at normal direction of the surface, which can be determined by summation of the dipole moments per volume of the crystal unit. Because each dipole is a vector, all the dipoles constitute a vector field. Dipoles close to each other may interplay and often be aligned in small regions. Such regions with aligned dipoles are called Weiss domains. All the domains can orient randomly, or be aligned upon polarizing process (similar but not the same as magnetic polarizing). Piezoelectricity takes place because of the occurrence of polarization when a stress/strain is applied. The piezoelectric mechanism can be attributed to the reconfiguration of the dipoles or the reorientation of the dipole moments under action of the external stress.

For a linear piezoelectric crystal, the charge, $Q$, induced by a force, $F$, applied on the surface is proportional to the force:

$$Q = d \cdot F \tag{10.67}$$

where $d$ indicates piezoelectric constant. Obviously, the greater the piezoelectric constant, the more pronounced the piezoelectric effect.

Considering that both force and electrical charges are vectors (the force in various directions and the charge accumulation on various surfaces are different), the linear piezoelectric constitutive equations can be expressed as

$$P^j_i = d_{ij} \cdot \sigma_j \qquad (10.68)$$

where $P$ and $\sigma$ indicate charge and force per unit area, respectively. Subscript $i$ indicates the polarization direction of the crystal, that is, the charged surface is perpendicular to the $X$ axis (or $Y$ axis, or $Z$ axis), marking $i = 1, 2, 3$. Superscript or subscript $j$ indicates three normal stresses and three shear stresses in the rectangular coordinates, that is, $j = 1, 2, \ldots, 6$. $P^j_i$ indicates the surface density of charge accumulation in $i$ direction when force is applied to $j$ direction (i.e., the polarization intensity along $i$ direction). $\sigma_j$ indicates stress along $j$ direction. $d_{ij}$ is the piezoelectric constant when stress is applied along $j$ direction and the charge is generated in $i$ direction.

In general, the piezoelectric characteristics of crystals (piezoelectric materials) can be represented by their piezoelectric constant matrix:

$$d_{ij} = \begin{bmatrix} d_{11} & d_{12} & d_{13} & d_{14} & d_{15} & d_{16} \\ d_{21} & d_{22} & d_{23} & d_{24} & d_{25} & d_{26} \\ d_{31} & d_{32} & d_{33} & d_{34} & d_{35} & d_{36} \end{bmatrix} \qquad (10.69)$$

For reverse piezoelectric effect, the linear piezoelectric constitutive equations can be expressed as

$$\varepsilon_h = d_{hk} \cdot E_k \qquad (10.70)$$

where $\varepsilon_h$ indicates the strain in $h$ direction, $h = 1, 2, \ldots, 6$. $E_k$ indicates the electric field applied in $k$ direction, $k = 1, 2, 3$. The piezoelectric constant of the reverse piezoelectric effect is equal to that of the direct piezoelectric effect, and corresponds to one by one.

$$d_{hk} = \begin{bmatrix} d_{11} & d_{12} & d_{13} \\ d_{21} & d_{22} & d_{23} \\ d_{31} & d_{32} & d_{33} \\ d_{41} & d_{42} & d_{43} \\ d_{51} & d_{52} & d_{53} \\ d_{61} & d_{62} & d_{63} \end{bmatrix} \qquad (10.71)$$

The piezoelectric constant matrix of the reverse piezoelectric effect is equivalent to the transposed matrix of the direct piezoelectric effect. For a specific crystal, the 18 components of the matrix can be simplified owing to the symmetry of the crystal. For example, the piezoelectric constant matrix of quartz crystal can be expressed as follows:

$$d_{ij} = \begin{bmatrix} d_{11} & -d_{11} & 0 & d_{14} & 0 & 0 \\ 0 & 0 & 0 & 0 & -d_{14} & -2d_{11} \\ 0 & 0 & 0 & 0 & 0 & 0 \end{bmatrix} \text{ for direct piezoelectric effect} \qquad (10.72)$$

$$d_{hk} = \begin{bmatrix} d_{11} & 0 & 0 \\ -d_{11} & 0 & 0 \\ 0 & 0 & 0 \\ d_{14} & 0 & 0 \\ 0 & -d_{14} & 0 \\ 0 & -2d_{11} & 0 \end{bmatrix} \quad \text{for reverse piezoelectric effect} \qquad (10.73)$$

Since the piezoelectric effect is related to the crystal symmetry, one can identify the piezoelectric crystals from the crystallographic point groups. In the 32 point groups of crystal symmetry, 11 are centrosymmetric (having symmetric centers), of which all components of the piezoelectric tensors are equal to zero. They are not the piezoelectric crystal. The crystals categorized in point group of 432 (not having symmetric centers) are also not piezoelectric because cubic system has high symmetry. Piezoelectric crystals can only exist in other 20 point groups as listed in Table 10.9. Of these 20 groups (all exhibiting direct piezoelectricity), ten exhibit spontaneous polarization without external stress. This is because their crystal cells always retain electric dipole moments. The crystals in this class also exhibit pyroelectricity. If the couple torque can be reversed by applying an electric field, the crystal is also ferroelectric.

**Table 10.9:** Dielectric crystal classes.[12, 13].

| | | |
|---|---|---|
| | Polar crystals (pyroelectric crystals) (10 kinds) | 1, 2, m, mm2, 3, 3m, 4, 4 mm, 6, 6 mm |
| 21 crystals not having symmetric centers, in which 20 are piezoelectric crystals | Nonpolar crystals (11 kinds) | 222, 32, 422, $\bar{4}$, $\bar{4}$2m, 622, $\bar{6}$, $\bar{6}$2m, 23, $\bar{4}$3m |
| | | 432 (without piezoelectric effect) |
| 11 crystals having symmetric centers | | $\bar{1}$, 2/m, mmm, $\bar{3}$, $\bar{3}$m, 4/m, 4/mmm, 6/m, 6/mmm, m3, m3m |

In the large category of piezoelectric materials, piezoelectric ceramics are of special interest due to their significant piezoelectric effect. A typical titanate, $BaTiO_3$, exhibits distinct piezoelectricity in any one of its four crystal structures (rhombohedral, orthorhombic, tetragonal and hexagonal) that present in turn from low temperature to high temperature. These four crystal structures also exhibit ferroelectric effect. The cubic structure of $BaTiO_3$, existing in the range of 130–1,460 °C, is not piezoelectric or ferroelectric, because of its highly structural symmetry without dipolar moment. Since barium titanate was firstly reported to be piezoelectric, many other ceramics with structures similar to perovskite and tungsten-bronze are also affirmed to be piezoelectric. Although barium titanate has been used for microphones and other transducers, lead zirconate titanate (PZT) with more superior piezoelectricity is the most widely used piezoelectric ceramic so far.

Lead zirconate titanate is a solid solution of $PbZrO_3$ and $PbTiO_3$. Its nominal composition can be expressed as $Pb[Zr_xTi_{1-x}]O_3$, $0 \leq x \leq 1$. It exhibits typical perovskite structure as shown in Fig. 10.34. Lead titanate is a ferroelectric, and its Curie temperature is 492 °C, while lead zirconate is antiferroelectric, and the Curie temperature is 232 °C. The combination of both results in the Curie temperature above 350 °C. The piezoelectric effect of PZT can be adjusted by changing composition and doping. Two kinds of doping can be applied in modifying PZT: acceptors or donors. The former brings oxygen (anion) vacancies that reduce piezoelectric constant. The doping also impede the motion of domain walls, decreasing the internal losses. This kind of PZT is usually able to bear stronger electrical excitation and larger stress, so is called "hard" PZT. The latter brings metal (cation) vacancies that can increase piezoelectric constant, leading to a "soft" PZT. The doping also promotes the motion of domain walls, so as to increase internal losses. PZT has better piezoelectric and dielectric properties than other ferroelectrics. Its $d_{15}$, $d_{31}$, and $d_{33}$ can be as high as $7 \times 10^{-10}$C/N, $-2 \times 10^{-10}$C/N, and $5 \times 10^{-10}$C/N, respectively. Its electromechanical coupling factor reaches 0.7.

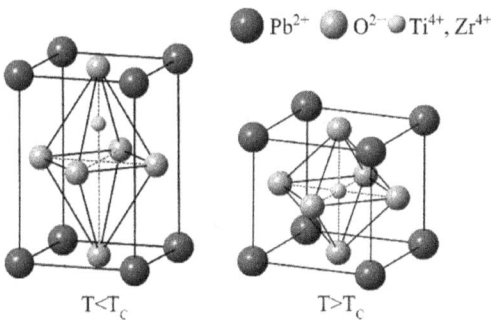

**Fig. 10.34:** Cell structure of lead zirconate titanate.

The other important piezoelectric materials that should be mentioned are lithium niobate ($LiNbO_3$) and polyvinylidene fluoride (PVDF). The former has a trigonal crystal system, which lacks inversion symmetry (as shown in Fig. 10.35) and thus exhibits obvious piezoelectric effect. The latter, a thermoplastic fluorine-containing polymer, is synthesized by vinylidene fluoride polymerization. Its chemical structure is shown in Fig. 10.36.

PVDF was found to exhibit strong piezoelectric effect in 1969. The piezoelectric constant of poled PVDF thin films reaches up to $6-7 \times 10^{-12}$ C/N (net dipole moment under a strong electric field), ten times larger than that searched in any other polymers. Compared with other commonly used piezoelectric materials, for example,

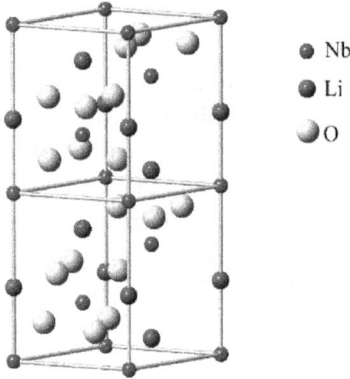

**Fig. 10.35:** Cell structure of lithium niobate.

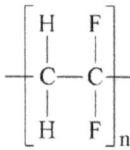

**Fig. 10.36:** Chemical structure of PVDF.

PZT, BaTiO$_3$, and LiNbO$_3$, PVDF has an opposite piezoelectricity, that is, PVDF compresses rather than expands (or vice versa) when they are exposed to the same electric field. This means that PVDF has a negative $d_{33}$ value.

As early as World War I, piezoelectric materials have been applied to the electroacoustic transducers. Following the way proposed by Langevin, quartz crystals were utilized to create sonic wave. This was the basis for the electroacoustic converters. Potassium sodium tartrate, the ferroelectric crystal that was first discovered, was extensively used in underwater acoustic transducers before World War II. Since then, piezoelectric materials have been widely used in technical fields including defense equipments, medical diagnostics, audio entertainment apparatus, and electronic communications. The typical applications can be classified into the following six subsidiary sets: (1) sensors, such as pressure and shock-wave sensors, flow sensors, mass sensitive sensors, and acceleration sensors, are used for automobile, automation, and medical devices. Suitable piezoelectric materials for these applications include rings, plates, and disks of soft PZT, LiNbO$_3$ substrates, PVDF foils, and quartz disks. (2) Actuators and motors, for example, transformers, bimorph actuators, multilayer actuators, printers, and injection systems, are used for automotive fuel valves, fine positioning and optics, inkjet printing head, micropumps, and minitype transformers and motors. Bars, plates, tubes, rings, thin films, as well as multilayer of PZT (including "soft" and "hard" PZT) are used for these applications. (3) Sound and ultrasound, that is, atomizer, air ultrasonic generator, buzzer, hydrophonics, ultrasonic imaging, microphones and speakers, shock wave generators and superpower transducers, and so on, are used for aerosols, telephone, distance meter, humidifiers, intrusion alarm, medical diagnostics, machining, ultrasonic cleaning, lithotripsy, oil atomizers, sonic alerts, sound sources and detectors. These applications are mainly based on the use of hard and soft PZT with different shapes and sizes, or even its thin films. (4) Frequency control and signal processing, that is, surface/bulk acoustic wave devices, mechanical frequency filters, frequency-time standards, are used for precise frequency control and filtering for wireless communication and signal identification. These applications involve not only PZT, but also other piezoelectric materials including quartz single crystal, $LiNbO_3$ and $LiTaO_3$. (5) Adaptronics (adaptive devices) are used for eliminating vibration and noise or adaptive controlling. Various shapes of soft PZT are best suited for this application due to the larger piezoelectric constant. (6) Gas and fuel ignition. The best suitable materials for these applications are hard PZT cylinders.

### 10.6.3 Magnetostriction

A ferromagnetic material extends (or shortens) under an external magnetic field, and returns to its original size after removing the external magnetic field. This phenomenon is called *magnetostrictive effect* (or magnetostriction). It was first discovered in iron metal by James Joule in 1842. Physically, when a magnetic field acts on a ferromagnetic material both its volume and length will change. But the length changes much more than its volume. The magnitude of the length change is about $10^{-5}$ to $10^{-6}$ for some transition metal alloys and ferrite ceramics, but reaches up to $10^{-3}$ for giant magnetostrictive materials such as terbium–dysprosium–iron compounds ($TbFe_2$, $TbFe_3$, and $Tb(Dy)Fe_2$).

Magnetostrictive effect is originated from the magnetization process of the material. Atomic magnetic moments under magnetic field coordinate and balance the elastic bond length, resulting in magnetostriction. For ferromagnetic materials, their microstructure is composed of magnetic domains (uniform magnetic polarizing region). Inside the regions, the magnetic moments are arranged toward specific crystal direction deferring to the local anisotropy due to the exchange interaction. The randomly distributed domains will rotate under magnetic field, resulting in the domain boundary shifts and the dimension changes due to the shape anisotropy of the domains.

Magnetic anisotropy refers to the phenomenon that the magnetic properties of matter change with direction. This involves magnetoelastic anisotropy due to magnetoelastic coupling, magnetic shape anisotropy for nonspherical symmetric objects, and magnetocrystalline anisotropy due to dipolar magnetic interaction and anisotropic exchange effect. They determine the easy axis in magnetic materials and the anisotropic magnetostrictive effect. When an external magnetic field is consistent with the direction of the easy axis of magnetization, the magnetic domains in the material rotate and rearrange along the direction of the external magnetic field to reduce the free energy of the system. Because of shape anisotropy, this process produces magnetostriction. Obviously, the magnitude of the magnetostrictive effect is related to the orientation of domains and the external magnetic field direction. For a polycrystalline material, the larger magnetostriction can only occur in the microstructure with preferential crystallographic orientation.

To measure magnetostriction, the magnetostriction coefficient, $\lambda$, equal to the ratio of elongation along the direction of magnetization and the total length of the sample tested, is frequently used:

$$\lambda = \frac{L_H - L_0}{L_0} \tag{10.74}$$

where $L_0$ denotes the initial length and $L_H$ denotes the length after extending (or shortening) under external magnetic field. Units of the magnetostriction coefficient generally take ppm. If $\lambda > 0$, the material extends along the magnetization direction, which is called positive magnetostriction, for example, iron. If $\lambda < 0$, the material shortens, known as negative magnetostriction, for example, nickel.

If the magnetic state in a material responses to an external stress, this phenomenon is called the inverse magnetostrictive effect (it is also called Villari effect or magnetoelastic effect). The inverse magnetostrictive materials can be utilized for stress and torsion sensors.

In addition, there are other magnetostrictive effects in ferromagnetic materials. A ferromagnetic rod placed in the direction of a longitudinal magnetic field will twist when the electrical current flows through it. This effect was observed in 1858 by Gustav Wiedemann (German physicist). The Wiedemann effect involves the combined magnetostriction that results from both the external magnetic field and the induced circular magnetic field generated by the current. If the applied magnetic field or the electric current is set as the signal source, the Wiedemann device can be configured as a torsional oscillator.

The torsional angle, $\theta$, of the rod-like material with magnetoelastic effect can be related to the electrical current flowed through the rod by using linear approach:

$$\theta = j\frac{h_{15}}{2G} \tag{10.75}$$

where $j$ indicates current density; $h_{15}$ is a component of magnetoelastic properties, which is proportional to the external magnetic field intensity; and $G$ is the shear modulus.

Matteucci effect is thermodynamically inverse to Wiedemann effect. This phenomenon can be demonstrated by using a high permeability rod that is pretorsioned and fixed. When it is magnetized with a parallel AC field, an AC pulse voltage is generated synchronously between the two ends of the rod.

In a more general expression, Matteucci effect indicates the spiral anisotropic magnetization of a magnetostrictive material under a torque. Accordingly, a spiral wire with ferromagnetic effect intends to magnetize preferentially toward the helical direction. When the energy density arisen from magnetic strain/stress is greater than that of magnetic anisotropy, the saturation torsion is achieved. In this situation, all spins rotate toward this direction. The torsion combines the longitudinal magnetic flux with the circumferential magnetic flux, resulting in a decrease in voltage. In practice, Matteucci effect will be significantly improved by pre-forming spiral domains in microstructure. This can be realized by twisting or annealing under twist. The Matteucci effect is used for mechanical sensors.

Besides, Guillemin effect is also related to magnetostriction. According to the effect, a prebent magnetostrictive rod will be straightened when it is placed in a longitudinal magnetic field.

It is obvious that the amplitude of magnetostriction of a specific ferromagnetic material depends on the intensity of the applied magnetic field in the initial stage of magnetization. The magnetostrictive strain (i.e., magnetostriction coefficient) increases as the applied magnetic field strength increases, until reaching its saturation value (i.e., maximal value), $\lambda_s$, as shown in Fig. 10.37. Besides, the magnetostriction coefficients of ferromagnetic materials are significantly affected by temperature, composition of alloys, crystallographic order and crystalline orientation of crystals. As an example, the saturation magnetostriction of polycrystalline nickel is significantly affected by temperature as shown in Fig. 10.38. Alloying generally alters the magnetostriction coefficient because the alloy elements may change interactions in the magnetic structure. Nickel alloy with the addition of a nonmagnetic metal exhibits the decrease of magnetostriction. But, adding with Co, Mn, and Fe may result in a positive effect on the magnetostriction, because these elements added contribute to the saturation moment of the alloy. Pd is a definite metal that improves the

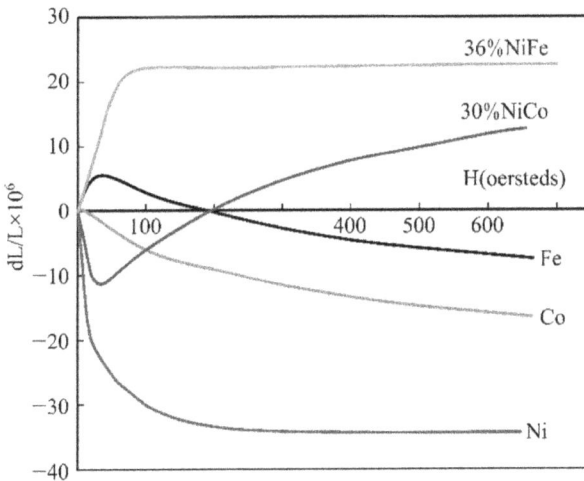

**Fig. 10.37:** Magnetostrictive strains of some metals as a function of magnetic field intensity[14].

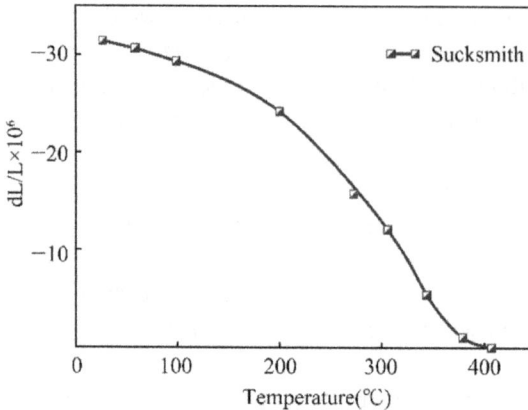

**Fig. 10.38:** Magnetostrictive strains of polycrystalline nickel at different temperature[15].

magnetostrictive effect of nickel. $Fe_{50}Co_{50}$ alloy with ordered structure exhibits much larger magnetostriction coefficient than it with disordered structure. When measuring a single crystal, one can find that magnetostrictive strain varies with crystalline orientation. Therefore, controlling grain orientation toward the preferred direction (e.g., unidirectional solidification) can facilitate greater magnetostriction.

As early as in the 1960s, rare earth elements, terbium and dysprosium, were found to exhibit large magnetostrictive effects at low temperature. However, in order to obtain the large magnetostrictive strain, it is necessary to utilize a strong magnetic field at low temperature. This is very inconvenient and costly. Fortunately, the similar large magnetostrictive effect at ambient temperature was discovered in terbium/dysprosium-iron compounds, $TbFe_2$ and $DyFe_2$. The drawback is that they all exhibit large anisotropy. A strong magnetic field must be applied to achieve saturation magnetization. These compounds are therefore still not suitable for use as practical applications. In order to avoid this disadvantage, a pseudobinary alloy $(Tb_{0.3}Dy_{0.7})Fe_2$ with small anisotropy constant at room temperature was developed. This compound has a greater permeability and a smaller coercive force. Early work on this material indicated that the magnetostrictive strain occurred along the [111] crystal orientation. Its values reached about $2,000 \times 10^{-6}$ at 240 K under a magnetic field of 15 kOe, and about $1,640 \times 10^{-6}$ at room temperature under the same magnetic field intensity.

Magnetostrictive effect can easily convert magnetic (electrical) signals into mechanical energy or action, and vice versa. The materials with giant magnetostrictive effect are very important for manufacturing high intensity electric-to-force transducers. Some successful applications include high power ultrasonic wave generators, highly stable electromechanical oscillators and filters, and electroacoustic transducers. In the past two decades, high-power underwater acoustic sonar transducer has been developed by using the rare earth-iron giant magnetostrictive materials, which plays a key role in the underwater communication and ocean exploration. They are also used for microdisplacement driving, vibration attenuation, noise reduction, intelligent wing, robot, valves, pumps, fuel injection, and fluctuating oil production.

# References

[1]  W. D. Callister Jr., Materials Science and Engineering: An Introduction, sixth edition, Wiley, New York, 2003.

[2]  D. A. Porter, K. E. Easterling, and M. Y. Sherif, Phase Transformations in Metals and Alloys, third edition, Taylor & Francis, Boca Raton, 2008.

[3]  M. E. Glicksman, Principles of Solidification: An Introduction to Modern Casting and Crystal Growth Concepts, Springer, Berlin, 2011.

[4]  E. J. Mittemeijer, Fundamentals of Materials Science, Springer, Berlin, 2011.

[5]  W. Kurz and D. J. Fisher, Fundamentals of Solidification, fourth revised edition, TTP, Zurich, 1998.

[6]  M. Ohring, Materials Science of Thin Films, second edition, Academic, London, 2001.

[7]  D. Smith, Thin-Film Decomposition: Principles and Practice, McGraw-Hill, New York, 1995.

[8]  D. I. Brower, An Introduction to Polymer Physics, Cambridge, London, 2002.

[9]  G. R. Strob, The Physics of Polymers: Concepts for Understanding Their Structures and Behavior, third edition, Springer, Berlin, 2007.

[10]  D. R. Askeland, P. P. Fulay, and W. J. Wright, The Science and Engineering of Materials, Six Edition, Cengage Learning, Stamford, 2011.

[11]  Y. Du, S. H. Liu, L. J. Zhang, H. H. Xu, D. D. Zhao, A. J. Wang, and L. C. Zhou. An overview on phase equilibria and thermodynamic modeling in multicomponent Al alloys: Focusing on the Al-Cu-Fe-Mg-Mn-Ni-Si-Zn system, CALPHAD: Computer Coupling of Phase Diagrams and Thermochemistry, 35, 2011, 427–445.

[12]  U. R. Kattner. The thermodynamic modeling of multicomponent phase equilibria, JOM, (12), 1997, 14–19.

[13]  A. A. Kodentsov, G. F. Bastin, and F. J. J. Loo. The diffusion couple technique in phase diagram determination, Journal of Alloys and Compounds, 320, 2001, 207–217.

[14]  R. W. Cahn and P. Haasen, Physical metallurgy, 4th Edition, Elsevier Science B.V., Amsterdam, 1996.

[15]  B. Predel, M. Hoch, and M. Pool, Phase Diagrams and Heterogeneous Equilibria, Springer, Berlin, 2004.

[16]  D. R. F. West and N. Saunders, Ternary Phase Diagrams in Materials Science, third edition, CRC, London, 2017.

[17]  A. D. Pelton, Phase Diagrams and Thermodynamic Modeling of Solutions, Elsevier, Amsterdam, 2018.

[18]  M. Hillert, Phase equilibria, Phase Diagram and Phase Transformations, Their Thermodynamic Basis, second edition, Cambridge, London, 2007.

[19]  A. Van der Ven and L. Delaey. Models for precipitate growth during the $\gamma \rightarrow \alpha + \gamma$ transformation in Fe-C and Fe-C-M alloys, Progress in Materials Science, 40, 1996, 181–264.

[20]  E. Pereloma and D. V. Edmonds, Phase Transformations in Steels, Elsevier, Amsterdam, 2012.

[21]  M. E. Gurtin, Phase Transformations and Material Instabilities in Solids, Elsevier, Amsterdam, 1984.

[22]  H. I. Aaronson, M. Enomoto, and J. K. Lee, Mechanisms of Diffusional Phase Transformations in Metals and Alloys, Taylor & Francis, Boca Raton, 2016.

[23]  J. S. Galsin, Solid State Physics: An Introduction to Theory, Academic Press, London, 2019.

[24]  A. O. Mohamed and E. K. Paleologos, Electrical Properties of Soils, in Fundamentals of Geoenvironmental Engineering, Butterworth-Heinemann, Oxford, 2018.

[25]  P. Schelling, Metallic Films for Electronic, Optical and Magnetic Applications: Structure, Processing and Properties, Woodhead Publishing Limited, Cambridge, 2014.

https://doi.org/10.1515/9783110495379-011

[26]  R. W. Collins and A. S. Ferlauto, Handbook of Ellipsometry, William Andrew, Inc, Norwich, 2005.

[27]  B. C. Chakraborty and D. Ratna, Polymers for Vibration Damping Applications, Elsevier Inc, Amsterdam, 2020.

[28]  Y. Poplavko and Y. Yakymenko, Functional Dielectrics for Electronics: Fundamentals of Conversion Properties, Woodhead Publishing, Cambridge, 2020.

[29]  Y. M. Poplavko, Electronic Materials: Principles and Applied Science, Elsevier, Amsterdam, 2019.

[30]  G. Engdahl, Handbook of Giant Magnetostrictive Materials, Academic Press, London, 2000.

# Index

https://doi.org/10.1515/9783110495379-012

www.ingramcontent.com/pod-product-compliance
Lightning Source LLC
Chambersburg PA
CBHW080911220326
41598CB00034B/5541